Hydrology: The Scientific Study of Water

Hydrology: The Scientific Study of Water

Editor: William Sobol

www.callistoreference.com

Callisto Reference,
118-35 Queens Blvd., Suite 400,
Forest Hills, NY 11375, USA

Visit us on the World Wide Web at:
www.callistoreference.com

ISBN: 978-1-63239-869-7 (Hardback)

Cataloging-in-publication Data

Hydrology : the scientific study of water / edited by William Sobol.
   p. cm.
Includes bibliographical references and index.
ISBN 978-1-63239-869-7
1. Hydrology. 2. Groundwater. 3. Earth sciences. 4. Water. I. Sobol, William.
GB661.2 .H93 2017
551.48--dc23

Table of Contents

Preface

The book presents researches and studies performed by experts across the globe in the field of hydrological sciences. Hydrological science is the study of the movement and distribution of water sources. With an increase in the world's demand for energy sources, there is a growing risk of depletion of our water resources. This book is a compilation of chapters that discuss the most vital concepts and emerging trends in the field of hydrology in recent times. While understanding the long-term perspectives of the topics, the book makes an effort in highlighting their impact as a modern tool for the growth of the discipline. It includes contributions of experts and scientists, which will provide innovative insights into this field.

After months of intensive research and writing, this book is the end result of all who devoted their time and efforts in the initiation and progress of this book. It will surely be a source of reference in enhancing the required knowledge of the new developments in the area. During the course of developing this book, certain measures such as accuracy, authenticity and research focused analytical studies were given preference in order to produce a comprehensive book in the area of study.

This book would not have been possible without the efforts of the authors and the publisher. I extend my sincere thanks to them. Secondly, I express my gratitude to my family and well-wishers. And most importantly, I thank my students for constantly expressing their willingness and curiosity in enhancing their knowledge in the field, which encourages me to take up further research projects for the advancement of the area.

Editor

Comprehensive experimental study on prevention of land subsidence caused by dewatering in deep foundation pit with hanging waterproof curtain

T. L. Yang[1,2,3], X. X. Yan[1,2,3], H. M. Wang[1,2,3], X. L. Huang[1,2,3], and G. H. Zhan[1,2,3]

[1]Shanghai Institute of Geological Survey, Shanghai, 200072, China
[2]Key Laboratory of Land Subsidence Monitoring and Prevention, Ministry of Land and Resources, Shanghai, 200072, China
[3]Shanghai Engineering Research Center of Land Subsidence, Shanghai, 200072, China

Correspondence to: T. L. Yang (sigs_ytl@163.com)

Abstract. Land subsidence caused by dewatering of deep foundations pit has currently become the focus of prevention and control of land subsidence in Shanghai. Because of the reliance on deep foundation dewatering pit projects, two comprehensive test sites were established to help prevent land subsidence. Through geological environmental monitoring during dewatering of a deep foundation pit, the analysis of the relation between artesian water level and soil subsidence, some basic features of land subsidence caused by dewatering of deep foundation pits are elucidated. The results provide a scientific basis for prevention and control of land subsidence caused by dewatering in deep foundation pits.

1 Introduction

The exploitation and utilization of the shallow subsurface in Shanghai keeps increasing with the demands of urban development deep foundation pits used in construction, leads to a series of environmental and geological problems (Yang and Gong, 2010; Wei et al., 2009). Among the problems, the most serious is the uneven subsidence caused by the compression of shallow strata, which arises from the dewatering of deep foundation pit engineering (Yang, 2010). Research on existing foundation pit engineering reveals that dewatering during excavation for the engineered hanging waterproof curtain has a great impact on the artesian water level at significant distances from the foundation pit (Yang et al., 2009). The large decline of artesian water level beyond the foundation pit leads directly to the uneven subsidence, which adversely affects the environment and important nearby structures. Furthermore, the subsidence is difficult to regulate and threatens the elevation safety in Shanghai (Yang et al., 2014).

Taking deep foundation pits of two construction projects in Shanghai as typical cases, we conducted a comprehensive dewatering-recharge test for prevention of land subsidence during the period of excavation and dewatering. The results can be applied elsewhere to prevent and manage subsidence caused by deep foundation pit engineering.

2 The testing design

2.1 Geological conditions of the testing field

The shallow strata of the testing field, which are comprised of clay, silt and sandy soil, are normal sedimentary strata. The objective of the foundation pit engineering is to decompress the first confined aquifer (the 7th soil layer). It is characterized by large depth, high water yield and permeability. According to the foundation pit retaining design, dewatering for decompression during the excavation is necessary. The underground continuous wall as a hanging waterproof curtain did not block the target aquifer – thus there is a directly hydraulic connection inside and outside the foundation pit.

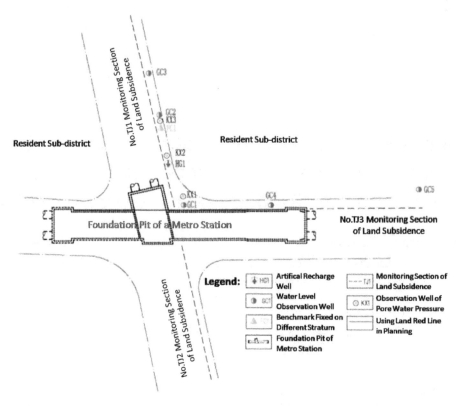

Figure 1. Layout chart of comprehensive testing site of land subsidence control in certain metro station foundation pit.

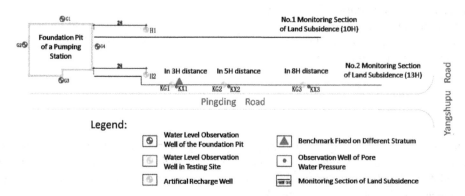

Figure 2. Layout chart of comprehensive testing site of land subsidence control in certain pumping station foundation pit.

2.2 Construction of the testing site

To monitor the geological and environmental effects caused by the pit excavation and dewatering, such as land subsidence and groundwater seepage, deep foundation pits of two construction projects in Shanghai were used as comprehensive testing sites to control engineering subsidence. The land subsidence monitoring section, groundwater observation wells for the confined aquifer, pore water pressure monitoring boreholes and extensometers for the layered subsidence of the soil were built inside the influence of the foundation pit and the dewatering (Yang et al., 2010). The length of monitoring section is more than 10 times the excavation

depths (H) of the foundation pits. The interval between two monitoring points is 5 to 10 m. Observation wells for the confined aquifer and pore water pressure monitor holes and extensometers (benchmarks fixed on different stratum) for the layered subsidence of soil and recharge wells were arranged in the main monitoring sections of the foundation pits (Wang et al., 2012), shown in Figs. 1 and 2.

3 Comprehensive test

According to the test objective and the facilities at the testing sites, the comprehensive test included 2 parts, monitoring and recharge. The monitoring test included the monitoring

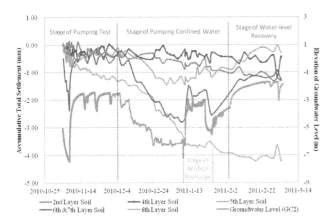

Figure 3. Duration graphical chart of soil laminar settlement and artesian head.

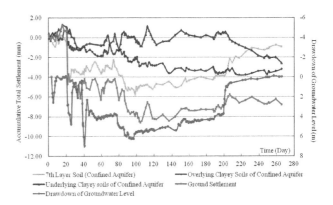

Figure 4. Duration graphical chart of soil laminar settlement and confined aquifer head drop depth at the point of 3 times the depth of foundation pit.

of the water level of the confined aquifer, the layered subsidence of the soil, the pore water pressure and the land subsidence. The monitoring spatial extent (10 times the excavation depth [H] of the foundation pit) covered most of the area influenced by dewatering. Through monitoring the groundwater seepage and the soil settlement during the dewatering period, the relation of the water-level change in the confined aquifer and the land subsidence caused by dewatering of the foundation pit was determined. The recharge test employed an artificial recharge well that penetrated the aquifer targeted for dewatering. A single-well pressurized recharge test was conducted to measure the effect of simultaneously recharging the shallow confined aquifer during the dewatering period on the control of land subsidence.

4   Comprehensive experimental analysis and study

The dewatering of the foundation pit during the construction period causes land subsidence in a wide area around the pit. Large scale water level declines in the confined aquifer that may cause environmental problems. About 70 m from the foundation pit, a group of extensometers, land subsidence monitor section and observation wells for the confined aquifer were set. The relationship between the water level in the confined aquifer and the soil displacement during the dewatering was studied.

4.1   Characteristics of confined aquifer level and different soil layer subsidence

A group of extensometers were arranged to measure the layered soil subsidence during the dewatering period. Figures 3 and 4 show the characteristic curves of subsidence of the soil layers. The results indicated that during the excavation and dewatering, with the change of water level, the characteristics of soil displacement were as follow:

1. During the initial stage of engineered dewatering, with the rapid decline of the confined aquifer water level, compression (settlement) is mainly focused on the soil layer of the confined aquifer. Furthermore, settlement gradually stabilized with stablization of the water level. Following engineered dewatering, the soil layer of the confined aquifer rebounded with recovery of the water level, while the overlying clayey soils sustained compression with gradual slow. At this time, settlement was mainly focused on the overlying clayey soil layer. In addition, during the engineered dewatering, compression of the underlying soil layer of the confined aquifer resulted in small settlement of that layer.

2. From the point of view of deformation characteristic caused by the engineered dewatering, initial settlement mainly showed elastic deformation of the confined aquifer, and overlying clayey soils were less affected. With the increase of pumping quantity and duration, elastic deformation of the confined aquifer was changed into elastic-plastic deformation of overlying clayey soils. After dewatering was over the soil layer of the confined aquifer rebounded, but compression occurred in the overlying clayey soils.

3. A short term of artificial recharge to the first confined aquifer was conducted during the dewatering period. The test result showed that the artificial recharge could increase the confined aquifer water level and reverse settlement of the confined aquifer soil layer, so that the land subsidence could be controlled.

4.2   Characteristics of confined aquifer level and land subsidence

Water level in the confined aquifer within the spatial extent of $10H$ around the foundation pit varied with distance from the

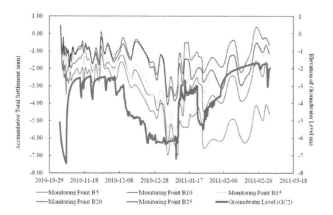

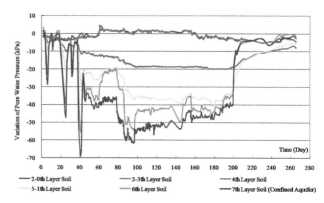

Figure 7. Duration graphical chart of pore water pressure at the point of 3 times the depth of foundation pit.

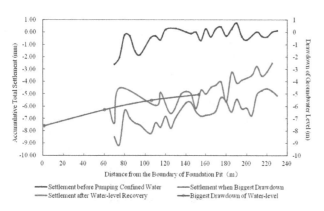

Figure 5. Duration graphical chart of land subsidence and artesian head in monitoring section.

Figure 6. Graphical chart of spatio-temporal features of land subsidence and artesian head in monitoring section.

pit during the excavation and dewatering period. The corresponding subsidence and rebound were monitored. Figures 5 and 6 show the relationship between land subsidence and the water-level change or 3 monitor sections (Fig. 6). The results show the following characteristics:

1. The drawdown and recovery of the confined water level was directly linked with the land subsidence and rebound. The former was the triggering factor and the later the result.

2. The confined water level around the foundation pit dropped a lot during the dewatering period. The spatial extent of the confined aquifer water level influenced by dewatering was larger than $10H$, which corresponded with the land subsidence influence. The magnitude of subsidence and the water-level decline gradually decreased with distance from the foundation pit. The subsidence was positively associated with the variation of the confined aquifer water level.

3. The excavation and dewatering were the main reason why land subsidence occurred. Within $3H$ around the foundation pit, the land subsidence was caused by the

superposition of excavation and dewatering, and the subsidence magnitude was very large. The subsidence profiles were spoon-shaped. The maximum cumulative subsidence occurred within the site of $0.5–1.0H$ from the foundation pit. Where the distance was larger than $3H$ from the foundation pit, the subsidence was caused mainly by dewatering. The subsidence magnitude there was relatively small, yet the scope was very large.

4. A short term of artificial recharge to the objective aquifer was conducted during the excavation and dewatering period. Within the influence range of recharging, soil rebound occurred and the subsidence magnitude was inversely proportion to the distance from the recharge well. So, the simultaneous artificial recharge to the confined aquifer during dewatering could help control land subsidence during the dewatering period.

5. After dewatering, the confined aquifer water level rose, but the subsidence kept developing. When the water level rose for a period of time, the land surface elevation recovered. Within $3H$ from the foundation pit, the final subsidence magnitude was large while the rebound value was small. Beyond $3H$ outside the scope from the foundation pit, the rebound magnitude was large while the final subsidence value was small.

4.3 Characteristics of pore water-pressure change

As shown in Fig. 7, during the stage of pumping test and engineered dewatering, the water level (expressed as an equivalent pore water pressure in Fig. 7) in the confined aquifer fell. During this time there was an obvious decline of pore water pressure among the 5th and 6th soil layers overlying the confined aquifer. However, pore water pressure of the 4th soil layer and above was only slightly affected and relatively stable. Following dewatering of the foundation pit ,the water level of the confined aquifer rapidly recovered, but the pore water pressure of the overlying cohesive soil layers recovered slowly.

5 Conclusions

1. The dewatering of foundation pits using hanging waterproof curtains cause land subsidence.

2. The decline and recovery of the confined aquifer water level is directly linked with the land subsidence and rebound. The former is the triggering factor, and the later the result.

3. Monitoring results of extensometers showed that land subsidence caused by dewatering of the foundation pit was mainly comprised of the compression of the targeted aquifer for dewatering – the confined aquifer; and the neighboring soft soil layers. The compression of the confined aquifer was mainly elastic and was in accord with the variation of the confined aquifer water level while the neighboring soft soil layers was visco-elasto-plastic and was irreversible.

4. The spatial extent of land subsidence caused by dewatering of the foundation pit using the hanging waterproof curtain was larger than $10H$. The magnitude of subsidence and the decline of the confined aquifer water level gradually decreased with the increasing distance from the foundation pit. The subsidence was positively associated with the variation of the water level.

5. Within $3H$ around the foundation pit, the land subsidence was caused by the superposition of excavation and dewatering, and the subsidence value was large. The subsidence profile was spoon-shaped. The maximum cumulative subsidence occurred within a distance of 0.5–$1.0H$ from the foundation pit. Beyond a distance of $3H$ from the foundation pit, the subsidence was caused mainly by the dewatering. The subsidence there was relatively small, yet the spatial extent was large.

6. A short term of artificial recharge to the confined aquifer was conducted during the excavation and dewatering period. Within the range of influence from recharging, soil rebound occurred and the magnitude was inversely proportion to the distance from the recharge well. So, simultaneously, artificially recharging the objective aquifer during its engineered dewatering could help control land subsidence during the dewatering period.

Acknowledgements. This work was supported by public welfare and scientific research subject projects funded by the Ministry of Land and Resources of China (201311045) and major scientific research subjects from the Science and Technology Commission of Shanghai Municipality (12231200700, 14DZ1207802).

References

Wang, J. X., Wu, Y. B., Zhang, X. S., Liu, Y., Yang, T. L., and Feng, B.: Filed experiments and numerical simulations of confined aquifer response to multi-cycle recharge-recovery process through a well, J. Hydrol., 464–465, 328–343, 2012.

Wei, Z. X., Wang, H. M., Wu, J. Z., Fang, Z. L., and Liu, G. B.: Land subsidence and its influences on urban security of Shanghai, Shanghai Geology, 30, 34–39, 2009.

Yang, T. L.: Analysis of the Land Subsidence Impact of Dewatering of Deep Foundation Pits, Shanghai Geology, 33, 41–44, 2010.

Yang, T. L. and Gong, S. L.: Microscopic analysis of the engineering geologic behavior of soft clay in Shanghai, China, Bull. Eng. Geol. Environ., 69, 607–615, 2010.

Yang, T. L., Yan, X. X., Wang, H. M., Xu, J. H., and Cheng, H. S.: Study on land subsidence induced by excavation engineering, Shanghai Geology, 30, 15–21, 2009.

Yang, T. L., Yan, X. X., Wang, H. M., Wu, J. Z, and Liu, J. B.: Experimental research of shallow aquifer pressure groundwater recharge based on dual control pattern of groundwater level and land subsidence, Shanghai Geology, 31, 12–17, 2010.

Yang, T. L., Wang, H. M., and Jiao, X.: Land subsidence zoning control in Shanghai, Shanghai Geology, 35, 105–109, 2014.

2

Fuzzy-logic assessment of failure hazard in pipelines due to mining activity

A. A. Malinowska and R. Hejmanowski

AGH University of Science and Technology, Cracow, Poland

Correspondence to: A. A. Malinowska (amalin@agh.edu.pl)

Abstract. The present research is aimed at a critical analysis of a method presently used for evaluating failure hazard in linear objects in mining areas. A fuzzy model of failure hazard of a linear object was created on the basis of the experience gathered so far. The rules of Mamdani fuzzy model have been used in the analyses. Finally the scaled model was integrated with a Geographic Information System (GIS), which was used to evaluate failure hazard in a water pipeline in a mining area.

1 Introduction

Water pipelines localized in active mining areas are particularly hazarded with factors which may potentially evoke additional strains. Previous work shows that such factors include: seismic tremors, landslides, and continuous and discontinuous deformations of terrain due to underground extraction (Kalisz, 2007; Knothe, 1953; Kowalski and Kwiatek, 1995; Kwiatek and Mokrosz, 1996; Mendec et al., 1996; Mokrosz and Zawora, 1997; Szadziul, unpublished data; Talesnick and Baker, 2008; Zhao et al., 2005). Therefore, the failure-prone zones hazarding the pipelines, need to be managed because of the potential risk they bring. Depending on the significance of a given object, various evaluation methods are used for assessing the risk in particular situations. In the case of objects, whose failure may threaten the general safety and well-being of people, the applied methods are both time-consuming and very costly. Among the most frequent methods of the strain modeling is Finite Element Method (FEM; Zhao et al., 2005). Another approach is applicable when the linear objects are numerous and their failure may be only noxious for the users. In this case the applied estimation methods allow for a quick but approximated estimation of hazard. Unfortunately, approximated methods are inaccurate, and still time consuming. Therefore an attempt to work out a fuzzy model for assessing hazards caused by continuous deformations acting on linear objects was made. This model is integrated with GIS which significantly accelerates

and simplifies the evaluation and reporting about the zones potentially threatened by underground extraction operations.

2 Background

2.1 Pipeline hazard estimation – approximated point method

In Poland, the strength of water pipelines in mining areas is evaluated with the point method. This method was created in the 1950s and has been commonly applied by the coal mine industry. This method offers an approximated evaluation of failure hazard of pipelines and consists of three stages:

1. resistance evaluation of water pipeline (expressed as resistance category (Table 1);

2. prediction of continuous strains hazard with Knothe function theory (expressed as terrain category) (Knothe, 1953);

3. comparison of terrain category with pipeline resistance category.

The resistance of the pipelines is assessed on the basis of four risk factors, and each of them is ascribed a certain number of points. On this basis the pipeline is given its category of resistance to horizontal strains. The higher is the number of points, the lower is the resistance category. A low resistance category means that a linear object is vulnerable and

Table 1. Resistance classification of water pipelines – point method.

Hazard factor		Number of points			
Matter	PE	0–10			
	Cast iron or steel	10–15			
	PCV	15–20			
	Asbestos	20–30			
Compensation	Pipe sleeve	0–10			
	Compensators	10–20			
	No compensators	20–30			
Type and number of connections	< 1 connection $100\,\mathrm{m}^{-1}$	0–10			
	1–3 connection $100\,\mathrm{m}^{-1}$	10–20			
	3–5 connection $100\,\mathrm{m}^{-1}$	20–30			
	> 5 connection $100\,\mathrm{m}^{-1}$	30–40			
Technical condition	Very good	0–10			
	Good	10–20			
	Acceptable	20–30			
	Poor	30–40			
Number of points	0–24		25–48	49–80	> 80
Resistance category	4		3	2	1
Acceptable horizontal strains $[\mathrm{mm\,m}^{-1}]$	9.0		6.0	3.0	1.5

prone to surface strains. In this case the probability of a failure occurrence is higher.

The terrain hazard is defined on the basis of predicted horizontal strains and tilts (Knothe, 1953). Then the terrain, where continuous deformations are possible, is categorized. Six categories have been distinguished depending on the intensity of predicted maximal horizontal strain ε max:

1. 0 (ε max $\in$ 0–0.3 $\mathrm{mm\,m}^{-1}$);

2. I (ε max $\in$ 0.3–1.5 $\mathrm{mm\,m}^{-1}$);

3. II (ε max $\in$ 1.5–3.0 $\mathrm{mm\,m}^{-1}$);

4. III (ε max $\in$ 3.0–6.0 $\mathrm{mm\,m}^{-1}$);

5. IV (ε max $\in$ 6.0–9.0 $\mathrm{mm\,m}^{-1}$);

6. V (ε max $>$ 9.0 $\mathrm{mm\,m}^{-1}$).

The terrain categories are compared with pipeline resistance categories and on this basis the ultimate assessment is made.

2.2 Estimation of continuous deformation hazard in linear objects – shortcomings

The method used in Poland for evaluating hazard with continuous deformations in linear objects is very simplified. The assessment lies in finding objects where, the resistance category is exceeded by the terrain category (Fig. 1) (Szadziul, unpublished data).

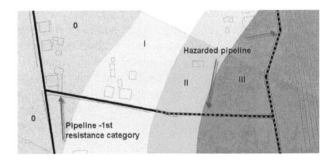

Figure 1. Example of the course of a linear object through terrains of belonging to different terrain categories (0, I, II, and III represent different terrain categories, with III being the most hazardous).

The practical application of this method has been exemplified below. A water pipeline belonging to the first resistance category is deposited in a ground of category 0–III (Fig. 1). Horizontal strains of terrain in that zone range between 0.1 $\mathrm{mm\,m}^{-1}$ (0 terrain category) to 4.5 $\mathrm{mm\,m}^{-1}$ (III terrain category). The pipeline can withstand horizontal strains under 1.5 $\mathrm{mm\,m}^{-1}$ (Table 1), therefore is hazarded in the terrain belonging to the II and III terrain category.

The major shortcoming of the presently applied method is discretisation of the described hazard. Moreover, the resistance of pipelines is defined in a very general way. Two sections of water pipeline (5 m from each other) can be evaluated as hazarded and not hazarded. Such discretisation creates serious problems to the organization responsible for safety in

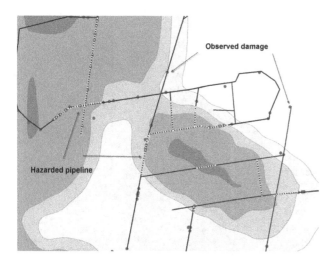

Figure 2. Hazard evaluation of water pipeline with point method (yellow dotted line: endangered part of the pipeline, red dot: observed damage).

mining areas, which have to make decisions about possible hazards and replacements of particular pipeline sections. As visualized in the presented example, the congruence between actual failures and places of potential hazard is low (Fig. 2).

Another problem lies in the lack of direct connection between terrain categories and pipeline resistance categories. Moreover, the present methodology is burdened with high uncertainty stemming from a number of factors including: incomplete information about the mining-geology conditions, incorrect modeling of continuous deformations, subjective evaluation of pipeline resistance; and, incomplete information about pipelines. This prompted the current study to identify other solutions, which would increase the efficiency of evaluation of pipelines hazard in areas staying under constant deformations of terrain.

3 Modeling of pipeline hazard with fuzzy logic

The analyses were focused on assuming additional risk factor for the assessment of pipeline hazard and developing a fuzzy-logic model based on risk factors.

Fuzzy logic has been used in analyses since the 1960s (Zadeh, 1965). Initially it was used in industry for image processing, complex processes control, and computer-aided decision making. In the subsequent years attempts were undertaken to implement fuzzy theory in social systems, economy and even medicine. In environmental sciences fuzzy theories were used for evaluating hazards which could be generated by selected natural elements (Adriaenssensa et al., 2009; Bojorquez-Tapia et al., 2002; Busch and Maas, 2006; Gheorghe et al., 2000; Liu et al., 2005), e.g. for evaluating the landslide hazard (Lee, 2007), and evaluation of environmental elements in the aspect of their degradation (e.g., hazard evaluation of a river ecosystem, Ioannidou et al., 2003). It

Table 2. Rule base.

			HSM/HSR					
			0	I	II	III	IV	V
			L	VL	M	H	VH	EH
	4	L	La	La	La	VL	L	M
V	3	VL	La	La	VL	L	M	A
	2	M	La	VL	L	M	A	H
	1	H	VL	L	M	A	H	VH

should be emphasized that in a majority of cases the fuzzy analyses were supported by GIS tools. The present work represent the first time fuzzy logic has been used for assessing hazard assessment in pipelines placed in mining areas.

The first stage of creating a reasoning fuzzy system was determining input and output variables of the model. Basing on the studies by Malinowska et al. (unpublished data) two variables regarding the linear objects hazard were defined: predicted maximum horizontal strains ($HSM \in (0,9)$) and the difference between predicted major strains in two directions ($HSR \in (0,18)$). Another linguistic variable was the sensitivity of linear objects (V), expressed as resistance points ($V \in (0,100)$). The failure hazard of a linear object expressed in the point scale ($R \in (0,100)$) was the output variable. This variable can be used for determining the failure hazard of a pipeline in points. The space in which these variables were analyzed was assumed on the basis of extreme values observed in the mining areas in Poland. Then the linguistic values were defined for each of the variables (Table 1): very low (VL), low (L), medium (M), high (H), very high (VH), extremely high (EH).

Pipeline vulnerability was described by 4 linguistic variables: very low (VL), low (L), medium (M), high (H), very high (VH).

Hazard of pipeline failure was characterized by 7 linguistic variables: lack (La), very low (VL), low (L), medium (M), appreciable (A), high (H), very high (VH).

Each linguistic value has a defined fuzzy set described by a characteristic function. The information which could describe the characteristic function is very limited, therefore authors assumed a triangle shape of the membership function for fuzzy sets. The membership functions were modeled based on the following assumptions:

1. characteristic points of the membership function are defined on the basis of limits of terrain category and resistance category,

2. membership functions meet the unity condition.

Then, a rule base was defined on the following assumptions (Table 2).

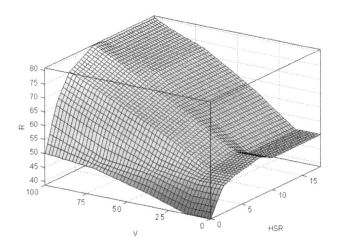

Figure 3. Surface of fuzzy model (V vs. HSR).

1. If the terrain category is higher than the resistance category by two or more grades, the failure hazard is serious.

2. If the terrain category is equal to the resistance category, the failure hazard is very low.

Fuzzy modelling was realized with the use of the Mandami dependence. The degree to which the premises in the inference block have been met was evaluated with the product operator (PROD). The successively activated premises allowed for determining the degree to which conclusions have been met. The last stage of fuzzy reasoning lied in determining the output function of rule base conclusions. The output membership function was defined on the basis of membership functions of particular conclusions of rules in the accumulation process. The accumulation in the model was performed for the SUM operator. The output membership function obtained in the course of inference was used for determining a sharp output value representing this set in the most reliable manner. Defuzzification was performed with the gravity center method. On this basis a surface was created, allowing for the evaluation of pipeline hazard with continuous strains (Fig. 3).

4 Model application

Modeling of pipeline failure hazard is supported by GIS programming. This is of special significance in predominantly urbanized areas, where the water network is extensive. It should be stressed that a sound evaluation of the failure hazard with GIS can be done only when reliable data about the resistance of the pipeline and the expected deformations which may occur in the study area are available (Fig. 4).

In the proposed solution based on a fuzzy model hazard in pipelines can be determined in a continuous manner, for each segment of the pipeline (the length to be defined arbitrarily). The pipeline can be ascribed a point value of failure

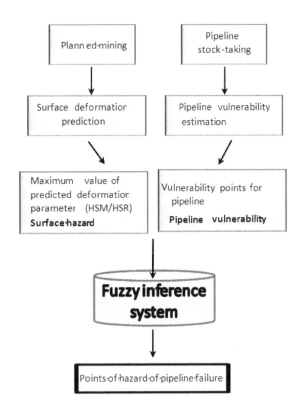

Figure 4. Pipeline damage risk assessment algorithm.

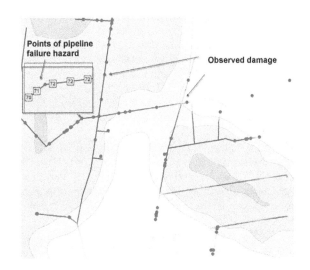

Figure 5. Evaluation of failure hazard of a pipeline performed with a fuzzy model (yellow dotted line: endangered part of the pipeline, red dot: observed damage, line with the blue intensive color-predicted high risk of pipeline damage).

risk ($R \in (0,100)$). The pipeline sections marked in intense colors are more endangered with failures (Fig. 5). The congruence between actual failures and places of potential hazard is high.

Organizations responsible for water pipelines are free to define the limits of risk points, for which the hazard is very high. The proposed model can be used for evaluating hazards in other linear objects, such as gas pipelines, when parameters have been adapted to the local conditions.

5 Summary

In a fuzzy-set based model the uncertainty resulting from the subjective evaluation of experts or incomplete data can be accounted for. The applied solution increases the efficiency of estimation of a water pipeline hazard. Moreover, thanks to the application of the inference block, the hazard can be presented as a continuous variable, which seems to be very advantageous compared to other methods. The comparison of modeling results with actual observations of damaged pipelines revealed a considerably higher accuracy using this new method. In addition, the presented method of a failure hazard evaluation for water pipelines seems to be readily and easily integrated with geographic information systems.

Acknowledgements. The research reported in this paper has been supported by a grant from the National Science Centre no. 2011/01/D/ST10/06958.

References

Adriaenssensa, V., Baetsb, B. D., Goethalsa, P. L. M., and Pauwa, N. D.: Fuzzy rule-based models for decision support in ecosystem management, Sci. Total Environ., 319, 1–12, 2009.

Bojorquez-Tapia, L. A., Juarez, L., and Cruz-Bello, G.: Integrating Fuzzy Logic, Optimization, and GIS for Ecological Impact Assessments, Environ. Manag., 30, 418–433, 2002.

Busch, W. and Maas, K.: Remarks to the risk assessment for abandoned mine sites, Acta Montanistica Slovaca Ročník 12, 3, 340–348, 2006.

Gheorghe, A. V., Mrock, R., and Kröger, W.: Risk assessment of regional systems, Reliab. Eng. Syst. Safe., 70, 141–156, 2000.

Ioannidou, I. A, Paraskevopoulos, S., and Tzionas, P.: Fuzzy modeling of Interactions among Environmental Stressors in the Ecosystem of Lake Koronia, Greece, Environ. Manag., 32, 624–638, 2003.

Kalisz, P.: The impact of the mining on the pipeline resiliency, Zeszyty Naukowe Politechnika Śląska S. Górnictwo, 278, 191–200, 2007 (in Polish).

Knothe, S.: Equation of the profile of finitely formed subsidence trough, Archiwum Górnictwa i Hutnictwa, 1(1), 1953 (in Polish).

Kowalski, A. and Kwiatek, J.: Surface and surface structure object protection on the mining area, Przegląd Górniczy, 4, 11–19, 1995 (in Polish).

Kwiatek, J. and Mokrosz, R.: Gas networks on the mining areas, WUG, 3, 1996 (in Polish).

Lee, S.: Application and verification of fuzzy algebraic operators to landslide susceptibility mapping, Environ. Geol., 52, 615–623, 2007.

Liu, H., Chen, W., and Kang, Z.: Fuzzy Multiple Attribute Decision Making for Evaluating Aggregate Risk in Green Manufacturing, J. Tsinghua Sci. Technol., 10, 627–632, 2005.

Mendec, J., Kliszczewicz, B., and Wytrychowska, M.: Surface and surface structure object protection on the mining area, Rules for the water pipeline protection against the effects of underground mining, Główny Instytut Górnictwa, Katowice, Poland, 53 pp., 1996 (in Polish).

Mokrosz, R. and Zawora, J.: The impact of the mining on the buried pipeline, Materiały konferencyjne: Ochrona powierzchni i o biektów budowlanych przed szkodami górniczymi, Główny Instytut Górnictwa, Katowice, Poland, 20-28, 1997 (in Polish).

Talesnick, M. and Baker, R.: Failure of flexible pipe with a concrete liner, Eng. Fail. Anal., 5, 247–259, 2008.

Zhao, W., Nassar, R., and Hall, D.: Design and reliability of pipeline rehabilitation liners, Tunn. Undergr. Sp. Tech., 20, 203–212, 2005.

Zadeh, L. A.: Fuzzy sets, Inform. Control, 38, 1–14, 1965.

3

New information on regional subsidence and soil fracturing in Mexico City Valley

G. Auvinet, E. Méndez-Sánchez, and M. Juárez-Camarena

Instituto de Ingeniería, UNAM, Mexico

Correspondence to: G. Auvinet (gauvinetg@iingen.unam.mx)

Abstract. In this paper, updated information about regional subsidence in Mexico City downtown area is presented. Data obtained by R. Gayol in 1891, are compared with information obtained recently from surveys using the reference points of Sistema de Aguas de la Ciudad de México (2008) and on the elevation of a cloud of points on the ground surface determined using Light Detection and Ranging (LiDAR) technology. In addition, this paper provides an overview of recent data obtained from systematic studies focused on understanding soil fracturing associated with regional land subsidence and mapping of areas susceptible to cracking in Mexico City Valley.

1 Introduction

Mexico Valley lacustrine subsoil has an exceptionally high compressibility and low resistance. Additionally, a regional subsidence phenomenon is affecting the urban area since the early Twentieth Century. One consequence of this phenomenon is the generation of cracks in the soil in many places. Both problems, regional subsidence and soil fracturing represent a risk for the stability of buildings and affect the urban infrastructure.

The demographic development of Mexico City has created an accelerated demand for services, especially supply of drinking water. One of the cheapest ways to respond to this demand has been the exploitation of the aquifer underneath the urban area by pumping water from deep wells. This activity has produced a regional subsidence phenomenon and the cracking of the soil in the lacustrine and alluvial-lacustrine areas of Mexico City. Due to the high cost of other water-supply alternatives, it is expected that extraction of water from the local aquifer will continue for many years.

The regional subsidence in Mexico City has severe consequences. It affects the drainage system, transport infrastructure, foundations of buildings and generates serious risks to the population, since it induces other problems such as flooding of low areas. Therefore, although the regional subsidence is an ancient phenomenon, its study and analysis remain a priority nowadays, inasmuch as it has not been possible to control its basic cause, which frequently leads to adopt emergency solutions.

More and more frequently, cracks appear in the soil of Mexico City causing alarm among the population and damaging buildings. Therefore, since 2005, the Geocomputing Laboratory group of the Geotechnical Section of Instituto de Ingeniería, UNAM in collaboration with the Mexican Society for Geotechnical Engineering and with the support of municipal authorities has undertaken a systematic study of the phenomenon of soil cracking. The occurrence of cracking may result from any condition that causes important tension stresses in the soil (Auvinet, 2008) and the occurrence of cracking has different causes, including contraction of the lacustrine clays by drying, existence of tension stresses associated with buildings weight, hydraulic fracturing in areas of flooding, etc. However, the most important and destructive cracks are a direct consequence of regional subsidence that occurs in Mexico City as a result of pumping water from the aquifer.

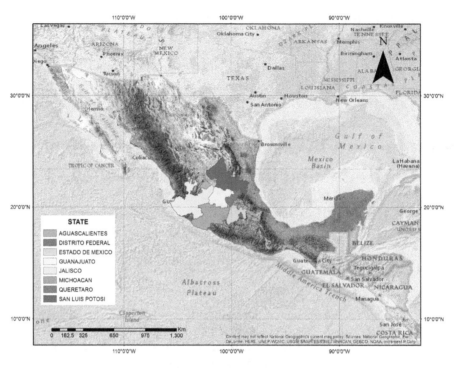

Figure 1. Spatial distribution of the phenomenon of soil cracking in Mexico.

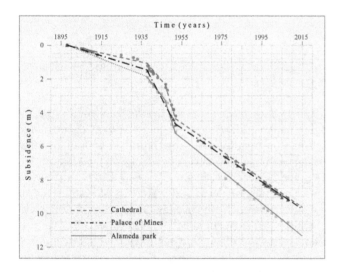

Figure 2. Evolution of subsidence for period 1898–2015.

2   Previous studies

The regional subsidence of Mexico basin was studied in the late nineteenth century by Téllez Pizarro (1899). In 1925, Roberto Gayol informed the Society of Engineers and Architects of Mexico that Mexico City was subsiding and that the probable cause was the "disturbance that the drainage of subsoil water was producing, in the bottom of Mexico basin". Gayol based his assertion on some surveys in Mexico City downtown and in the Texcoco Lake (1891).

Between 1920 and 1930, José A. Cuevas gave his support to the ideas of Gayol and asked Nabor Carrillo to study the influence of the pumping from water wells on subsidence. Carrillo (1948) explained the subsidence using newly developed techniques of Soil Mechanics and established that the cause of this phenomenon was the consolidation of clays due to the increase in effective stress caused by the drawdown of interstitial water pressure.

In 1952, the systematic study of subsoil and the first piezometric measurements made by Sandoval, Hiriart and Marsal, corroborated the findings of Nabor Carrillo. Also very meaningful were the investigations by Zeevaert and the periodical surveys and piezometric measurements performed since 1953 by Hydrological Commission of Mexico Basin Valley (CHCVM, SRH) and subsequently by the Water System of Mexico City (SACMEX).

Soil cracking is not exclusive of Mexico City, this phenomenon has spread to other states of the country (Estado de Mexico, Querétaro, Morelia, Silao, Aguascalientes, between others). In Fig. 1, the states of the country where the cracking phenomenon has been detected are indicated.

In Mexico City, due to increasing subsidence, soil fissuring that previously only occurred in the dry zone of former Lake of Texcoco has extended to the transition geotechnical zone of the city (0 to 20 m of lacustrine clay). This can be attributed to differential settlements between zones of soft and hard soils, damaging buildings and urban services. The problem of soil cracking in Mexico City Valley is of large magnitude and will require continued attention in the future.

3 Subsidence evaluation

From the available historical documents, it has been possible to reconstruct the history of subsidence of the old downtown area of Mexico City. Analyzing graphs in Fig. 2 and in particular, those corresponding to the Cathedral and Alameda Central, it can be concluded that the history of subsidence has gone through four stages during the twentieth century and in the first decade of the current century while, in the Palace of Mines, the subsidence history has gone through only three stages. These stages correspond to different rates of water extraction.

Figure 3a shows the shaded relief model (SRM) built from topographical data obtained in 1891 by Gayol. The model corresponds to the area of the old "traza" (center) of Mexico City. The so-called "corographic" map of Federal District by Manuel Fernandez Leal (1899) was superimposed on the model. Figure 3b shows the shaded relief model (SRM) constructed from a dense cloud of points obtained by scanning using the Light Detection and Ranging (LiDAR) technique by the National Institute of Geography Statistics and Computing (INEGI, 2010), corresponding to the old central area of Mexico City. Figure 4 shows the remnants of the buildings of the ceremonial precinct of the Mexica Empire and the trace of the Historic center of Mexico City superimposed on a shaded relief model, obtained by INEGI (2010).

Figure 5 shows an assessment of the magnitude of regional subsidence, both in the west-east direction along Tacuba Avenue and in the North-South direction along José María Pino Suárez Avenue, in the historical center of Mexico City for the 1891–2010 period (119 years).

4 Soil cracking associated to regional subsidence

Cracks originating in areas of abrupt transition between compressible and hard soils, are a direct consequence of the regional subsidence of Mexico Valley as a result of pumping water from the local aquifer.

These cracks are very destructive; they present a step towards the area of greatest settlement and are generally parallel to the elevation contours at the foot of the mountains, hills or outcrops of rock or rigid materials (Fig. 6).

The cracking phenomenon associated with regional subsidence in areas of abrupt transition takes a particular importance in the area surrounding Santa Catarina range, Peñón Viejo-Peñón Márquez hill, Guerrero hill, Peñón de los Baños hill, Chimalhuacán hill, Xico hill and in the northern limit of the Chichinautzin range. Figure 7 shows the principal zones where soil is most suceptible to cracking as defined from the gradient of variation of clay layers thickness and recorded differential settlements.

Photographs of Fig. 8 show the destructive capacity of soil cracking associated with regional subsidence.

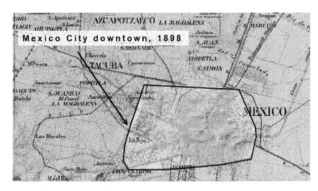

a) Old map (1898)

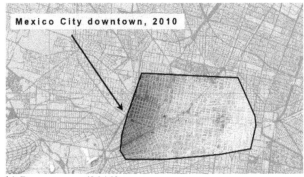

b) Recent map (2010)

Figure 3. Maps of Mexico City downtown area over a shading relief model.

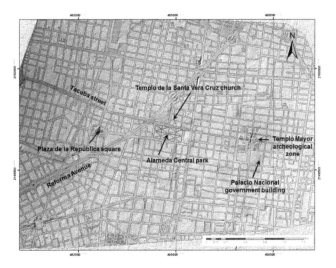

Figure 4. Mexica constructions remnants in Mexico City downtown area over a raster model obtained from LiDAR data.

5 Conclusions

The effort undertaken by different groups and in particular by the Geocomputing Laboratory of the Engineering Institute, UNAM to achieve a satisfactory evaluation of the phenomenon of subsidence, as well as other aspects of geotechnical problems such as soil fissuring has given useful and

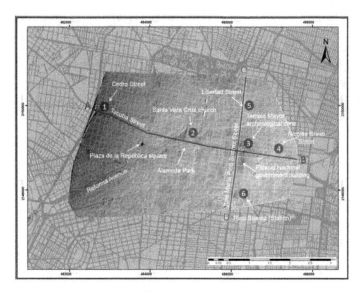

a) Location of profiles

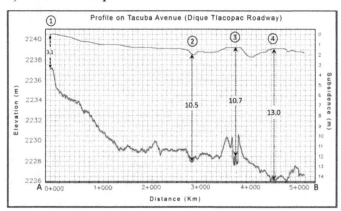

b) Profile along Tacuba Avenue (A-B)

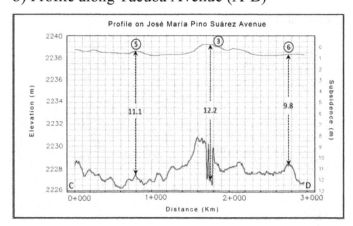

c) Profile along José María Pino Suárez Avenue (C-D)

| ——— Surface Profile, 1891 | ——— Surface Profile, 2010 |

Figure 5. Location map and topographic profiles along Tacuba Avenue and J. M. Pino Suárez Avenue.

Figure 6. Crack with vertical step associated with the subsidence in abrupt transition zone.

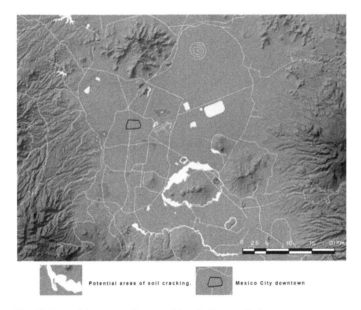

Figure 7. Main zones of Mexico City Valley subject to surface cracking (yellow strips).

Figure 8. Effects of differential soil subsidence and cracks on constructions.

promising results, but this is only the first stage of a huge job to be performed consistently in the future.

In a short period, it has been possible to update and expand the database of sites where cracks are found and to categorize them according to their mechanism of generation and/or propagation; further efforts will be necessary to define and extend the techniques to control the cracking phenomenon and mitigate its harmful effects.

Acknowledgements. The authors acknowledge the support of different institutions: Mapoteca Manuel Orozco y Berra, Palacio de Minería library, CONACYT, DGAPA-UNAM, IPN, GDF, CNA SACMEX and other institutions and agencies for the realization of this work. The valuable contributions from companies and geotechnical consultants who kindly gave to the authors access to their collections of subsoil explorations are also recognized.

References

Auvinet, G.: Agrietamiento de suelos, Proceedings, XXIV Reunión Nacional de Mecánica de Suelos, special volume, 299–313, Mexico, 2008.

Carrillo, N.: Influence of Artesian Wells on the Sinking of Mexico City, Proceedings of 2th Congress of International Soil Mechanics and Foundation Engineering, Vol. VI, 156–159, Rotterdam, 1948.

Fernandez Leal, M.: Carta Corográfica del Distrito Federal, Secretaría de Fomento, Official map, Mexico, 1899.

Gayol, R.: Plano General de las atarjeas conforme al proyecto de desagüe y saneamiento de la Ciudad de México, official map, Mexico, 1891.

Gayol, R.: A study of disturbances caused by draining out water from the subsoil of the Valley of Mexico, First part, Revista Mexicana de Ingenieria y Arquitectura, Vol. III, No. 3, 96–132; Second part, Revista Mexicana de Ingenieria y Arquitectura, Vol. III, No. 8, 507–559, Mexico, 1925 (in Spanish).

Instituto Nacional de Estadística, Geografía e Informática (INEGI): Data of LiDAR technology in Mexico City, data bases, Mexico, 2010.

Sistema de Aguas de la Ciudad de México: Set of topographic data of Mexico City, from 1977 to 2008, Mexico, 2008.

Téllez-Pizarro, A.: Apuntes acerca de los cimientos de los edificios en la ciudad de México, Reprinted in the commemorative volume of the XXV anniversary of the Mexican Society for Soil Mechanics (1982), Mexico, 1899.

4

Combination with precise leveling and PSInSAR observations to quantify pumping-induced land subsidence in central Taiwan

C. H. Lu[1], C. F. Ni[1], C. P. Chang[2,3], J. Y. Yen[4], and W. C. Hung[5]

[1]Graduate Institute of Applied Geology, National Central University, Zhongli District, Taoyuan City, Taiwan
[2]Center for Space and Remote Sensing Research, National Central University,
Zhongli District, Taoyuan City, Taiwan
[3]Institute of Geophysics, National Central University, Zhongli District, Taoyuan City, Taiwan
[4]Department of Natural Resources and Environmental Studies, National Dong Hwa University,
Shoufeng, Hualien, Taiwan
[5]Green Environmental Engineering Consultant Co. Ltd., Hsinchu, Taiwan

Correspondence to: C. H. Lu (fox52600@gmail.com)

Abstract. Choushui River Fluvial Plain (CRFP) is located in the western central Taiwan, where the geomaterials are composed of alluvial deposits. Because the CRFP area receives highly variable rainfall in wet and dry seasons, the groundwater becomes the main resource of residential water. The precise leveling monitoring from 1970s indicated that the coastal areas of CRFP had been threatened by serious pumping-induced land subsidence. On the basis of relatively accurate measurements of precise leveling measurements, we used cokriging technique to incorporate a number of InSAR images to quantify the surface deformation in CRFP. More specifically, the well-developed Persistent Scatterer InSAR (PSI) was employed to process 34 Envisat images (2005–2008) and the results of PSI was then used for improving the spatial resolution of data from precise leveling. The results of cokriging estimation indicate whether the rate or the area of the land subsidence slows down gradually from 2005 to 2008. The subsidence in the northern part of CRFP was influenced by the groundwater decline in aquifer III, and the southern part was influenced by groundwater decline in aquifer II and III. The cokriging estimation was also comparable with continuous GPS data, and their correlation coefficient is 0.9603 and the root mean square is $10.56\,\mathrm{mm\,yr}^{-1}$.

1 Introduction

Groundwater resource management is a notable issue due to growing population and rapid urban development, especially in areas where surface water is limited or shortage. Many cities, such as Shanghai, Las Vegas, and Bangkok had faced serious land subsidence because of unlimited groundwater withdrawal. To identify the land subsidence range effectively the geodetic techniques with different scales record the behaviors of surface deformation. Precise leveling, Global Positioning System (GPS) and Interferometric Synthetic Aperture Radar (InSAR) have been the most powerful techniques

in geodesy (Galloway et al., 1998; Massonnet and Feigl, 1998; Galloway and Burbey, 2011; Hung et al., 2010).

In the central Taiwan, Choushui River fluvial plain (CRFP) faced the serious problem of land subsidence by poignant groundwater drawdown since 1970's. Central Geological Survey (CGS) has investigated several hydrologic and geologic studies. Thereafter, Water Resources Agency (WRA) have been collecting geodetic data such as precise leveling, GPS and compaction monitoring well (CMW) to monitoring the rate of land subsidence in CRFP (Central Geological Survey, 1999; Liu et al., 2001, Liu et al., 2004; Hwang et al., 2008). Hung et al. (2011) apply the Persistent Scatter InSAR (PSI), which was developed by Hooper et al. (2004,

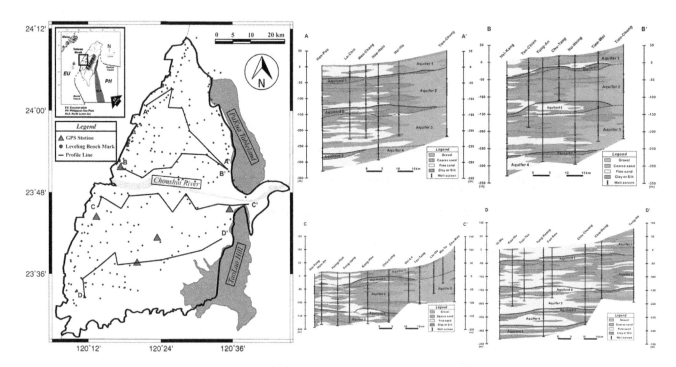

Figure 1. The study area. The black circle is the leveling benchmarks. The red triangle is continuous GPS station. The black lines indicate the location of hydrogeological profile. The right side of figure shows the hydrogeological profiles. The hydrogeological profiles were modified from Central Geological Survey (1999).

2007) and Hooper (2008), for identifying PS pixels and extracting more than 100 pixels km^{-2} in CRFP from 2006 to 2008. In this paper, the main aim is to integrate precise leveling data and PSI result with geostatistical method, variogram and cokriging, for retaining the accuracy of precise leveling and the spatial resolution of PSI and estimating the surface deformation, then comparing cokriging estimation with the difference of yearly groundwater level, at last verifying the cokriging estimation with the observed GPS data from 2005 to 2008 in CRFP.

2 Background of study area

2.1 Geological setting

Taiwan is located at the boundary between the Philippine Sea Plate (PSP) and Eurasian Plate (EUP), where the still ongoing collision started about 5–7 Ma. The PSP move toward the northwest to EUP at a rate of 82 mm yr^{-1}. Choushui River Fluvial Plain (CRFP) is one part of the coastal plain in the western Taiwan. This area covers about 2000 km^2 and the boundary is enclosed by Pakua tableland and Taolao hill to the east, Taiwan Strait to the west, Wu River to the north and Putzu River to the south. Choushui River is the longest river in Taiwan and flows cross Central Range, Western Foothill and Coastal Plain then in to Taiwan Strait. The range of the sedimentary thickness is from 750 to 3000 m and the grain size become finer from the east to the west. The materials

of the sedimentary contain clay, fine sand, coarse sand, and gravel. According the drilling core in the depth of 300 m, the subsurface hydrogeology can be divided into eight overlapping sequences. There are four marine sequences (aquitard I–IV) and four non-marine sequences (aquifer I–IV). The right side of Fig. 1 shows four hydrogeological profiles in CRFP (Central Geological Survey, 1999; Liu et al., 2001; Liu et al., 2004).

2.2 Observation data

2.2.1 Precise leveling

The land subsiding surveys determined by precise leveling in CRFP has recorded since 1975. These data acquired from Water Resource Agency (WRA) are based on the first or second order standard of leveling procedures proposed by Ministry of the Interior (MOI). The specifications demand that any loop misclosure should be below 3 mm $\sqrt{K}$, where K is the distance between two neighboring benchmarks in km. In this study 298 precise leveling benchmarks were be used from 2005 to 2008 (Fig. 1).

2.2.2 Persistent Scatterers InSAR (PSI)

Due to the area of CRFP covers two SAR image modes (Track 232, Frame 3123 and Track 232, Frame 3141), 34 Envisat radar images (16 images from Track 232, Frame 3123; 18 images from Track 232, Frame: 3141) from European

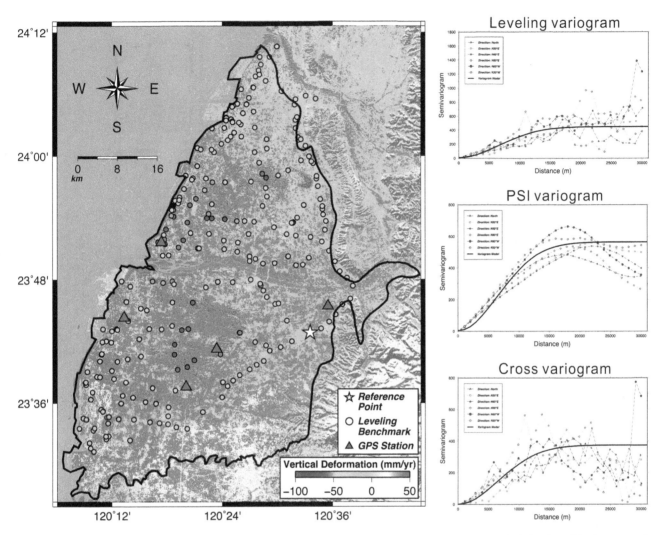

Figure 2. The left side of the figure is the vertical deformation of precise leveling and PSI between 2005 and 2008. The cold color means the land subsidence. The right side of the figure from top to bottom are precise leveling variogram, PSI variogram and cross-variogram between precise leveling and PSI. Every variogram has 6 directions and a Gaussian model.

Space Agency (ESA) were used in this study. Envisat radar images were taken from May 2005 to September 2008. The PSI was processed with StaMPS/MTI (Hooper et al., 2007; Hooper, 2008). The topographic effect was removed by the DEM from NASA's SRTM mission and the precise orbit parameters from Delft Institute for Earth-Oriented Space Research were used to correct the orbit errors.

2.2.3 Groundwater level

Since 1992 WRA installed hydrologic monitoring well evenly, and dug the depths ranging from 200 to 300 m in CRFP. According to the hydrogeology profiles (Fig. 1) CRFP can be divided into four aquifers in depth of 300 m. In this study we collect 98 monitoring well screens in the aquifer I, 97 screens in the aquifer II, 44 screens in the aquifer III, and 21 screens in the aquifer IV. The yearly average of groundwa-

ter level from 2005 to 2008 was interpolated by using kriging method.

3 Methodology

The grid of 9867 active cells is set up in the study area and the size of each cell is 500 m × 500 m. Two geostatistical methods, variogram and cokriging, are adopted for analyzing and combining the precise leveling and PSI result. The FORTRAN codes gamv and cokb3d in GSLIB (Deutsch and Journel, 1998) make use of generating experimental variograms and cokriging respectively.

3.1 Variogram analysis

An experimental semivariogram, $\gamma'(h)$, is defined as half of the average squared difference between two values of an uni-

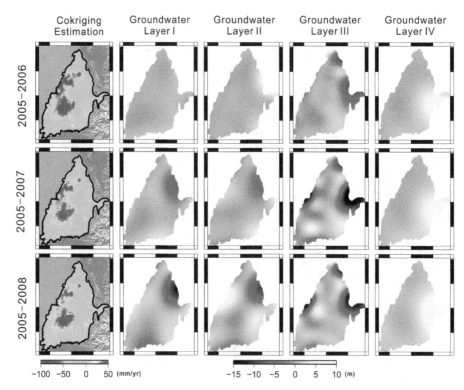

Figure 3. First column is the cokriging estimation in the three intervals (2005–2006, 2005–2007 and 2005–2008). The rest columns are the difference of yearly average of groundwater level in four aquifers.

variate approximately separated by vector h and can be written as:

$$\gamma'(h) = \frac{1}{2N(h)} \left\{ \sum_{i=1}^{N(h)} [z(u_i + h) - z(u_i)]^2 \right\}. \tag{1}$$

where $N(h)$ is the number of pairs; $z(u_i)$ and $z(u_i + h)$ are the value of precise leveling or PSI at the location u_i and $u_i + h$ of the pair i; and the vector h can be specified with particular directions and lag distance.

3.2 Cokriging interpolation

The cokriging estimate is a linear combination of both primary and secondary data value and minimizes the variance of the estimation error by exploiting the cross-correlation between two variables (Isaaks and Srivastava, 1989). The precise leveling is set to the primary variable, $L(u_{a_1})$, the PSI results is sets to the secondary variable, $P(u'_{a_2})$, and the cokring estimation, $Z_{\text{COK}}(u)$, can be written as:

$$Z_{\text{COK}}(u) = \sum_{a_1=1}^{n_1} \lambda_{a_1}(u) \cdot L(u_{a_1}) + \sum_{a_2=1}^{n_2} \lambda'_{a_2}(u) \cdot P(u'_{a_2}), \tag{2}$$

where λ_{a_1} are the weights applied to the n_1 precise leveling data, u_{a_1} are the n_1 locations of precise leveling benchmarks, λ'_{a_2} are the weights applied to the n_2 PSI results, u'_{a_2} are the n_2 locations of PSI pixel.

4 Results and discussion

4.1 Variogram analysis

15 Envisat interferograms from Track 232, Frame 3123 and 17 Envisat interferogram form Track 232, Frame 3141 were used to estimate the displacement of the radar line-of-sight (LOS). At the same period 298 precise leveling data recorded the vertical deformation in CRFP. The left side of Fig. 2 shows the vertical displacement rate of precise leveling and PSI from 2005 to 2008. The cold color represents the land subsidence relative to the reference point and the warm color mean surface uplift. The most subsidence rate of PSI result almost reaches $60 \, \text{mm yr}^{-1}$ and the most subsidence rate of precise leveling data reaches $80 \, \text{mm yr}^{-1}$ in the center of southern CRFP. The lag distance of 1 km was used to analyze the semivariograms and cross semiveriogram in 6 horizontal directions (right side of Fig. 2). The red triangle is north direction, the green square is N–30° E direction, the blue star is N–60° E, the yellow diamond is N–90° E, the purple circle is N–60° W, the gray hexagon is N–30° W, and the black solid line is the omnidirectional semivariogram fitted by Gaussian model.

4.2 Cokriging estimation and groundwater level

The cokriging estimations in the three intervals display on first column of Fig. 3. The negative values of the rate (cold

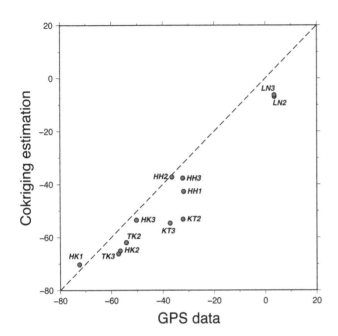

Figure 4. The comparison of cokriging estimation with GPS data. The code name of GPS station is written in text.

color) mean the area of land subsidence. In 2005–2006 the highest rate of land subsidence reaches $-99.98\,\mathrm{mm\,yr^{-1}}$, in 2005–2007 is $-93.17\,\mathrm{mm\,yr^{-1}}$, and in 2005–2008 is $80.97\,\mathrm{mm\,yr^{-1}}$. There are $440\,\mathrm{km^2}$ (1760 cells) where the subsidence rate is higher than $-50\,\mathrm{mm\,yr^{-1}}$ in 2005–2006, $340\,\mathrm{km^2}$ (1360 cells) in 2005–2007, and $316.5\,\mathrm{km^2}$ (1266 cells) in 2005–2008. Whether the rate or the area of the land subsidence slows down gradually from 2005 to 2008.

In general the subsidence in the northern part of CRFP was influenced by groundwater decline in aquifer III, and the southern part was influenced by groundwater decline in aquifer II and III. In 2005–2006 the land subsidence is most serious however the groundwater decline is slightest. It reveals that the elastic compaction would not be a major factor to induce the land subsidence during this interval. In 2005–2007 and 2005–2008, the regional groundwater decline corresponds the same location of land subsidence. Especially in aquifer III the depth is about 175–300 m, and the regional decline of groundwater level distributes in CRFP obviously. It may reflect the pumping-water location or behavior in aquifer III. The geo-material near eastern CRFP contains the gravel higher than 90 % and the permeability of the gravel is between 1.0×10^{-4} and 9.9×10^{-4} (Central Geological Survey, 1999). It gave a good path for groundwater flow and supply. Therefor even the groundwater level near Pakua Tableland draws down almost 15 m, the surface deformation reveals slight.

The five continuous CPS stations in CRFP were acquired from WRA. Two stations, Hsi-Kang (HK) and Hsin-Hsing (HH) recorded the vertical displacement from 2005 to 2008 and the others stations, Tu-Ku (TK), Lin-Nei (LN) and Ko-

Tso (KT) recorded the vertical displacement from 2006 to 2008. Figure 4 shows the direct comparisons of cokriking estimation with GPS data. The correlation coefficient between the cokriging estimation and GPS data within the same cell is 0.9603 and the RMS is $10.56\,\mathrm{mm\,yr^{-1}}$. Thus the cokriging estimation is a reliable data for detecting the land subsidence in this study.

To sum above results, the mechanism of land subsidence in CRFP is quite complication. The models not only include elastic and inelastic compaction behaviors but also should consider the characteristics of different aquifers.

This research was partially supported by the Ministry of Science and Technology of the Republic of China (contracts MOST 103-2221-E-008 -049-MY3 and MOST 102-2116-M-008-010).

References

Central Geological Survey: Project of groundwater monitoring network in Taiwan during first stage – research report of Choushui River alluvial fan, Water Resources Bureau press, Taipei County, 383 pp., 1999.

Deutsch, C. V. and Journel, A. G.: Geostatistical software library and user's guide, Oxford University Press, Oxford, UK, New York, USA, 369 pp., 1998.

Galloway, D. L. and Burbey, T. J.: Review: Regional land subsidence accompanying groundwater extraction, Hydrogeol. J., 19, 1459–1486, doi:10.1007/s10040-011-0775-5, 2011.

Galloway, D. L., Hudnut, K., Ingebritsen, S., Phillips, S., Peltzer, G., Rogez, F., and Rosen, P.: Detection of aquifer system compaction and land subsidence using interferometric synthetic aperture radar, Antelope Valley, Mojave Desert, California, Water Resour. Res., 34, 2573–2585, 1998.

Hooper, A.: A multi-temporal InSAR method incorporating both persistent scatterer and small baseline approaches, Geophys. Res. Lett., 35, L16302, doi:10.1029/2008GL034654, 2008.

Hooper, A., Zebker, H., Segall, P., and Kampes, B.: A new method for measuring deformation on volcanoes and other natural terrains using InSAR persistent scatterers, Geophys. Res. Lett., 31, 1–5, 2004.

Hooper, A., Segall, P., and Zebker, H.: Persistent scatterer interferometric synthetic aperture radar for crustal deformation analysis, with application to Volcán Alcedo, Galápagos, J. Geophys. Res., 11, B07407, doi:10.1029/2006JB004763, 2007.

Hung, W. C., Hwang, C., Chang, C. P., Yen, J. Y., Liu, C. H., and Yang, W. H.: Monitoring severe aquifer-system compaction and land subsidence in Taiwan using multiple sensors: Yunlin, the southern Choushui River Alluvial Fan, Environ. Earth Sci., 59, 1535–1548, 2010.

Hung, W. C., Hwang, C., Chen, Y. A., Chang, C. P., Yen, J. Y., Hooper, A., and Yang, C. Y.: Surface deformation from persistent scatterers SAR interferometry and fusion with leveling data: A case study over the Choushui River Alluvial Fan, Taiwan, Remote Sens. Environ., 115, 957–967, 2011.

Hwang, C., Hung, W. C., and Liu, C. H.: Results of geodetic and geotechnical monitoring of subsidence for Taiwan High Speed Rail operation, Nat. Hazards, 47, 1–16, doi:10.1007/s11069-007-9211-5, 2008.

Isaaks, E. H. and Srivastava, R. M.: An introduction to applied geo-statistics, Oxford University press, Oxford, UK, 561 pp., 1989.

Liu, C. W., Lin, W. S., Shang, C., and Liu, S. H.: The effect of clay dehydration on land subsidence in the Yun-Lin coastal area, Taiwan, Environ. Geol., 40, 518–527, 2001.

Liu, C. H., Pan, Y. W., Liao, J. J., Huang, C. T., and Ouyang, S.: Characterization of land subsidence in the Choshui River alluvial fan, Taiwan, Environ. Geol., 45, 1154–1166, 2004.

Massonnet, D. and Feigl, K.: Radar interferometry and its application to changes in the Earth's surface, Rev. Geophys., 36, 441–500, 1998.

5

Surface deformation on the west portion of the Chapala lake basin: uncertainties and facts

M. Hernandez-Marin[1], J. Pacheco-Martinez[1], J. A. Ortiz-Lozano[1], G. Araiza-Garaygordobil[1], and
A. Ramirez-Cortes[2]

[1]Departamento de Geotecnia e Hidráulica, Universidad Autónoma de Aguascalientes, Aguascalientes, Mexico
[2]Doctorado en Ciencias de los Ámbitos Antrópicos, Universidad Autónoma de Aguascalientes, Aguascalientes,
México

Correspondence to: M. Hernandez-Marin (mhernandez@correo.uaa.mx)

Abstract. In this study we investigate different aspects of land subsidence and ground failures occurring in the west portion of Chapala lake basin. Currently, surface discontinuities seem to be associated with subsiding bowls. In an effort to understand some of the conditioning factors to surface deformation, two sounding cores from the upper sequence (11 m depth) were extracted for analyzing physical and mechanical properties. The upper subsoil showed a predominant silty composition and several lenses of pumice pyroclastic sand. Despite the relative predominance of fine soil, the subsoil shows mechanical properties with low clay content, variable water content, low plasticity and variable compressibility index, amongst some others. Some of these properties seem to be influenced by the sandy pyroclastic lenses, therefore, a potential source of the ground failure could be heterogeneities in the upper soil.

1 Introduction

Mexico is one of the most affected countries by anthropogenic subsidence. Here, conditioning factors such as topography, hydrostratigraphy, depositional environments, and others are combined with large periods of groundwater withdrawal, resulting in surface deformation in the form of subsidence (vertical deformation), or faulting and fissuring (horizontal or combined horizontal and vertical deformation). The zone known as the Transmexican Volcanic Belt is perhaps the most affected area in the country, mostly due to the postdepositional processes resulting in fine sediments that potentially contain smectitic clays, usually derived from volcanic materials deposited in lacustrine and currently urbanized valleys such as Mexico City, Queretaro, Chapala, and others.

In this work, new findings regarding surface deformation on the west portion of the Chapala lake basin are analyzed and discussed, and previous measured land subsidence previously measured and mapped faults are recalled. Novel information includes the description and geotechnical characterization of sediments from two cores of 11 m from of the upper sedimentary sequence. Additionally, uncertainties regarding

the factors that may condition or trigger surface deformation in this particular area are discussed.

2 Conditions of Chapala lake basin

The lake of Chapala and its potential flooding area is bounded by ranges. The elevation difference between the top of the mountains and the lacustrine plain is close to 500 m. This difference is more significant along the western portion of the basin. Due to the potential for tourism on the lake shore, several urban centers around the lake have experienced significant growth in the past few decades, one of them is the community of Jocotepec on the west with close to 37 700 inhabitants (INEGI, 2010). The increasing population in all communities around the lake has triggered the demand of water to satisfy agricultural and municipal needs in the last decades. Usually people from the valley prefer groundwater because is cleaner and more accessible than the water from the lake.

In the basin of Chapala lake, as most inside the Transmexican Volcanic Belt, tectonism, volcanism and sedimen-

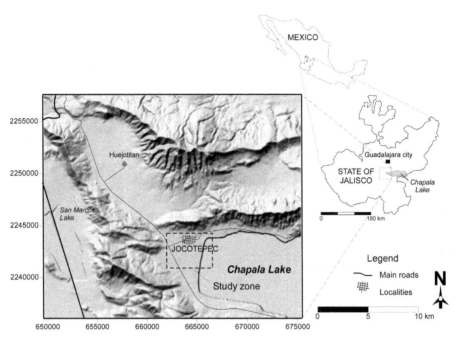

Figure 1. Location of the study area in the west part of the Chapala lake. From Hernandez-Marin et al. (2014).

tation occurred simultaneously during tertiary an quaternary ages. Regionally, this basin is seismically active with recent shallow earthquakes. This basin presents its longest side in an east-to-west direction, which is collinear to the main regional fault system. The ranges bounding the lake plain are mainly volcanic and plutonic, formed by basalts, rhyolite and pyroclastic material (Rosas-Elguera and Urrutia-Fucugauchi, 1998); and the sedimentary sequence of the plain presents alluvial, lacustrine and volcanic deposits (ash) including sands, silts and clays (Zarate-del-Valle and Simoneit, 2005). Figure 1 depicts the location and main characteristics of the study area.

Land subsidence in this part of Mexico is recent since the first reports of surface deformation are from the early 90's, however, it was not documented until 2007 by Castillo-Aja and Valdivia-Ornelas (2007). Locally, the first cores for physically analyzing sediment samples were used to determine mineralogy of the fine sediments and the rate of sedimentation (Fernex et al., 2001). Through the soil samples from these cores it was observed that the main minerals of the clay portion are halloysite, hectorite and illite, and the sand grains are mainly composed of vitreous volcanic ash (pumice) (Fernex et al., 2001).

3 Monitoring campaign of surface deformation

Land subsidence and surface faulting have recently been a recurrent problem in the urban area of Jocotepec and suburbs. The problems associated to ground deformation include damage to buildings, roads and other constructions. In 2012, as part of a project to quantify subsidence, identify

ground failures and determine damage to infrastructures, it was found that for a period of 8 months (April–November), a maximum deformation of 7.16 cm was measured with a calculated rate of 0.89 cm month^{-1}. Additionally, two zones of the surveyed area showed uplifting close to 2 cm. Four alignments of surface faults with preferential direction SW-NE are affecting several constructions (Hernandez-Marin et al., 2014). Land subsidence, fault alignments and other characteristics are shown in Fig. 2.

4 Sampling and analyses

In order to determine the stratigraphy of the upper sequence (< 11 m) inside the subsiding zone, and determine its main geotechnical properties, two cores of 11 m depth were sampled. Analyses included stratigraphy determination, sediment description and geotechnical testing in accordance with ASTM standards.

4.1 Stratigraphy of the analyzed sequence

Observations from samples in both cores indicate a predominance of green-olive fine sediments with minor layers of brown fine sediments and alternating gray-light and gray-dark pyroclastic sand. The thickness of the sandy layers is variable in both cores, in core 1 (northernmost) is more than 2 m, but in the core 2 (drilled at 287 m to the southeast from the core 1) the thickness is close to 20 cm only. Figure 3 shows the observed stratigraphy with descriptions of the sediments from the sampled cores.

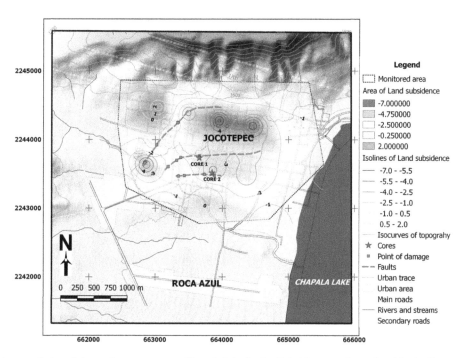

Figure 2. General characteristics of the study zone at west Chapala lake basin. Subsidence registered and fault alignments observed in 2012 are present. Also the two cores for geotechnical analysis of the subsoil are indicated. Units of area and isolines of land subsidence are in centimeters.

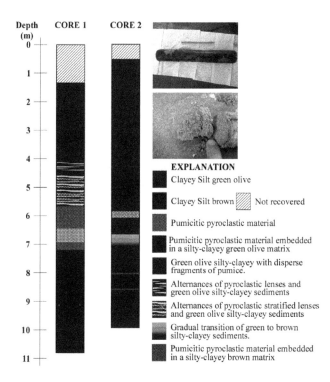

Figure 3. Described stratigraphy of both cores. Sample photographs of the predominant green-olive sediments, along with an example of a pumice sandy layer are shown.

4.2 Grain-size, water content, consistency limits and compressibility

Even though the sequence appears to be macroscopically dominated by fine sediments, the grain-size of sediments passing sieve # 10 (≤ 2 mm) using the hydrometer technique, demonstrated the presence of fine sands, particularly in core 1. As observed in Fig. 4 the clay content is very low in core 1, in comparison with the high content found at some depths in core 2, were proportion is as high as 37 % between 1 and 2 m (Fig. 5). Gravimetric water content is low compared with soil samples from other volcano-lacustrine basins such as Mexico City. Maximum water content is close to 215 %, although most of the water content measurements are no higher than 150 %, (in Chalco basin, adjacent to Mexico City, for instance, water content can reach 350 %, Hernandez-Marin, 2003). Consistency limits are low with liquid limits lower than 50 %. Finally, to obtain the compressibility properties, consolidation test were performed from unaltered samples. In general, compressibility index is lower than 0.2 (only one determination was close to 0.4), as shown in Figs. 4 and 5.

5 Uncertainties and discussion

Many questions remain unanswered regarding the process of subsidence in the locality of Jocotepec and in general in the Chapala lake basin. Some of these questions are for instance: is the groundwater depletion the unique process causing land subsidence? As demonstrated in Hernandez-Marin

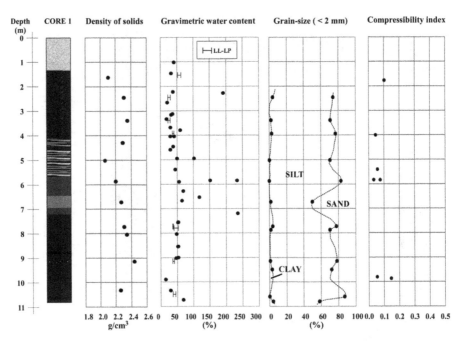

Figure 4. Core one and some of its geotechnical properties. LL-LP represents the values of liquid and plastic Atteberg limits.

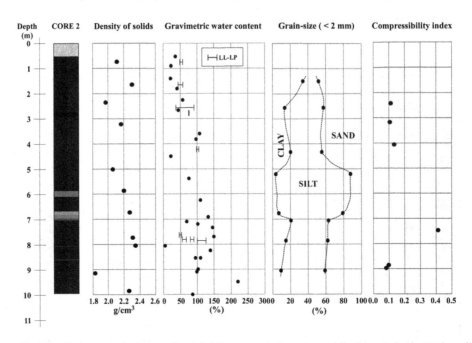

Figure 5. Core two and some of its geotechnical properties. LL-LP represents the values of liquid and plastic Atteberg limits.

et al. (2014), subsidence bowls do not coincide with groundwater cones of depression, even though some cones of depression are remarkable due to intensive pumping for irrigation at the south of the study area. Even though groundwater levels are shallow in most of the area, since as observed during drilling of cores, the groundwater table was no deeper than 2 m. Another question is regarding the tectonic component: is this stress contributing to surface deformation? The zone is tectonically active and is regionally dominated by a

triple junction rift–rift–rift (Pacheco et al., 1999). The last registered earthquake occurred in 17 May 2012, of magnitude 4.4 and the epicenter was located 6 km from the study zone, as reported by the National Seismological Service, indicating tectonic activity and potential stressing. Fault alignments tend to follow isolines of subsidence, as presented in Fig. 2, however that tendency is only suggestive; therefore, the real relationship between faults and subsidence is still under investigation. The origin of faults is also unknown.

A first hypothesis is that differential subsidence due to heterogeneities in the lacustrine sequence is the possible cause. However, future work must corroborate this hypothesis.

6 Conclusions

New information regarding the geotechnical properties of the upper sequence of the west portion of Chapala lake basin is presented and discussed, and the previous study to quantify subsidence and delineate ground failures is revisited. Two cores for the geotechnical analysis of sediments show predominance of silt and fine sand, whereas geotechnical properties such as water content, consistency limits, and compressibility index are low compared with samples from other volcano-lacustrine basins such as Mexico basin. Differences in the stratigraphy, in particular those related to pumice layers, less compressible than silty and clayey layers, are a potential cause to differential subsidence and surface faulting.

Acknowledgements. This work was financed by the Autonomous University of Aguascalientes with the internal project: PIIC13-4N. The authors are also thankful for the reviewers comments, which improved the quality of this manuscript.

References

Castillo-Aja, M. R. and Valdivia-Ornelas, L.: Amenazas por agrietamiento en el Valle de Tesistán [Risks due to cracking in the valley of Tesistan], Universidad de Guadalajara, 2007

Fernex, F., Zarate-del-Valle, P., Ramirez-Sanchez, H., Michaud, F., Parron, C., Dalmasso, J., Barci-Funel, G., and Guzman-Arroyo, M.: Sedimentation rates in Lake Chapala (western Mexico): possible active tectonic control, Chem. Geol., 177, 213–228, 2001.

Hernandez-Marin, M.: Estudio del comportamiento hidromecánico de sedimentos vulcanolacustres durante el proceso de consolidación [Study of the hydromechanic behavior of the vulcanolacustrine sediments during the process of consolidation], Master Dissertation, National Autonomous University of Mexico, Mexico, 2003.

Hernandez-Marín, M., Pacheco-Martínez, J., Ramírez-Cortés, A., Burbey, T. J., Ortíz-Lozano, J. A., Zermeño-De-Leon, M. E., Guinzberg-Velmont, J., and Pinto-Aceves, G.: Evaluation and analysis of surface deformation in west Chapala basin, central Mexico, Environ. Earth Sci., 72, 1491–1501, 2014.

INEGI: Censo de población y vivienda [Census of population and housing], Mexico, 2010.

Pacheco, J. F., Montera-Gutierrez, C. A., Delgado, H., Singh, S. K., Valenzuela, W., Shapiro, N. M., Santoyo, M. A., Hurtado, A., Barron, R., and Gutierrez-Miguel, E.: Tectonic significance of an earthquake in the Zacoalco half-graben, Jalisco, Mexico, J. S. Am. Earth Sci., 12, 557–565, 1999.

Rosas-Elguera, J. G. and Urrutia-Fucugauchi, J.: Tectonic control of the volcano-sedimentary sequence of the Chapala graben, western Mexico, Int. Geol. Rev., 40, 350–362, 1998.

Zarate-del-Valle, P. and Simoneit, B.: La generación de petróleo hidrotermal en sedimentos del Lago Chapala y su relación con la actividad geotérmica del rift Citala en el estado de Jalisco, México [Generation of hidrotermal oil into the Chapala Lake sediments and its relationship with the geotermal activity of the Citala rift in the state of Jalisco, Mexico], Revista Mexicana de Ciencias Geológicas, 22, 358–370, 2005.

6

Groundwater-abstraction induced land subsidence and groundwater regulation in the North China Plain

H. Guo[1], L. Wang[2], G. Cheng[1], and Z. Zhang[1]

[1]China Institute of Geo-Environment Monitoring, Beijing, 100081, China
[2]Beijing Geo-Environmental Monitoring Station, Beijing, 100195, China

Correspondence to: H. Guo (hpguo79@gmail.com)

Abstract. Land subsidence can be induced when various factors such as geological, and hydrogeological conditions and intensive groundwater abstraction combine. The development and utilization of groundwater in the North China Plain (NCP) bring great benefits, and at the same time have led to a series of environmental and geological problems accompanying groundwater-level declines and land subsidence. Subsidence occurs commonly in the NCP and analyses show that multi-layer aquifer systems with deep confined aquifers and thick compressible clay layers are the key geological and hydrogeological conditions responsible for its development in this region. Groundwater overdraft results in aquifer-system compaction, resulting in subsidence. A calibrated, transient groundwater-flow numerical model of the Beijing plain portion of the NCP was developed using MODFLOW. According to available water supply and demand in Beijing plain, several groundwater regulation scenarios were designed. These different regulation scenarios were simulated with the groundwater model, and assessed using a multi-criteria fuzzy pattern recognition model. This approach is proven to be very useful for scientific analysis of sustainable development and utilization of groundwater resources. The evaluation results show that sustainable development of groundwater resources may be achieved in Beijing plain when various measures such as control of groundwater abstraction and increase of artificial recharge combine favourably.

1 Introduction

Land subsidence can be defined as the sinking of the ground surface with respect to surrounding terrain or sea level. Land subsidence is the result of consolidation of subsurface strata caused by natural causes such as tectonic motion or man-induced causes such as the withdrawal of groundwater, oil and gas (Abidin et al., 2013; Zeitoun and Wakshal, 2013; Galgaro et al., 2014; Papadaki, 2014). Due to increased groundwater abstractions to supply rapid industrial and urban development, subsidence is prevalent in the NCP (Fig. 1), and critically impacts sustainable economic and social development. The first occurrence of land subsidence in the NCP took place in Tianjin in the 1920s, and by the 1960s it became a severe hazard in this city. Since the 1980s, the land affected by subsidence extended from cities to rural areas. Ground cracks or fissures are observed in areas where large differential subsidence occurs. Commonly subsidence in the

NCP is slow, accumulative, and irreversible. Economic losses due to subsidence in the NCP have exceeded Yuan 200 billion (Ye and He, 2006). Due to groundwater overexploitation, the maximum subsidence reached up to 1.1 m in Beijing plain during the period 1955–2007 (Zhang et al., 2014). To control subsidence in the Beijing plain, reducing groundwater abstraction and increasing artificial recharge will be important goals for groundwater resources management.

This paper presents a brief summary of the subsidence mechanism in the NCP. A transient 3-D groundwater flow model of the Beijing plain is used to simulate 10 proposed future groundwater regulation scenarios. A multi-criteria fuzzy pattern recognition method is used to evaluate the simulated scenarios based on a set of specified groundwater management goals and constraints. Evaluation of the proposed regulation scenarios is discussed and one scenario is recommended.

Figure 1. Distribution of the depression cones of shallow and deep groundwater, and major land subsidence area in the NCP.

2 Regional hydrogeology

The NCP is located in the eastern part of China, has a total area of about 140 000 km². This region comprises the plain area of Beijing, Tianjin and Hebei Province and the plain area of Henan and Shandong provinces to the north of the Yellow River, which is a typical plain landscape with elevation less than 100 m a.s.l. The NCP comprises three distinct hydrogeological settings within the Quaternary sediments (Fig. 1): the piedmont plain and associated major alluvial fans, the alluvial plains with many abandoned river channels, and the coastal plain strip around the margin of the Bohai Sea (Foster et al., 2004). The main stratigraphy of this region consists of unconsolidated Quaternary sediments with thickness ranging from 200 m to more than 600 m, comprising unconsolidated pebble, gravel, sand, silt, and clay. Groundwater occurs principally in the pores of the unconsolidated Quaternary sediments. From the piedmont plain to alluvial and coastal plains in the middle-east portion of the NCP, the aquifer systems typically transistion from a single aquifer of sandy gravel to multiple aquifers of sand separated by silt or clay layers (Fig. 2).

3 Influencing factors of land subsidence

The factors affecting subsidence in the NCP are complex. Geologically, the aquifer system comprises aquifers and aquitards in the multi-layered Quaternary sediments, consisting of layers of principally coarse-grained materials, mainly unconsolidated sands, as well as layers of principally fine-grained deposits, which are predominantly clays and silts that have medium to high compressibility. The alternating sequences of sand and compressible clay layers are very susceptible to deformation due to increases in effective stress accompanying groundwater-level declines. Excessive withdrawal of underground fluids such as groundwater, oil, and gas is the important driving condition for subsidence. Shallow groundwater abstraction may be the main driving condition for subsidence in the piedmont area and associated alluvial fans (e.g. Beijing and Handan). Figure 1 shows that most serious subsidence areas in the NCP lie in the alluvial plain and in the coastal strip around the margin of the Bohai Bay, which are well correlated with the distribution of depression cones of deep groundwater. In the alluvial and coastal plains, subsidence is commonly caused by withdrawal of deep groundwater, but in a few areas where large volumes of shallow groundwater, and oil and gas are extracted, subsidence is also observed.

4 Groundwater regulation in Beijing plain

A transient, calibrated regional groundwater flow model was developed using MODFLOW for a simulation period of 1995–2011. The model consists of 9 model layers: 5 aquifers separated by 4 aquitards. The top aquifer is unconfined and the underlying four aquifers are semi-confined.

4.1 Formulation of regulation scenarios

During the past 30 years, overexploitation of the groundwater resource has led to severe land subsidence, which seriously affects economically and socially sustainable development in Beijing municipality. Some actions such as artificial recharge and reduction of groundwater abstraction could be taken to overcome the subsidence crisis. The major regulation measures include: reduce or stop the abstraction from emergency groundwater well fields, reduce the abstraction for irrigation and from well fields in the downtown area and self-supplied water supply wells, artificially increase groundwater recharge to aquifers in alluvial fans of rivers. Groundwater demand will increase with the growth of population. Ten groundwater regulation scenarios (Table 1) were designed based on the water demand and the capacity of water supply in Beijing plain. In scenario 1, the total groundwater abstraction of 2600×10^6 m³ a^{-1} would remain unchanged as before. The groundwater flow model described above was used to simulate effects of proposed regulation scenarios. The sce-

Table 1. Groundwater regulation scenarios (10^6 m^3 a^{-1}).

Abstraction and recharge	Case 1	Case 2	Case 3	Case 4	Case 5	Case 6	Case 7	Case 8	Case 9	Case 10
Domestic and industrial	847	508	847	508	847	847	508	847	847	847
Agriculture	1023	1023	1023	1023	1023	1023	818	1023	1023	1023
Well fields	354	354	354	501	364	331	354	402	354	354
Emergency water source	291		291			204				291
Self-water supply wells	113	113	113				113			113
Artificial recharge			405			150			215	520
Total abstraction	2628	1999	2628	2033	2234	2415	1795	2270	2234	2628

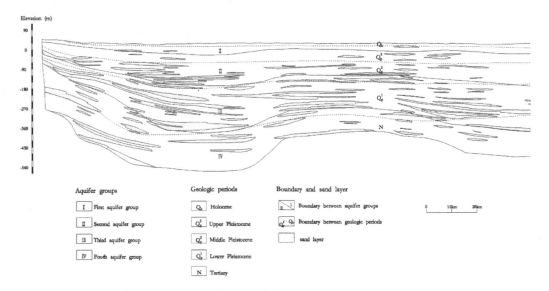

Figure 2. Hydrogeological cross section A-A' (location shown in Fig. 1).

narios were simulated for the 50 year period from 2012 to 2061.

4.2 Multi-criteria fuzzy pattern recognition (MFPR) approach for quantitative estimation of regulation scenarios

Assume $X = (x_1, x_2, \ldots, x_n)$ be a finite set of n groundwater regulation scenarios. Each regulation scenario can be valued by m quantitative factors, and then the decision-making matrix can be denoted as

$$X = \left(x_{ij}\right)_{m \times n} \tag{1}$$

where x_{ij} is the value of the ith quantitative factor for the jth regulation scenarios.

Generally, there exist three types of estimation factors: larger-the-better, smaller-the-better and nominal-the-better by characteristics. These three types of estimation factors can, respectively, be normalized by the following equations (Chen, 2002):

$$r_{ij} = \frac{x_{ij} - \min_j x_{ij}}{\max_j x_{ij} - \min_j x_{ij}}, \quad i \tag{2}$$

$$r_{ij} = \frac{\max_j x_{ij} - x_{ij}}{\max_j x_{ij} - \min_j x_{ij}}, \quad i \tag{3}$$

$$r_{ij} = 1 - \frac{|x_{ij} - x_i|}{\max_j |x_{ij} - x_i|}, \quad i \tag{4}$$

where r_{ij} is the relative membership degree of the jth scenario regarding the ith factor, $\max_j x_{ij}$ and $\min_j x_{ij}$ are the maximum and minimum value regarding the ith factor, respectively; and x_i is the optimum target value of the ith factor.

Each groundwater regulation scenario can be evaluated by c levels, where the relative membership degree of the "superior" level is assigned to be 1 and relative membership degree of "inferior" level is 0. With $c = 2$, the relative membership matrix of each groundwater regulation scenario belonging to each level can be established (Chen, 2002):

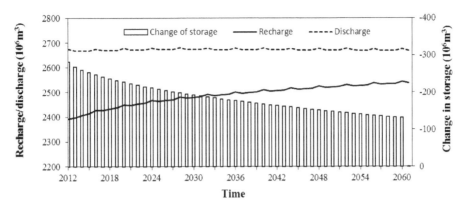

Figure 3. Change of groundwater balance under scenario 1.

5 Quantitative estimation of regulation scenarios

$$u_{hj} = \left(\sum_{k=1}^{2} \frac{\sum_{i=1}^{m} \left[w_i \left(r_{ij} - s_{ih} \right) \right]^2}{\sum_{i=1}^{m} \left[w_i \left(r_{ij} - s_{ik} \right) \right]^2} \right)^{-1} \quad (5)$$

where u_{hj} is the relative membership grade of the jth scenario belonging to the hth ranking, $j = 1, 2, \ldots, n$; $h = 1\text{--}2$; and w_i is the weight vector of factor set.

The ranking feature of each groundwater regulation scenario is,

$$H = (1, 2, \ldots, c) \cdot \left(u_{hj} \right) = (H_1, H_2, \ldots, H_n) \quad (6)$$

where H_j is ranking feature value of the jth scenario. Ranking feature values for the groundwater development scenarios are ordered lowest to highest representing best to worst, respectively.

5 Quantitative estimation of regulation scenarios

Seven main factors are chosen to quantatively estimate different groundwater regulation scenarios, which include amount of groundwater exploitation, evaporation from unconfined aquifer, groundwater outflows to rivers, regional average groundwater-level depth below the ground, drawdowns in depression cone centers and the ratio of groundwater storage to total recharge. The amount of groundwater exploitation needed to meet the requirements for everyday life, city construction, agriculture and industry, and sustain groundwater levels above critical thresholds should be balanced so that severe land subsidence cannot be induced. Regional average groundwater levels should be controlled in a reasonable range so that the ecosystems dependent on groundwater can be sustained and nurtured. Sufficient groundwater discharge to rivers is required to sustain baseflow for riverine ecosystems. Evaporation from unconfined aquier is unproductive and should be reduced as much as possible. The ratio of groundwater storage to total recharge ought to keep in a reasonable range.

The ranking feature value vector for the 10 regulation scenarios (1 : 10) subject to the set of criteria described above was obtained from Eq. (6):

$$H = (1.906 \ 1.174 \ 1.140 \ 1.174 \ 1.187 \ 1.137 \ 1.140$$
$$1.199 \ 1.128 \ 1.119).$$

Scenarios 9 and 10 clearly have the lowest ranking feature values and are therefore considered as the best choices for managing the groundwater resource in the Beijing plain. Scenario 1 has the highest ranking feature value representing the worst choice for managing the groundwater resource. From a water balance perspective, the groundwater system in all scenarios approach a new equilibrium at the end of the 50 year simulation period, except scenario 1 where groundwater levels continue to decline significantly (Fig. 3), indicating poor near-future sustainability of the groundwater resources. The simulation results indicate that two major groundwater depression cones are developed, which are located in the southeast of Beijing city (Fig. 1).

In scenario 9, production from emergency well fields in the plain and most deep self-supply wells would be stopped, and reclaimed wastewater and imported surface water would be artificially recharged to aquifers in alluvial fans of major rivers. The annual total artificial recharge reaches $215 \times 10^6 \, \text{m}^3 \, \text{a}^{-1}$ (Table 1). The total reduction of annual abstraction rates amounts to $394 \times 10^6 \, \text{m}^3 \, \text{a}^{-1}$ compared with scenario 1. Implementing scenario 9 would result in immediate, rapid recovery of groundwater levels, especially at the depression cone centers. As a result, the continuous decline in groundwater levels and subsidence would be greatly mitigated. Figure 4 shows the change of groundwater balance under scenario 9. Here, groundwater recharge exceeds abstraction, and groundwater storage is gradually increasing as the change in groundwater flowing out of storage is decreasing. Sustainable groundwater resources development may be achieved after many decades under this scenario.

In scenario 10, sustainable groundwater resources development may be achieved sooner. However, in the Bejing plain with limited water supply, it would be difficult to ensure

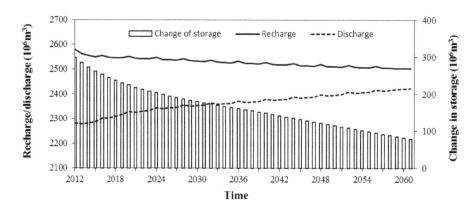

Figure 4. Change of groundwater balance under scenario 9.

an alternative water source of $520 \times 10^6\,\text{m}^3\,\text{a}^{-1}$ for artificial recharge specified in this scenario. In scenario 9, one of the best evaluated scenarios, a reasonable combination of artificial recharge and reduced abstraction would be implemented, which is more feasible and easier to implement. Thus, scenario 9 may be the best groundwater regulation scenario for the Beijing plain.

6 Summary and conclusions

The main factors leading to land subsidence in the NCP are represented by geological conditions such as the presence of highly compressible, thick, unconsolidated sediments and dynamic conditions of the groundwater system, e.g. groundwater abstraction. The best method to control subsidence is to recover groundwater levels especially in the known subsidence areas. Beijing is facing challenges in managing scarce water resources and controling subsidence. A multi-criteria fuzzy pattern recognition (MFPR) approach is introduced and applied to quantitatively evaluate 10 proposed groundwater regulation scenarios simulated using a regional groundwater flow model for the Beijing plain. The evaluation model will be helpful for determining the most reasonable and feasible scenarios for sustainable development and utilization of groundwater resources.

Currently, in the NCP The areal extent and accumulative magnitude of subsidence is increasing though some controlling measures have been taken. Subsidence in the NCP has a direct relationship with groundwater exploitation, so rationally reducing and controlling groundwater abstraction should be carried out to decrease groundwater production. However, because both population and economic activities are heavily dependent upon the groundwater resource, it is important to seek a balance between controlling subsidence and ensuring adequate water supply so that the NCP can be developed in a sustainable and harmonious manner.

Acknowledgements. This work was supported by the China Geological Survey (No. 12120113011700), and the National Basic Research Program of China (973 Program, 2010CB428806).

References

Abidin, H. Z., Gumilar, I., Andreas, H., Murdohardono, D., and Fukuda, Y.: On causes and impacts of land subsidence in Bandung Basin, Indonesia, Environ. Earth Sci., 68, 1545–1553, 2013.

Chen, S. Y.: Optimal Fuzzy Recognition Theory and Application in Complex Water Resources System, JiLin University Press, Jilin, China, 2002.

Foster, S., Garduno, H., Evant, R., Olson, D., Tian, Y., Zhang, W., and Han, Z.: Quaternary Aquifer of the North China Plain – assessing and achieving groundwater resource sustainability, Hydrogeol. J., 12, 81–93, 2004.

Galgaro, A., Boaga, J., and Rocca, M.: HVSR technique as tool for thermal-basin characterization: a field example in N-E Italy, Environ. Earth Sci., 71, 4433–4446, 2014.

Papadaki, E. S.: Monitoring subsidence at Messara basin using radar interferometry, Environ. Earth Sci., 72, 1965–1977, 2014.

Ye, X. B. and He, Q. C.: Estimation of economic loss due to land subsidence in the North China Plain, China Land Press, Beijing, 2006.

Zeitoun, D. G. and Wakshal, E.: Land Subsidence Analysis in Urban Areas: the Bangkok Metropolitan Area Case Study, Springer Environmental Science and Engineering, Springer, Dordrecht, 2013.

Zhang, Y. Q., Gong, H. L., Gu, Z. Q., Wang, R., Li, X. J., and Zhao, W. J.: Characterization of land subsidence induced by groundwater withdrawals in the plain of Beijing city, China, Hydrogeol. J., 22, 397–409, 2014.

7

Study on the risk and impacts of land subsidence in Jakarta

H. Z. Abidin[1], H. Andreas[1], I. Gumilar[1], and J. J. Brinkman[2]

[1]Geodesy Research Group, Faculty of Earth Science and Technology, Institute of Technology Bandung, Bandung 40132, Indonesia
[2]Deltares, Delft, Netherlands

Correspondence to: H. Z. Abidin (hzabidin@gmail.com)

Abstract. Jakarta is the capital city of Indonesia located in the west-northern coast of Java island, within a deltaic plain and passes by 13 natural and artificial rivers. This megapolitan has a population of about 10.2 million people inhabiting an area of about $660\,km^2$, with relatively rapid urban development. It has been reported for many years that several places in Jakarta are subsiding at different rates. The main causative factors of land subsidence in Jakarta are most probably excessive groundwater extraction, load of constructions (i.e., settlement of high compressibility soil), and natural consolidation of alluvial soil. Land subsidence in Jakarta has been studied using leveling surveys, GPS surveys, InSAR and Geometric-Historic techniques. The results obtained from leveling surveys, GPS surveys and InSAR technique over the period between 1974 and 2010 show that land subsidence in Jakarta has spatial and temporal variations with typical rates of about 3–$10\,cm\,year^{-1}$. Rapid urban development, relatively young alluvium soil, and relatively weak mitigation and adapatation initiatives, are risk increasing factors of land subsidence in Jakarta. The subsidence impacts can be seen already in the field in forms of cracking and damage of housing, buildings and infrastructure; wider expansion of (riverine and coastal) flooding areas, malfunction of drainage system, changes in river canal and drain flow systems and increased inland sea water intrusion. These impacts can be categorized into infrastructural, environmental, economic and social impacts. The risk and impacts of land subsidence in Jakarta and their related aspects are discussed in this paper.

1 Introduction

Land subsidence is a natural-anthropogenic hazard affecting quite many large urban areas (cities) in the world, including Jakarta, the capital city in Indonesia. It is located on the lowland area in the northern coast of West Java, centered at the coordinates of about $6°15'\,S$ and $+106°50'\,E$ (see Fig. 1), within a deltaic plain and passes by 13 natural and two canals. This megapolitan has a population of about 10.1 million people in 2014, inhabiting an area of about $662\,km^2$, with relatively rapid urban development. Topographically, the area of Jakarta has slopes ranging between 0 and 2° in the northern and central parts, between 0 and 5° in the southern part, and its southern-most area has an altitude of about 50 m above mean sea level.

Land subsidence is a well known phenomena affecting Jakarta, and it has been observed using several geodetic techniques and has been widely reported for many years (Rismianto and Mak, 1993; Murdohardono and Sudarsono, 1998; Purnomo et al., 1999; Rajiyowiryono, 1999; Abidin et al., 2001, 2004, 2008, 2010, 2011, 2013; Koudogbo et al., 2012; Ng et al., 2012; Chaussard et al., 2013). According to those studies, land subsidence in Jakarta has spatial and temporal variations with typical rates of about 3–$10\,cm\,year^{-1}$.

Land subsidence impacts in Jakarta can be seen already in the field in forms of cracking and damage of housing, buildings and infrastructure; wider expansion of (riverine and coastal) flooding areas, malfunction of drainage system, changes in river canal and drain flow systems and in-

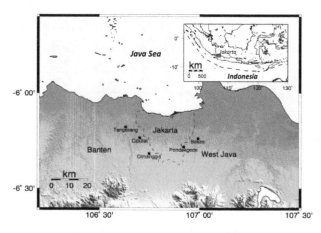

Figure 1. Location of Jakarta, the capital city of Indonesia.

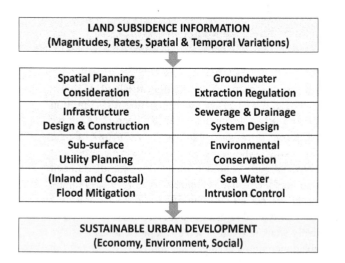

Figure 2. Importance of land subsidence information for urban development.

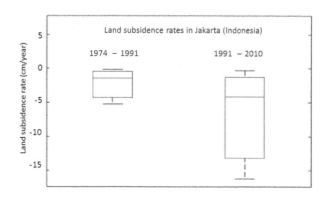

Figure 3. Box plots of land subsidence rates in Jakarta during the period of 1974–1991 (left) and 1991–2010 (right). Subsidence rate unit is cm year^{-1}.

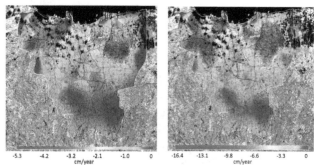

Figure 4. Rates of land subsidence in Jakarta during the period of 1974–1991 (left) and 1991–2010 (right); courtesy of Irwan Gumilar of Geodesy Research Group of ITB.

creased inland sea water intrusion. In general, these impacts can be categorized into infrastructural, environmental, economic and social impacts.

Since information of land subsidence characteristics is important for sustainable urban development of Jakarta (see Fig. 2), then understanding the risk and impacts of land subsidence in Jakarta and their related aspects is very important and strategic, and land subsidence charactristics should be continuously monitored.

2 Land subsidence characteristics in Jakarta

Land subsidence phenomena in Jakarta has actually been observed using several geodetic methods, such as Leveling survey, GPS survey, InSAR, Microgravity survey, and Geometric-Historic method. In general, the observed land subsidence in Jakarta has spatial and temporal variations, with the typical rates of about 3–10 cm year^{-1} (see Figs. 3, 4). These subsidence rates are obtained from the in-

tegration of results obtained by the Leveling surveys (1974–1997), GPS surveys (1997–2010), and InSAR technique (2006–2010). In this period of about 37 years (1974–2010), the total subsidence of up to about 4 m have been observed in several locations in Jakarta. The observed land subsidences along the coastal areas of Jakarta are relatively have larger rates than the inland areas. Figures 3 and 4 also show the existence of spatial variation in subsidence rates along the coastal zone of Jakarta, in which the central and western parts have relatively larger rates. These larger rates of subsidence are mainly due to higher volumes of groundwater extraction, combined with relatively younger alluvium soil composition. More information on land subsidence characteristics in Jakarta can be seen in Abidin et al. (2001, 2004, 2008, 2010, 2011, 2013, 2014, 2015a, b), Ng et al. (2012), Chaussard et al. (2013).

The studies from Murdohardono and Sudarsono (1998), Rismianto and Mak (1993), Harsolumakso (2001), and Hutasoit (2001) suggested that the possible causative factors of land subsidence in Jakarta are most probably: excessive groundwater extraction, load of constructions (i.e., settlement of high compressibility soil), natural consolidation of alluvial soil, and tectonic activities. Land subsidence in certain area is usually caused by combination of those factors. Considering the spatial variation of land subsidence rates in Jakarta area, then it can be expected that the contribution of each factor on land subsidence at each location also will has spatial variation. In Jakarta, tectonic activities seem to be the least dominant factor, while excessive groundwater extraction is considered to be one of dominant factor for causing land subsidence.

3 Land subsidence risk in Jakarta

Considering the aforementioned causative factors of land subsidence in Jakarta, actually there are three main risk increasing factors of land subsidence in Jakarta that can be considered, namely: relatively young alluvium soil in which Jakarta is located, rapid urban development, and relatively weak imposement of land subsidence mitigation and adaptation measures (see Fig. 5).

Jakarta is located in a lowland coastal areas, composed by relatively young and soft alluvium soil. According to Rimbaman and Suparan (1999), there are five main landforms of Jakarta, namely: alluvial landforms (southern part), landforms of marine-origin (northern part adjacent to the coastline), beach ridge landforms (northwest and northeast parts), swamp and mangrove swamp landforms (coastal fringe), and former channels (perpendicular to the coastline). There are also 13 rivers and 2 canals flowing across Jakarta from its southern part and have estuaries in Java sea along an approximately 35 km of coastline. Jakarta basin, according to Yong et al. (1995), its consisted of a 200–300 m thick sequence of quaternary deposits which overlies tertiary sediments, in which the top sequence is thought to be the base of the groundwater basin. The quaternary sequence itself can be further subdivided into three major units, which, in ascending order are: a sequence of pleistocene marine and non-marine sediments, a late pleistocene volcanic fan deposit, and holocene marine and floodplain deposits.

The above geological and geomorphological conditions make Jakarta has a higher risk (prone) toward land subsidence phenomena, especially in cases of excessive groundwater extraction and massive loading from buildings and infrastructures.

Rapid urban development in Jakarta also contributes to increasing risk of land subsidence occurences in Jakarta (Abidin et al., 2011). In this case, the increases in built-up areas, population, economic and industrial activities, will then increase groundwater extraction and also buildings and infrastructure loadings; which usually in turn lead to land subsidence phenomena. The relatively very rapid urban development of Jakarta as a megapolitan city is mainly in the sectors of industry, trade, transportation, housing, hotel and apartment, and many others (Firman, 1999, 2004; Hudalah et al., 2013); and they have introduced several negative environmental problems (Firman and Dharmapatni, 1994; Hu-

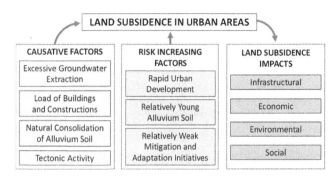

Figure 5. The causative factors, risk increasing factors and impacts of land subsidence in urban areas of Indonesia.

dalah and Firman, 2012), such as: extensive conversion of agricultural areas into residential and industrial areas, significant disturbance to ecological and hydrological functions of the upland of Jakarta area and river catchment areas, and increase in groundwater extraction due to development of industrial activities and the high population increase. Coastal area of Jakarta which is mainly composed by relatively young and soft alluvium soil, has also experienced extensive urban development, such as establishment of sea port, coastal resort, golf course, residential areas, industries, apartments, malls, hotels, and commercials and office buildings. Coastal reclamation has also been conducted to accomodate more coastal development initiatives. These extensive urban development activities will have contribution into increasing land subsidence risk in several places in Jakarta (Abidin et al., 2011).

The other risk increasing factor of land subsidence in Jakarta is a relatively weak mitigation and adaptation initiatives (see Table 1) being implemented in Jakarta. At present, land subsidence hazard is not yet properly considered in urban development and spatial planning, groundwater extraction regulation system, and building codes in Jakarta. Although the regulation has been introduced to limit groundwater extraction in several subsidence prone areas in Jakarta; however, its enforcement is still need to be proven.

4 Land subsidence impacts in Jakarta

In general, the impacts of land subsidence in Jakarta can be seen in the field in various forms as shown in Fig. 6, such as cracking of permanent constructions and roads, tilting of houses and buildings, "sinking" of houses and buildings, changes in river canal and drain flow systems, wider expansion of coastal and/or inland flooding areas, and increased inland sea water intrusion. These subsidence impacts can be categorized into infrastructure, environmental, economic, and social impacts (see Table 2). Most of these impacts are indirectly generated due to land subsidence in the areas, and several of them are directly caused by subsidence. More over, these subsidence impacts have also rela-

Table 1. Examples of mitigation and adaptation initiatives for decreasing land subsidence risk in urban areas.

Causative factors	Mitigation	Adaptation
Excessive groundwater extraction	– Limit or prohibit groundwater extraction in subsidence prone areas. – Enforcement of strict regulation and punishment for groundwater extraction in subsidence prone areas.	– Increase the surface groundwater supply and resources. – Continuous monitoring of subsidence characteristics.
Load of buildings and infrastructures	– Urban development planning takes into account the land subsidence characteristics in the areas. – Urban development rate in subsidence prone areas is properly controlled.	– Implementation of special building codes for land subsidence prone areas. – Continuous monitoring of subsidence characteristics.
Natural consolidatium of alluvium soil	None	– Implementation of subsidence-adaptive urban developmen and spatial planning.
Tectonic activities		– Continuous monitoring of subsidence characteristics.

Table 2. Characteristics of land subsidence impacts; after Abidin et al. (2015c).

No.	Category	Representation of impact	Level of impact
1.	Infrastructural	cracking of permanent constructions and roads	direct
		tilting of houses and buildings	direct
		"sinking" of houses and buildings	direct
		breaking of underground pipelines and utilities	direct
		malfunction of sewerage and drainage system	indirect
		deterioration in function of building and infrastructures	indirect
2.	Environmental	changes in river canal and drain flow systems frequent coastal flooding wider expansion of flooding areas inundated areas and infrastructures increased inland sea water intrusion deterioration in quality of environmental condition	indirect
3.	Economic	increase in maintenance cost of infrastructure decrease in land and property values abandoned buildings and facilities disruption to economic activities	indirect
4.	Social	deterioration in quality of living environment and life (e.g. health and sanitation condition) disruprion to daily activities of people	indirect

tion among each other, and its connection system is simplified by Fig. 7.

The potential losses due to land subsidence in urban areas such as Jakarta are actually quite significant (Ward et al., 2011; Viets, 2010). Related infrastructural, social and environmental costs due to direct and indirect impacts of land subsidence are economically quite significant, and can not be under estimated in sustainable urban development. In this regard for example, the planning, development and maintenance costs of building and infrastructures in the affected areas are usually much higher than the normal situation. The

collateral impact of coastal subsidence in Jakarta, in the form of coastal flooding during high tides is also quite damaging (Abidin et al., 2011, 2015b). This repeated coastal flooding in several areas along the coast, will deteriorates the structure and function of building and infrastructures, badly influences the quality of living environment and life (e.g. health and sanitation condition), and also disrupts economic and social activities in the affected areas. Inland subsidence should also has an impact on inland flooding phenomena in Jakarta (Texier, 2008), since it will theoretically lead to expanded coverage and deeper water depth of flooded (inundated) areas

Figure 6. Examples of representation in the field of land subsidence impacts in Jakarta.

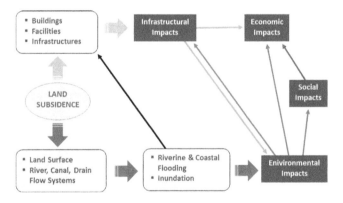

Figure 7. Land subsidence impacts in urban areas and its connection system, from Abidin et al. (2015c).

(Abidin et al., 2015b, c). The losses due to inland flooding in Jakarta are also significant since it affects many economic related activities in the city and surrounding regions.

5 Closing remarks

Land subsidence is a natural-anthropogenic hazard affecting Jakarta, with typical rates of about 3–10 cm year^{-1}. Its direct and indirect impacts can be easily seen nowadays in the field. Without good and effective mitigation and adaptation initiatives, land subsidence hazard in Jakarta can have more disastrous impacts in the future. In this case, the infrastructural, economic, environmental and social impacts of land subsidence can be quite a heavy burden for sustainable development of Jakarta.

At present, detail characteristics and mechanisms of land subsidence in Jakarta, both in spatial and temporal domains, are still not yet fully established. Contribution of each causative factors on observed land subsidence in certain areas still needs to be estimated and modelled. In this regard, the implementation of continuous subsidence monitoring system (e.g. GNSS CORS system) in Jakarta is necessary. Since, there is quite a strong linkage between geological and geomorphological condition, and rapid urban development in Jakarta with land subsidence characteristics, then subsidence mitigation and adaptation initiatives should be integrated in urban development program and spatial planning of Jakarta.

Finally, it should be pointed out that based on preliminary studies (Abidien et al., 2015b), it is found that there is some spatial correlation between land subsidence affected areas with flooded (inundated) areas in Jakarta. However, since the exact relation mechanism between the two phenomena is not yet established; then more quantitative characteristics of this correlation cannot be estimated. In order to establish a quantitative relation between land subsidence and flooding phenomena in Jakarta, then the following activities should be conducted, namely: detail mapping of the spatial and temporal rates and impacts of land subsidence, detail mapping of flooded (inundated) area during the flooding events, and detail flood risk modelling for Jakarta.

Acknowledgements. Land subsidence study in Jakarta using space geodetic techniques has been conducted by the Geodesy Research group of ITB since 1997, by using several research grants from Ministry of Science, Technology, and Higher Education of Indonesia, from the Provincial Government of Jakarta, and from ITB Research programs. The GPS surveys were conducted by the Geodesy Research Group of ITB, the Geospatial Agency of Indonesia, and the staffs and students from the Department of Geodesy and Geomatics Engineering of ITB.

References

Abidin, H. Z., Djaja, R., Darmawan, D., Hadi, S., Akbar, A., Rajiyowiryono, H., Sudibyo, Y., Meilano, I., Kusuma, M. A., Kahar, J., and Subarya, C.: Land Subsidence of Jakarta (Indonesia) and its Geodetic-Based Monitoring System, Nat. Hazards, 23, 365–387, doi:10.1023/A:1011144602064, 2001.

Abidin, H. Z., Djaja, R., Andreas, H., Gamal, M., Hirose, K., and Maruyama, Y.: Capabilities and Constraints of Geodetic Techniques for Monitoring Land Subsidence in the Urban Areas of Indonesia, Geomatics Research Australia, 81, 45–58, 2004.

Abidin, H. Z., Andreas, H., Djaja, R., Darmawan, D., and Gamal, M.: Land subsidence characteristics of Jakarta between 1997 and 2005, as estimated using GPS surveys, GPS Solutions, Springer Berlin/Heidelberg, 12, 23–32, 2008.

Abidin, H. Z., Andreas, H., Gamal, M., Gumilar, I., Napitupulu, M., Fukuda, Y., Deguchi, T., Maruyama, Y., and Riawan, E.: Land

Subsidence Characteristics of the Jakarta Basin (Indonesia) and its Relation with Groundwater Extraxtion and Sea Level Rise, in: Groundwater Response to Changing Climate, IAH Selected Papers on Hydrogeology No. 16, edited by: Taniguchi, M. and Holman, I. P., CRC Press, London, Chapter 10, 113–130, 2010.

Abidin, H. Z., Andreas, H., Gumilar, I., Fukuda, Y., Pohan, Y. E., and Deguchi, T.: Land subsidence of Jakarta (Indonesia) and its relation with urban development, Nat. Hazards, 59, 1753–1771, 2011.

Abidin, H. Z., Andreas, H., Gumilar, I., Sidiq, T. P., and Fukuda, Y.: On the Roles of Geospatial Information for Risk Assessment of Land Subsidence in Urban Areas of Indonesia, in: Intelligent Systems for Crisis Management, Lecture Notes in Geoinformation and Cartography, edited by: Zlatanova, S., Peters, R., Dilo, A., and Scholten, H., Springer-Verlag Berlin Heidelberg, 277–288, 2013.

Abidin, H. Z., Andreas, H., Gumilar, I., and Yuwono, B. D.: Performance of the Geometric-Historic Method for Estimating Land Subsidence in Urban Areas of Indonesia, Proceedings of the XXV FIG Congress, TS 10B Session – Earth Geodynamics and Monitoring 2, 16–21 June 2014, Kuala Lumpur, Malaysia, 2014.

Abidin, H. Z., Andreas, H., Gumilar, I., Yuwono, B. D., Murdohardono, D., and Supriyadi, S.: On integration of geodetic observation results for assessment of land subsidence hazard risk in urban areas of Indonesia, in: International Association of Geodesy Symposia Series, Springer Berlin Heidelberg, 143, 1–8, 2015a.

Abidin, H. Z., Andreas, H., Gumilar, I., and Wibowo, I. R. R.: On correlation between urban development, land subsidence and flooding phenomena in Jakarta, Proc. IAHS, 370, 15–20, doi:10.5194/piahs-370-15-2015, 2015.

Abidin, H. Z., Andreas, H., Gumilar, I., Sidiq, T. P., and Gamal, M.: Environmental Impacts of Land Subsidence in Urban Areas of Indonesia, in: Proceedings of the FIG Working Week 2015, TS 3 – Positioning and Measurement, Sofia, Bulgaria, Paper no. 7568, 17–21 May 2015, 2015c.

Chaussard, E., Amelung, F., Abidin, H. Z., and Hong, S.-H.: Sinking cities in Indonesia: ALOS PALSAR detects rapid subsidence due to groundwater and gas extraction, Remote Sens. Environ., 128, 150–161, 2013.

Firman, T.: From Global City to City of Crisis: Jakarta Metropolitan Region Under Economic Turmoil, Habitat Int., 23, 447–466, 1999.

Firman, T.: New town development in Jakarta Metropolitan Region: a perspective of spatial segregation, Habitat Int., 28, 349–368, 2004.

Firman, T. and Dharmapatni, I. A. I.: The challenges to suistanaible development in Jakarta metropolitan region, Habitat Int., 18, 79–94, 1994.

Harsolumakso, A. H.: Struktur Geologi dan Daerah Genangan, Buletin Geologi, 33, 29–45, 2001.

Hudalah, D. and Firman, T.: Beyond property: Industrial estates and post-suburban transformation in Jakarta Metropolitan Region, Cities, 29, 40–48, 2012.

Hudalah, D., Viantari, D., Firman, T., and Woltjer, J.: Industrial land development and manufacturing deconcentration in Greater Jakarta, Urban Geogr., 34, 950–971, 2013.

Hutasoit, L. M.: Kemungkinan Hubungan antara Kompaksi Alamiah Dengan Daerah Genangan Air di DKI Jakarta, Buletin Geologi, 33, 21–28, 2001.

Koudogbo, F. N., Duro, J., Arnaud, A., Bally, P., Abidin, H. Z., and Andreas, H.: Combined X- and L-band PSI analyses for assessment of land subsidence in Jakarta, Proc. SPIE, Remote Sensing for Agriculture, Ecosystems, and Hydrology XIV, 8531, 853107, doi:10.1117/12.974821, 2012.

Murdohardono, D. and Sudarsono, U.: Land subsidence monitoring system in Jakarta, Proceedings of Symposium on Japan-Indonesia IDNDR Project: Volcanology, Tectonics, Flood and Sediment Hazards, 21–23 September 1998, Bandung, Indonesia, 243–256, 1998.

Ng, A. H.-M., Ge, L., Li, X., Abidin, H. Z., Andreas, H., and Zhang, K.: Mapping land subsidence in Jakarta, Indonesia using persistent scatterer interferometry (PSI) technique with ALOS PALSAR, Int. J. Appl. Earth Obs., 18, 232–242, 2012.

Purnomo, H., Murdohardono, D., and Pindratno, H.: Land Subsidence Study in Jakarta, Proceedings of Indonesian Association of Geologists, 30 November–1 December 1999, Jakarta, Volume IV: Development in Engineering, Environment, and Numerical Geology, 53–72, 1999.

Rajiyowiryono, H.: Groundwater and Landsubsidence Monitoring along the North Coastal Plain of Java Island, CCOP Newsletter, 24, 19 pp., 1999.

Rimbaman and Suparan, P.: Geomorphology, in: Coastplan Jakarta Bay Project, Coastal Environmental Geology of the Jakarta Reclamation Project and Adjacent Areas, Jakarta, Bangkok, CCOP COASTPLAN Case Study Report No. 2, 21–25, 1999.

Rismianto, D. and Mak, W.: Environmental aspects of groundwater extraction in DKI Jakarta: Changing views, Proceedings of the 22nd Annual Convention of the Indonesian Association of Geologists, 6–9 December 1993, Bandung, Indonesia, 327–345, 1993.

Texier, P.: Floods in Jakarta: when the extreme reveals daily structural constraints and mismanagement, Disaster Prevention and Management, 17, 358–372, 2008.

Viets, V. F.: Environmental and Economic Effects of Subsidence, Publication of Lawrence Berkeley National Laboratory, LBNL Paper LBL-8615, 251 pp., available at: http://escholarship.org/uc/item/1sb4c8vh, last access: 20 October 2015, 2010.

Ward, P. J., Marfai, M. A., Yulianto, F., Hizbaron, D. R., and Aerts, J. C. J. H.: Coastal inundation and damage exposure estimation: a case study for Jakarta, Nat. Hazards, 56, 899–916, 2011.

Yong, R. N., Turcott, E., and Maathuis, H.: Groundwater extraction-induced land subsidence prediction: Bangkok and Jakarta case studies, Proceedings of the Fifth International Symposium on Land Subsidence, IAHS Publication no. 234, 89–97, 1995.

8

Towards a global land subsidence map

G. Erkens[1,2] and E. H. Sutanudjaja[1,2]

[1]Deltares Research Institute, Utrecht, the Netherlands
[2]Utrecht University, Utrecht, the Netherlands

Correspondence to: G. Erkens (gilles.erkens@deltares.nl)

Abstract. Land subsidence is a global problem, but a global land subsidence map is not available yet. Such map is crucial to raise global awareness of land subsidence, as land subsidence causes extensive damage (probably in the order of billions of dollars annually). With the global land subsidence map relative sea level rise predictions may be improved, contributing to global flood risk calculations.

In this paper, we discuss the approach and progress we have made so far in making a global land subsidence map. Initial results will be presented and discussed, and we give an outlook on the work needed to derive a global land subsidence map.

1 Introduction

Although its impact on flood risk is locally outranging the impact of absolute sea level rise, over the last decades land subsidence retrieved much less attention in terms of research. One of the reasons for this is the unknown extent of land subsidence around the world, in contrast to absolute sea level rise. For the latter, there are models that predict global absolute sea level rise (Church et al., 2013). For land subsidence there are many examples around the world, but a comprehensive picture of the global extent is currently unavailable. This originates from the heterogeneity of the land subsidence signal: an array of potential drivers and the heterogeneous subsurface hydrology and geology, makes land subsidence mainly a local phenomenon on a global scale.

A map showing the extent of land subsidence around the world, the global land subsidence map, would be very useful. It would raise awareness among scientist and policymakers alike. The land subsidence rates could be used – together with regional absolute sea level rise predictions – to estimate relative sea level rise in coastal areas. This serves as input for flood risk calculations. Lastly, a global land subsidence map would not only locate current land subsidence hotspots but also help to identify future sinking areas under predicted socio-economic development scenarios.

The maps that come close to a global land subsidence map are those based on a collection of globally spread case studies. However, these have two major limitations:

i. They are static and show the current land subsidence rates, but not necessarily at a single moment in time. They also have no predictive power.

ii. They are biased towards well-studied areas. Large cities or classic study sites in western countries show up most prominently.

Considering these limitations, we set out to produce a global land subsidence map that is derived from numerical model calculations. In this way, we are able to introduce a temporal component showing both historical and predicted future land subsidence under different development scenarios.

In this paper, we discuss the approach and progress we have made so far. Initial results will be presented and discussed, and we give an outlook on the work needed to derive a global land subsidence map.

2 The approach

Land subsidence is caused my many different drivers, both natural and anthropogenic-induced. Ideally, a global land subsidence map would cover all known drivers, but for sake of simplicity, in first instance we restricted ourselves

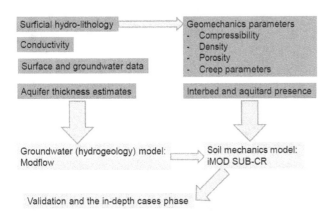

Figure 1. The proposed model suite to derive a global land subsidence map consists of three coupled models: the surface water model, the ground water model and a soil mechanics model.

to one driver. One of the most prominent causes for land subsidence is aquifer compaction as a result of excessive groundwater extraction for domestic, agricultural and industrial use (Galloway and Burbey, 2011). For instance, the Vietnamese Mekong Delta sinks on average $1.6\,\mathrm{cm\,yr^{-1}}$, attributed to groundwater extraction (Erban et al., 2014). Crucially, in many coastal mega-cities, land subsidence is accelerated by ongoing urbanization and related increased demands for (ground) water. In Jakarta land subsidence is up to $20\,\mathrm{cm\,yr^{-1}}$ (Abidin et al., 2008). With ongoing economic development in the near future the expectation is that groundwater abstraction and, consequently, land subsidence rates, will accelerate and areas affected will expand as well as. Therefore, in this paper, we chose to focus on aquifer abstraction induced land subsidence and present our numerical approach on constructing a global land subsidence map.

Figure 1 illustrates different the model components used in this study. The integrated global hydrological and water resources model, PCR-GLOBWB (Van Beek et al., 2011; Sutanudjaja et al., 2015; Van Beek, 2008; Van Beek and Bierkens, 2009), serves as the starting point. The model is fed by global meteorological forcing data (i.e. precipitation, temperature and reference potential evaporation) and parameterized based on only global datasets (for an extensive list, see e.g. Sutanudjaja et al., 2011). PCR-GLOBWB simulates daily river discharge and groundwater recharge, as well as surface water and groundwater abstraction rates. The latter are estimated internally within the model based on the simulation of their availabilities and water demands for irrigation and other sectors (De Graaf et al., 2014; Wada et al., 2014). The daily output of PCR-GLOBWB would then be aggregated to the monthly resolution and used to force the MODFLOW groundwater model (McDonald and Harbaugh, 1988; Sutanudjaja et al., 2011, 2014) resolving spatio-temporal groundwater head dynamics, incorporating the simulated groundwater abstraction of PCR-GLOBWB. Subsequently, the simulated monthly groundwater head changes are fed into

a land subsidence module, iMOD-SUB-CR (Bakr, 2015), which is an extension of the subsidence and aquifer-system compaction package (SUB-WT) of MODFLOW (Leake and Galloway, 2007). For this study all aforementioned models are simulated at the spatial resolution of 5 arc minutes (approximately $10\,\mathrm{km}$ at the equator).

3 First results

In this study we perform all aforementioned model simulations for the period 1958–2010. Our results are focused on the simulation results of PCR-GLOBWB that are fed in the MODFLOW model, particularly the rates of groundwater depletion, i.e. the abstraction of groundwater stores that are not being replenished by groundwater recharge. More specifically, we present and summarize the groundwater depletion rates in urban areas.

3.1 First attempt: depletion in urban area missed

Figure 2 shows first results for the island of Java, Indonesia. This is an area that is well known for its extensive land subsidence as a result ground water extraction. The map shows the depletion of ground water as calculated by the PCR-GLOBWB hydrology model. Most of the areas that show depletion are irrigation areas, where the model suggests that groundwater is excessively used. Large urban areas, such as Jakarta, with well-known depletion of groundwater and related land subsidence, are completely missed.

This misconception is attributed to the previous assumption in the PCR-GLOBWB model that simplifies that the fraction of water demands that is satisfied by groundwater resource (GW_{frac}) as a ratio between simulated baseflow and river discharge. Agricultural irrigation results in a high water demand, and because the river discharge in those areas is dominated by baseflow, the model assumes the use of groundwater. While in the urban area of Jakarta where river flows are not baseflow dominant, water demands are lower and as a result, surface water availability as simulated for Jakarta is sufficient for the domestic and industrial water demand. However, in reality, the urban areas in Java rely on ground water as a clean and reliable source of water. In turn, it is foremost surface water that is used for irrigation in Java.

3.2 Second attempt: including water resources uses

To improve the model, we included (global) databases on the relative contribution of ground water and surface water to the total water usage, for both urban agglomerations (McDonald et al., 2014: Dataset of water infrastructure) and agricultural areas (Siebert et al., 2010: Groundwater use for irrigation). Figure 3 shows the result of the second attempt in which the contribution of actual groundwater and surface water is included. The fraction of water demands that is satisfied by groundwater resource (GW_{frac}) is prescribed based on the

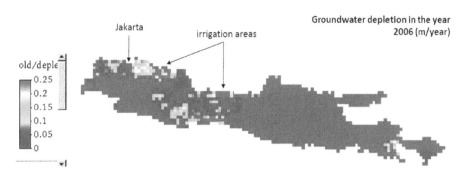

Figure 2. Screen shot of initial depletion $(\mathrm{m\,yr}^{-1})$ results as calculated by the global hydrology model for the island of Java, Indonesia. It shows the depletion of the groundwater as a result of abstraction for irrigation of agricultural areas, but misses known depleted urban areas, such as Jakarta.

Table 1. Calculated depletion $(\mathrm{m\,yr}^{-1}$ for the period 2000–2010) for a selection of cities that are known to subside.

Country	City	Depletion $(\mathrm{m\,yr}^{-1})$
United States	San Jose	3.27
Mexico	Mexico City	1.30
United States	Tucson	1.29
United States	San Francisco	1.25
Iran	Tehran	1.23
United States	Sacramento	1.18

aforementioned inventories. This outcome resembles much closer the expected outcome: dominant depletion in the urban areas.

In Fig. 4, the results of the same model run for the South China Sea area are depicted (depletion in $\mathrm{m\,yr}^{-1}$). This matches known areas of subsidence and groundwater extraction, such as the western half of Taiwan, the cities of Hong Kong, Shanghai, Qingdao and Beijing. It also captured the North China Plain, which is an agricultural area with strongly depleted aquifers. These results give confidence in the validity of the model.

3.3 A global city database of groundwater abstractions and discussion

Based on the model calculations, we are now able provide depletion per year for different urban areas. Table 1 lists some examples of calculated depletion rates for a selected number of cities for the period 2000–2010.

Most of the cities that are known to have depletion of the aquifers are found on the list. Some cities, such as Ho-Chi-Minh City (Vietnam) or Albuquerque (New Mexico, USA) are, however, missed. One of the possible explanations for this is the unknown contribution of irrigation. This may be overestimated in urban areas, which suffer from land surface sealing as a result of building. The rates as calculated by the model are on the lower end of reported depletion rates. This

may be the consequence of the scale that we use: depletion rates are averaged over a 10×10 km gridcel. Groundwater extraction and depletion, specifically for industrial sites, may however be much more local than that. Another interesting outcome is the prominent position of the city of San Jose (USA) with $3.27\,\mathrm{m\,yr}^{-1}$ depletion calculated for the 2000–2010 period (Table 1). In reality, depletion in the San Jose area has essentially stopped after dramatic subsidence forced the local government to reduce groundwater abstraction (Ingebritsen and Jones, 1999). Surface water is now being imported from elsewhere and the imported water is partly used for recharging the aquifer systems. The model however, still calculates the consumption of ground water in San Jose, as in the model local surface water availability is insufficient.

4 Towards a global land subsidence map

The PCR-GLOBWB outcomes will be used to force a two-layer global Modflow model, consisting of an upper unit with low permeability and a deeper confined aquifer system. The groundwater model will yield hydraulic heads. A decrease in hydraulic head will not automatically imply that subsidence will occur. If the deposits consist of sand, or are (over)consolidated, the hydraulic pressure decrease and effective stress increase will not lead to inelastic volume reduction. There is enough global information on the surface geology (e.g. Dürr, et al., 2005) to be able to focus only on those areas that are susceptible to subside. The required subsurface information to parameterize the geotechnical model is however unavailable on a global scale at this time, but will be approached by using different scenarios of subsurface build-up.

The outcomes will be compared to measured or modeled land level lowering in well-known case study areas, such as Jakarta and the Vietnamese Mekong Delta. The final map will include also future land subsidence rates under different development scenarios for the entire earth, and includes a sensitivity test for different subsurface build-up. The entire map will be used as input for a global flood risk model. This

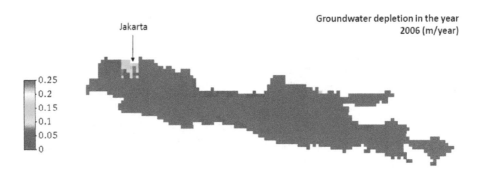

Figure 3. Screen shot of revised depletion (m yr^{-1}) results as calculated by the global hydrology model for the island of Java, Indonesia. It shows the depletion of the groundwater as a result of ground water abstraction in the urban area of Jakarta. The calculated depletion in the irrigation areas is now minor.

Figure 4. Screen shot of depletion (m yr^{-1}) results as calculated by the global hydrology model for the east coast of China and Taiwan. The cities of Hong Kong, Shanghai, Qingdao and Beijing can be recognised. The large area in the north is the North China Plain, which is an agricultural area.

will be done within the concept of a dynamic DEM (digital elevation model). Within this concept, the currently available height information (DEM) for a certain area is updated with the cumulative subsidence (or uplift) over a certain period of time. The predicted global land subsidence effects will be included in flood hazard and storm surge assessments within a global flood risk model.

Acknowledgements. The authors thank their colleagues Marijn Kuijper (Deltares Research Institute), Hessel Winsemius (Deltares Research Institute), Inge de Graaf (Utrecht University) and Marc Bierkens (Utrecht University, Deltares Research Institute) for data, support and discussion. This paper is a part of the World Resources Institute project Flood Risk and Intervention Assessment for Global Cities funded by the Netherlands Ministry of Infrastructure and Environment and the Netherlands Ministry of Foreign Affairs.

References

Abidin, H. Z., Andreas, H., Djaja, R., Darmawan, D., and Gamal, M.: Land subsidence characteristics of Jakarta between 1997 and 2005, as estimated using GPS surveys, GPS Solut., 12, 23–32, 2008.

Bakr, M.: Influence of Groundwater Management on Land Subsidence in Deltas A Case Study of Jakarta (Indonesia), Water Resour. Manag., 29, 1541–1555, doi:10.1007/s11269-014-0893-7, 2015.

Church, J. A., Clark, P. U., Cazenave, A., Gregory, J. M., Jevrejeva, S., Levermann, A., Merrifield, M. A., Milne, G. A., Nerem, R. S., Nunn, P. D., Payne, A. J., Pfeffer, W. T., Stammer, D., and Unnikrishnan, A. S.: Sea Level Change, in: Climate Change 2013: The Physical Science Basis. Contribution of Working Group I to the Fifth Assessment Report of the Intergovernmental Panel on Climate Change, edited by: Stocker, T. F., Qin, D., Plattner, G.-K., Tignor, M., Allen, S. K., Boschung, J., Nauels, A., Xia, Y., Bex, V., and Midgley, P. M., Cambridge University Press, Cambridge, United Kingdom and New York, NY, USA, 2013.

De Graaf, I. E. M., van Beek, L. P. H., Wada, Y., and Bierkens, M. F. P.: Dynamic attribution of global water demand to surface water and groundwater resources: Effects of abstractions and return flows on river discharges, Adv. Water Resour., 64, 21–33, 2014.

Dürr, H. H., Meybeck, M., and Dürr, S. H.: Lithologic composition of the Earth's continental surfaces derived from a new digital map emphasizing riverine material transfer, Global Biogeochem. Cy., 19, GB4S10, doi:10.1029/2005GB002515, 2005.

Erban, L. E., Gorelick, S. M., and Zebker, H. A.: Groundwater extraction, land subsidence, and sea-level rise in the Mekong Delta, Vietnam, Environ. Res. Lett., 9, 084010, doi:10.1088/1748-9326/9/8/084010, 2014.

Galloway, D. L. and Burbey, T. J.: Review: Regional land subsidence accompanying groundwater extraction, Hydrogeol. J., 19, 1459–1486, 2011.

Ingebritsen, S. E. and Jones, D. R.: Santa Clara Valley, California – A case of arrested subsidence, in: Land Subsidence in the United States by Galloway, D., Jones, D. R., and Ingebritsen, S. E., U.S. Geological Survey Circular 1182, 177 pp., 15–22, 1999.

Leake, S. A. and Galloway, D. L.: MODFLOW ground-water model – User guide to the Subsidence and Aquifer-System Compaction Package (SUB-WT) for water-table aquifers, U.S. Geological Survey, Techniques and Methods, 6-A23, 42 pp., 2007.

McDonald, M. G. and Harbaugh, A. W.: A modular three-dimensional finite-difference ground-water flow model: Techniques of Water-Resources Investigations of the United States Geological Survey, Book 6, Chapter A1, 586 pp., 1988.

McDonald, R. I., Weber, K., Padowski, J., Flörke, M., Schneider, C., Green, P. A., Gleeson, T., Eckman, S., Lehner, B., Balk, D., Boucher, T., Grill, G., and Montgomery, M.: Water on an urban planet: Urbanization and the reach of urban water infrastructure, Global Environ. Chang., 27, 96–105, 2014.

Siebert, S., Burke, J., Faures, J. M., Frenken, K., Hoogeveen, J., Döll, P., and Portmann, F. T.: Groundwater use for irrigation – a global inventory, Hydrol. Earth Syst. Sci., 14, 1863–1880, doi:10.5194/hess-14-1863-2010, 2010.

Sutanudjaja, E. H., van Beek, L. P. H., de Jong, S. M., van Geer, F. C., and Bierkens, M. F. P.: Large-scale groundwater modeling using global datasets: a test case for the Rhine-Meuse basin, Hydrol. Earth Syst. Sci., 15, 2913–2935, doi:10.5194/hess-15-2913-2011, 2011.

Sutanudjaja, E. H., van Beek, L. P. H., de Jong, S. M., van Geer, F. C., and Bierkens, M. F. P.: Calibrating a large-extent high-resolution coupled groundwater-land surface model using soil moisture and discharge data, Water Resour. Res., 50, 687–705, 2014.

Sutanudjaja, E. H., van Beek, L. P. H., Bosmans, J. H. C., Drost, N., de Graaf, I. E. M., de Jong, K., Peßenteiner, S., Straatsma, M. W., Wada, Y., Wanders, N., Wisser, D., and Bierkens, M. F. P.: PCR-GLOBWB 2.0: a 5 arc-minute global hydrological and water resources model, Geosci. Model Dev. Discuss., in preparation, 2015.

Van Beek, L. P. H.: Forcing PCR-GLOBWB with CRU data, [online] available at: http://vanbeek.geo.uu.nl/suppinfo/vanbeek2008.pdf (last access: 2 June 2015), 2008.

Van Beek, L. P. H. and Bierkens, M. F. P.: The Global Hydrological Model PCR-GLOBWB: Conceptualization, Parameterization and Verification, [online] available at: http://vanbeek.geo.uu.nl/suppinfo/vanbeekbierkens2009.pdf (last access: 2 June 2015), 2009.

Van Beek, L. P. H., Wada, Y., and Bierkens, M. F. P.: Global monthly water stress: 1. Water balance and water availability, Water Resour. Res., 47, W07517, doi:10.1029/2010WR009791, 2011.

Wada, Y., Wisser, D., and Bierkens, M. F. P.: Global modeling of withdrawal, allocation and consumptive use of surface water and groundwater resources, Earth Syst. Dynam., 5, 15–40, doi:10.5194/esd-5-15-2014, 2014.

9

Application of InSAR and gravimetric surveys for developing construction codes in zones of land subsidence induced by groundwater extraction: case study of Aguascalientes, Mexico

J. Pacheco-Martínez[1,a], S. Wdowinski[2], E. Cabral-Cano[3], M. Hernández-Marín[4], J. A. Ortiz-Lozano[1],
T. Oliver-Cabrera[2], D. Solano-Rojas[2], and E. Havazli[2]

[1]Departamento de Construcción y Estructuras, Universidad Autónoma de Aguascalientes, Aguascalientes,
Mexico
[2]Department of Marine Geosciences, University of Miami, Miami, Florida, USA
[3]Unidad de Geomagnetismo y Exploración, Universidad Nacional Autónoma de Mexico, D.F., Mexico, USA
[4]Departamento de Geotécnia e Hidráulica, Universidad Autónoma de Aguascalientes, Aguascalientes, Mexico
[a]now at: Department of Marine Geosciences, University of Miami, Miami, USA

Correspondence to: J. Pacheco-Martínez (jesus.pacheco@edu.uaa.mx)

Abstract. Interferometric Synthetic Aperture Radar (InSAR) has become a valuable tool for surface deformation monitoring, including land subsidence associated with groundwater extraction. Another useful tools for studying Earth's surface processes are geophysical methods such as Gravimetry. In this work we present the application of InSAR analysis and gravimetric surveying to generate valuable information for risk management related to land subsidence and surface faulting. Subsidence of the city of Aguascalientes, Mexico is presented as study case. Aguascalientes local governments have addressed land subsidence issues by including new requirements for new constructions projects in the State Urban Construction Code. Nevertheless, the resulting zoning proposed in the code is still subjective and not clearly defined. Our work based on gravimetric and InSAR surveys is aimed for improving the subsidence hazard zoning proposed in the State Urban Code in a more comprehensive way. The study includes a 2007–2011 ALOS InSAR time-series analysis of the Aguascalientes valley, an interpretation of the compete Bouguer gravimetric anomaly of the Aguascalientes urban area, and the application of time series and gravimetric anomaly maps for improve the subsidence hazard zoning of Aguascalientes City.

1 Introduction

The city of Aguascalientes (725 000 inhabitants), as well as other medium-sized cities (with a total of 275 000 inhabitants), is located within the Aguascalientes Valley, 430 km NW of Mexico City (Fig. 1). The growing population and increased agricultural and industrial activities along the valley have resulted in an intensive groundwater extraction, leading to develop land subsidence since earlies 1970's (Pacheco-Martínez et al., 2013).

In Aguascalientes City, surface faulting was first observed in the late of 1970's, when a group of houses developed cracks and fissures in walls and floors. Because all damaged properties were aligned along a linear feature, several studies concluded that the damage was caused by an active geological fault (Aranda-Gómez, 1989; Lermo et al., 1996).

From the discovery of this fault in the late 1970's until present, new surface faults have developed along the entire valley affecting the urban areas of Cosio, San Francisco del Rincón, San Francisco de los Romos, Jesús María and Aguascalientes City, as well as other smaller towns and rural communities (Fig. 1). As a consequence of this process, several legal disputes concerning land and housing selling affected by faulting were presented in local courts. Thus, sur-

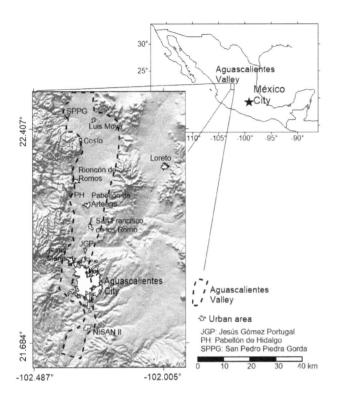

Figure 1. Location of Aguascalientes Valley. The names of minor cities and small towns are provided in the map legend.

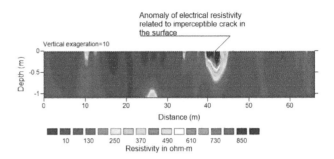

Figure 2. An example of resistivity anomaly related to a hidden or incipient surface fault surveyed in Aguascalientes City (After Pacheco-Martínez et al., 2013). Location of profile is shown in Fig. 3.

face faulting became a major concern for the local government, specially for those bureaus responsible for providing regulatory construction codes and for bureaus dealing with planning and growing of urban infrastructure.

In this work, we present geophysical data and geodetic observations, which provide information to improve the risk zoning defined in the current construction code of Aguascalientes. These observations include ALOS InSAR time-series the Aguascalientes valley, and gravimetric measurements in the city of Aguascalientes.

2   Land subsidence and related faulting in Aguascalientes City

The first reports of surface faults in the city of Aguascalientes were due to damage that a surface fault caused in several buildings in the late 1970's (Lermo et al., 1996; Aranda-Gómez, 1989), but there were no estimates about the subsidence magnitude at that time. However, currently the affected area by subsidence is around 942 km^2 with rates reaching 10 cm yr^{-1} as shown in Sect. 4.

Ground faulting within the city is a great concern for local population, because of the structural damage that it induces to buildings, as well as to public infrastructure (Pacheco-Martínez et al., 2011; Romero-Navarro et al., 2010). Table 1 shows that there are at least 208 ground faults locations, which affect over 1860 properties (SIFAGG, 2015).

A new and bigger concern to the local government is the damage to heritage buildings in the city caused by ground faulting. In recent years, three historical buildings in the valley have been affected by surface faulting; two of them, the Cathedral of Aguascalientes and the Museum of the Insurgency are listed in the national catalogue of historic buildings.

3   Current geotechnical practices and the Urban Code for the Aguascalientes State

The local government responded to the ground faulting problem by implementing several actions, including the recommendations and requirements for new construction projects which have been incorporated in the Urban Code for the Aguascalientes State. The most relevant statements in the code concerning ground failure damage prevention are:

1. Certification of proficient professionals in geology or geophysics, who have to prove a background in ground faulting studies in order to be certified.

2. A requirement for municipalities to prepare and publish ground faulting maps.

3. Proposal of a zoning risk based on the shortest distance to mapped ground faults.

4. A statement that requires constructors to present geophysical or geological studies prepared by the certificated geologist or geophysics in order to obtain a construction license.

Some of the weaknesses of the Urban Code concerning the ground faulting are:

1. The code does not recognize the phenomenon of regional subsidence of the valley. The ordinance is focused solely to prevent the effects of previously mapped surface faults.

Table 1. Ground faults length in municipalities and number of affected properties. AGS: Aguascalientes, JM: Jesús María, COS: Cosio, PAB: Pabellón, RIN: Rincón de Romos, SFCO: San Fancisco de los Romo.

	AGS	JM	COS	PAB	RIN	SFCO	TOTAL
Number of mapped surface fault locations	66	49	15	37	28	13	208
Accumulated length of ground faults (km)	83.575	60.473	30.620	43.534	47.275	25.231	290.678
Number of damaged properties	1438	183	35	86	35	88	1865

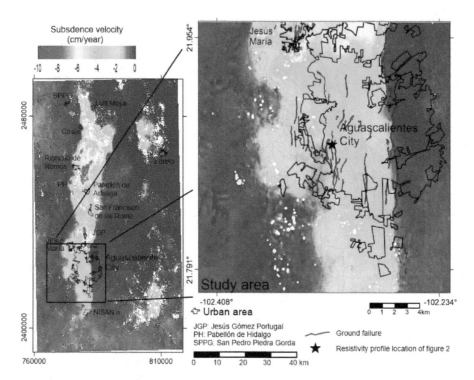

Figure 3. Subsidence velocity map of Aguascalientes Valley as determine from InSAR time series analysis of 2007–2011 ALOS data.

2. The proposed zoning in the code only points out the potential seismic risks to buildings related to the ground fault motion. The zoning does not include considerations about the hazard related to future ground fault evolution nor about the hazard associated to the differential subsidence in progress. The code recognizes three risk zones:

Zone I. High risk: terrain located within a 5 m wide buffer zone (or that determined by a municipality technician) along the axis of the ground failure. The code states that this zone has a high risk for urban development because the zone is under the influence of seismic activity and differential settlements caused by the motion of ground faults.

Zone II. Medium risk: terrain located within a 200-m wide buffer zone on each side from the zone I. The code states that in zone II the buildings or structures are still under the influence of seismic activity, which are originated along ground faults.

Zone III. All the terrain outside of zone I and II. The code states that the constructions in this zone are not influenced by ground faults.

3. The Urban Code does not clearly define what type of geological or geophysical study must be carried out depending the risk zone.

4. The code compels constructors to present geological and geophysical studies for any new constructions authorization, even if new constructions are outside of the subsiding area, for example at the eastern part of Aguascalientes City (see Fig. 3) where neither subsidence nor ground failure have been documented.

Regarding to the ground fault assessments, all studies that certificated geologist and geophysics develop for construction permissions focus on providing information that can lead to determine if the studied area is affected by ground faulting. In affirmative case, the studies have to include the determination of the area affected by the fault.

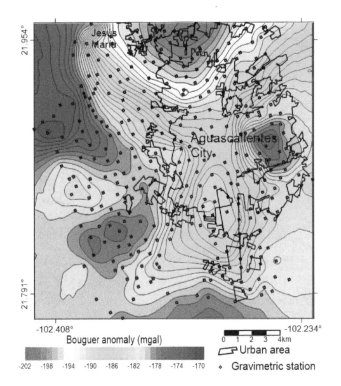

Figure 4. Complete Bouguer anomaly of the Aguascalientes city and surroundings.

The more difficult case is detecting faults at an early stage of formation. In this stage, faults may not be recognizable on the surface because either offset is very small and can be hard to detect by the erosion or by "in situ" pre-construction activities. To overcome this issue, measurements of electrical resistivity profiles of subsoil has been used with relative success to detect ground faulting in an early stage. Fractured subsoil under favorable conditions for the technique application (homogeneous stratigraphy and undisturbed soil) appears as an anomaly of high resistivity (Fig. 2).

However, under unfavourable conditions, such as a subsoil with a complex stratigraphy, deposits of anthropogenic fillings, or in zones with strong lateral water content variations, the interpretation of electrical resistivity data can be ambiguous. In such cases, geophysical measurements are complemented by trench excavation in order to directly explore where resistivity anomalies are observed. Exposed trenches help to verify the existence of incipient ground failures.

4 The role of InSAR and gravimetric methods in subsidence studies

InSAR techniques have been used successfully to study land subsiding areas (Ketelaar, 2009; Galloway and Burbey, 2011). The large-scale coverage of the technique and the relatively wide-spread availability of InSAR imagery have fostered its application for land subsidence studies. In some

studies, InSAR was used to detect the extent of subsiding areas and subsidence rates (Chaussard et al., 2014; Galloway et al., 1998; Bell et al., 2008). Some authors have reported the use of InSAR for hazard evaluation of ground failures related to subsidence (Cabral-Cano et al., 2010, 2011).

The large-scale coverage of SAR data enabled to demonstrate that the large spatial extent of land subsidence is controlled by geologic structures and stratigraphy of the subsiding area (Amelung et al., 1999; Cigna et al., 2012). These studies concluded that groundwater level decrease is the triggering factor of the subsidence process, whereas the presence of deformable sediments determines its occurrence.

In areas where the process of subsidence began long ago, the decrease in groundwater level has taken place along the entire aquifer system, suggesting that subsidence has affected the deformable sediments in greater or lesser extent. In these circumstances, a subsidence map should show the limits of the area which have potential to develop land subsidence. The map should also show those areas in which subsidence is not likely to develop.

In the case of Aguascalientes Valley, the deformation-prone aquifer system is composed of an alluvial sediments sequence with abundant silty sand and gravel contained within a regional graben; thus the limits of the subsiding-prone area are also the limits of the graben. Those limits are clearly identifiable in the subsidence map of Aguascalientes Valley (Fig. 3).

The velocity map of Fig. 3 was elaborated from the processing of 34 SAR images acquired by the ALOS satellite from 2007 to 2011. Images cover two frames (420, 430) from the ascending track 191. Imagery was processed with the differential InSAR technique (D-InSAR) to obtain 28 interferograms and the small baseline (SB) to generate the InSAR time-series (Rosen et al., 2000; Berardino et al., 2002; Lanari et al., 2004). We use for processing ROI_PACK software developed by NASA's Jet Propulsion Laboratory (Rosen et al., 2004). Topographic correction was applied according to Fattahi and Amelung (2013). Pixels with temporal coherence lesser than 0.7 were dismissed for elaborating the velocity map.

In Mexico, the subsurface geology of many subsiding areas consists of sequences of unconsolidated granular strata lying over a rigid bedrock, usually volcanic or carbonated rocks. Consequently, the density contrast between these units is significant, and very favorable for gravimetric analysis. Gravimetry provide in this way valuable information about the configuration of the bedrock in sedimentary basins (Pacheco-Martínez et al., 2006, 2010; Yutsis et al., 2014).

Furthermore, some studies have shown that in geological settings susceptible to develop subsidence by groundwater extraction, there is an inverse correlation between sediment thickness and the gravimetric anomaly (Jachens and Holzer, 1979; Pacheco-Martínez et al., 2006, 2010). In other words, the higher values of the gravimetric anomaly correlate with

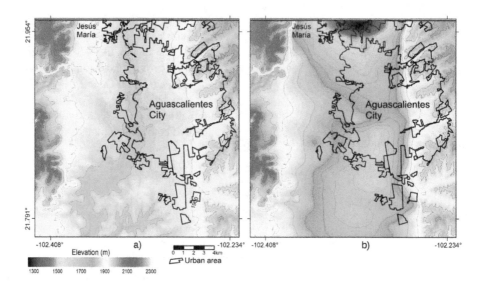

Figure 5. Surface (**a**) and bedrock (**b**) topography in the Aguascalientes City study area.

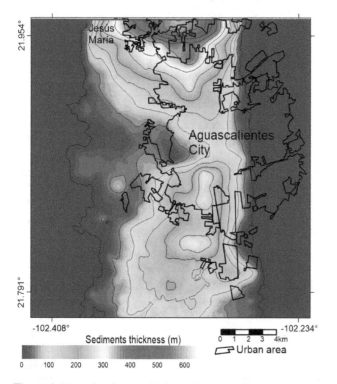

Figure 6. Map of sediments thickness in Aguascalientes City.

small sediment thicknesses. Inversely, lower anomaly values correspond to thicker sedimentary sequences.

Figure 4 shows the complete Bouguer's anomaly of Aguascalientes City subsidence area, which was elaborated from a surveying of 339 ground-based gravimetric measurements. Gravity field data were surveyed using a CG5 gravimeter from Scintrex. Standard procedures for fieldworks and data processing where followed (Telford et al.,

1990) . The gravity data were inverted according to Singh and Guptasarma (2001) using PyGMI software developed by Cole (2015) in order to determine a model of the rocky stratum (Fig. 5). The Rocky stratum model was restrained in several points in which rocky stratum depth is known through lithology logs wells and outcrops in both sides of the graben. Finally, an isopach map (Fig. 6), which is a sediment thickness variations map, was calculated from the difference in elevations of the rocky stratum and the terrain surface.

5   Subsidence hazard zoning map elaboration

The hazard level related to subsidence depends on the existence of deformable sediments. It depends also on the advance degree of the subsidence process, whether superficial sinking and ground failure have been observed. Hence, three hazard factors were considered in order to complete a hazard factors array, from which the hazard level was calculated according to the existence of such factors (Table 2).

The zoning map (Fig. 7) was prepared according to a spatial analysis of the hazard factors presented in Table 2. Each hazard zone and the suggested requirements for geotechnical studies needed to evaluate the existence of ground failure in the terrain of new projects are described bellow from low to high level.

Zone 4 non-existent hazard: neither subsidence nor ground failure have been detected. Furthermore deformable filling thicknesses are negligible in this zone. Standard geotechnical studies should be enough.

Zone 3 low hazard: neither subsidence nor ground failure have been detected. Nevertheless the isopach map shows thicknesses of deformable sediments which can be significant, then subsidence could develop in the future. Standard geotechnical studies plus a superficial geological study

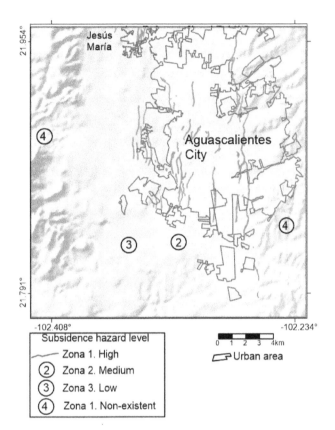

Figure 7. Subsidence hazard zoning map of Aguascalientes City.

Table 2. Hazard factor array and resulting hazard levels. H: High level, M: Medium level, L: Low level, N: non-existent level.

Hazard factors	Hazard levels			
	H	M	L	N
Deformable sediments existence	Yes	Yes	Yes	No
Detected surface sinkings	Yes	Yes	No	No
Ground failure precense	Yes	No	No	No

ning to better define the needed technical studies according to the hazard level of the zone in which new constructions are planned.

Due to the dynamic nature of the subsidence, the hazard map have to be updated with new information generated through the geotechnical studies required for construction permissions of new projects. New SAR-images acquisitions and their processing to obtain updated velocity maps will allow regular updates of the hazard map.

Acknowledgements. J. Pacheco-Martínez thanks CONACYT for financial support: "Investigación apoyada por el CONACYT". S. Wdowinski and E. Cabral-Cano acknowledge support through NASA-ROSES grant NNX12AQ08G. E. Cabral-Cano is supported by UNAM-PAPIIT projects IN104213-2, and IN109315-3. This material is partly based on data provided by the UNAVCO Facility with support from the National Science Foundation (NSF) under award EAR-1338091.

References

Amelung, F., Galloway, D. L., Bell, J. W., Zebker, H. A., and Laczniac, R. J.: Sensing the Ups and Downs of Las Vegas: InSAR Revelas Structural Control of Land Subsidence and Aquifer-System Deformation, Geology, 27, 483–486, 1999.

Aranda-Gómez, J. J.: Geología Preliminar Del Graben de Aguascalientes, Revista Mexicana de Ciencias Geológicas, 8, 22–32, 1989.

Bell, J. W., Amelung, F., Ferretti, A., Bianchi, M., and Novali, F.: Permanent Scatterer InSAR Reveals Seasonal and Long-Term Aquifer-System Response to Groundwater Pumping and Artificial Recharge, Water Resour. Res., 44, W02407, doi:10.1029/2007WR006152, 2008.

Berardino, P., Fornaro, G., Lanari, R., and Sansosti, E.: A New Algorithm for Surface Deformation Monitoring Based on Small Baseline Differential SAR Interferograms, IEEE T. Geosci. Remote, 40, 2375–2383, 2002.

Cabral-Cano, E., Osmanoglu, B., Dixon, T., Wdowinski, S., Demets, C., Cigna, F., and Díaz-Molina, O.: Subsidence and Fault Hazard Maps Using PSI and Permanent GPS Networks in Central Mexico, Proceedings of Eighth International Symposium on Land Subsidence, 255–59, IAHS Red Books, Querétaro, Mexico: IAHS Press, Publication no. 339, 2010.

Cabral-Cano, E., Díaz-Molina, O., and Delgado-Granados, H.: Subsidencia Y Sus Mapas de Peligro: Un Ejemplo En El Área Noror-

searching for evidences of pre-subsidence cracks should be enough.

Zone 2 medium hazard: subsidence is developing, but ground failure has not been detected. Same requirements of level 3 plus a geophysical study to find out incipient ground failure.

Zone 1 high hazard: this zone corresponds to the existent ground failure traces. Same requirements of level 2 plus a detailed study to determine the trace width over which differential subsidence could cause damage on new constructions.

6 Conclusions

The zoning map of subsidence and surface faulting prone area is an important element in risk management of land subsidence and associated surface faults. In this vein, Gravimetric measurements and InSAR-derived velocity maps provide valuable information for its generation. The first allows determining sediment thickness prone to deform, and the second one helps determining accurately where subsidence is present.

The combination of both information sources, along with reliable subsurface geological information, allows us to determine in a comprehensible and sustainable way the subsidence hazard map for Aguascalientes. This map will allow government agencies responsible for urban growth plan-

iental de La Zona Metropolitana de La Ciudad de Mexico, Bo-letín de La Sociedad Geológica Mexicana, 63, 53–60, 2011.

Chaussard, E., Wdowinski, S., Cabral-Cano, E., and Amelung, F.: Land Subsidence in Central Mexico Detected by ALOS InSAR Time-Series, Remote Sens. Environ., 140, 94–106, 2014.

Cigna, F., Osmanoğlu, B., Cabral-Cano, E., Dixon, T. H., Ávila-Olivera, J. A., Garduño-Monroy, V. H., DeMets, C., and Wdowinski, S.: Monitoring Land Subsidence and Its Induced Geological Hazard with Synthetic Aperture Radar Interferometry: A Case Study in Morelia, Mexico, Remote Sens. Urban Environ., 117, 146–161, 2012.

Cole, P.: PyGMI-Python Geophysical Modeling and Interpretation (version 2.2.6). South Africa: Council for Geosciences, Geological Survey of South Africa, available at: http://patrick-cole.github.io/pygmi/index.html, 2015.

Fattahi, H. and Amelung, F.: DEM Error Correction in InSAR Time Series, IEEE T. Geosci. Remote, 51, 4249–4259, 2013.

Galloway, D. L. and Burbey, T. J.: Review: Regional Land Subsidence Accompanying Groundwater Extraction, Hydrogeol. J., 19, 1459–1486, 2011.

Galloway, D. L., Hudnut, K. W., Ingebritsen, S. E., Phillips, S. P., Peltzer, G., Rogez, F., and Rosen, P. A.: Detection of Aquifer System Compaction and Land Subsidence Using Interferometric Synthetic Aperture Radar, Antelope Valley, Mojave Desert, California, Water Resour. Res., 34, 2573–2585, 1998.

Jachens, R. C. and Holzer, T. L.: Geophysical Investigation of Ground Failure Related to Ground Water Withdrawal-Picacho Basin, Arizona, Ground Water, 17, 574–585, 1979.

Ketelaar, V. B. H.: Satellite Radar Interferometry – Subsidence Monitoring Techniques, Springer, 2009.

Lanari, R., Mora, O., Manunta, M., Mallorqui, J. J., Berardino, P., and Sansosti, E.: A Small-Baseline Approach for Investigating Deformations on Full-Resolution Differential SAR Interferograms, IEEE T. Geosci. Remote, 42, 1377–1386, 2004.

Lermo, J., Nieto-Obregón, J., and Zermeño, M.: Faults and Fractures in the Valley of Aguascalientes. Preliminary Microzonification, Proceedings of the Eleventh World Conference on Earthquake Engineering, Acapulco Mexico, 1996.

Pacheco-Martínez, J., Arzate-Flores, J., Rojas, E., Arroyo, M., Yutsis, V., and Ochoa, G.: Delimitation of Ground Failure Zones due to Land Subsidence Using Gravity Data and Finite Element Modeling in the Querétaro Valley, Mexico, Eng. Geol., 84, 143–160, 2006.

Pacheco-Martínez, J., Arzate-Flores, J. A., López-Doncel, R., Barboza-Gudiño, R., Mata-Segura, J. L., Del-Rosal-Pardo, A., and Aranda-Gómez, J. J.: Zoning Map of Ground Failure Risk due to Land Susbidence of San Luís Potosí, Mexico, Proceedings of Eighth International Symposium on Land Subsidence, 179–184, Querétaro, Mexico: IAHS Press, Publication no. 339, 2010.

Pacheco-Martínez, J., Zermeño-De-León, M. E., Ortiz-Lozano, J. A., Solís-Pinto, A., Romero-Navarro, M. A., Aguilar-Valdés, F. J., and Fuente-López, J. A.: Soil cracks related to land subsidence. The main geotechnical hazard affecting to construction in Aguascalientes City, Proceedings of the 14th Pan-American Conference on Soil Mechanics and Geotechnical Engineering, 8, 1–8, Toronto, Canada, 2011.

Pacheco-Martínez, J., Hernandez-Marín, M., Burbey, T. J., González-Cervantes, N., Ortíz-Lozano, J. A., Zermeño-De-Leon, M. E., and Solís-Pinto, A.: Land Subsidence and Ground Failure Associated to Groundwater Exploitation in the Aguascalientes Valley, Mexico, Eng. Geol., 164, 172–186, 2013.

Romero-Navarro, M. A., Pacheco-Martínez, J., Ortiz-Lozano, J. A., Zermeño-De-León, M. E., Araiza-Garaygordobil, G., and Mendoza-Otero, E.: Land Subsidence in Aguascalientes Valley, Mexico. Historical Review and Present Situation, Proceedings of Eighth International Symposium on Land Subsidence, 207–9, IAHS Red Books, Querétaro, Mexico: IAHS Press, Publication no. 339, 2010.

Rosen, P. A., Hensley, S., Joughin, I. R., Li, F. K., Madsen, S. N., Rodríguez, E., and Goldstein, R. M.: Synthetic Aperure Radar Interferometry, Proceedings of the IEEE, 88, 333–382, 2000.

Rosen, P. A., Hensley, S., and Peltzer, G.: Updated Repeated Orbit Interferometry Package Released, Eos, Transactions American Geophysical Union, 85, 47, 2004.

SIFAGG: Sistema de Información de Fallas Geológicas Y Grietas on Line: http://www.aguascalientes.gob.mx/sop/sifagg/web/mapa.asp, Cartografía Digital Interactiva, Aguascalientes, Mexico: Secretaría de Obras Públicas del Estado de Aguascalientes, 2015.

Singh, B. and Guptasarma, D.: New Method for Fast Computation of Gravity and Magnetic Anomalies from Arbitrary Polyhedral, Geophysics, 66, 521–526, 2001.

Telford, W. M., Geldart, L. P., and Sheriff, R. E.: Applied Geophysics, New York, USA: Cambridge University Press, 1990.

Yutsis, V., Aranda-Gómez, J. J., Arzate-Flores, J. A., Bohnel, H., Pacheco Martínez, J., and López-Loera, H.: Maar Geophysics: Valle de Santiago Study, Proceedings of The 5th International MAAR Conference, 32–33, Querétaro, Mexico, 2014.

10

Groundwater management based on monitoring of land subsidence and groundwater levels in the Kanto Groundwater Basin, Central Japan

K. Furuno[1], A. Kagawa[2], O. Kazaoka[2], T. Kusuda[3], and H. Nirei[4]

[1]IUGS-GEM Japan branch, 6-41-8, Kotehashidai, Hanamigawa-ku, Chiba 262-0005, Japan
[2]Research Institute of Environmental Geology, Chiba, Japan
[3]Chiba Environmental Foundation, Chiba, Japan
[4]The Geo-pollution Control Agency, Chiba, Japan

Correspondence to: K. Furuno (kuniofurunojp@gmail.com)

Abstract. Over 40 million people live on and exploit the groundwater resources of the Kanto Plain. The Plain encompasses metropolitan Tokyo and much of Chiba Prefecture. Useable groundwater extends to the base of the Kanto Plain, some 2500 to 3000 m below sea level. Much of the Kanto Plain surface is at sea level. By the early 1970s, with increasing urbanization and industrial expansion, local overdraft of groundwater resources caused major ground subsidence and damage to commercial and residential structures as well as to local and regional infrastructure. Parts of the lowlands around Tokyo subsided to 4.0 m below sea level; particularly affected were the suburbs of Funabashi and Gyotoku in western Chiba. In the southern Kanto Plain, regulations, mainly by local government and later by regional agencies, led to installation of about 500 monitoring wells and almost 5000 bench marks by the 1990's. Many of them are still working with new monitoring system. Long-term monitoring is important. The monitoring systems are costly, but the resulting data provide continuous measurement of the "health" of the Kanto Groundwater Basin, and thus permit sustainable use of the groundwater resource.

1 Introduction

The Kanto plain is also called "the Kanto fore-arc submarine basin" or "the paleo-Kanto-submarine basin", based on surrounding geologic and physiographic features such as Nasu Volcanic Zone, Fuji Volcanic Zone, Japan trench and Izu-Ogasawara Trench (Nirei et al., 1990). Over 40 million people live on the Kanto plain which includes the Tokyo metropolis. The waterfront area around Tokyo Bay mostly includes both alluvium and man-made strata such as reclaimed areas and is densely covered with houses and factories. Sediments of the Kanto basin are 2500–3000 m thick and are of Miocene to Holocene age. The basin is also called the Kanto groundwater basin from the standpoint of the production of groundwater, natural gas and iodine which are contained in fossil sea water. Uncontrolled use of groundwater has caused land subsidence. To ensure effective use of groundwater, a monitoring system for their management has been developed. The system has about 500 self-recording observation wells, almost 5000 bench marks (leveled every year). Over-pumping of the groundwater (fresh water and brine groundwater) resulted in a serious land subsidence problem in various areas, especially alluvial plains and reclaimed area so that it was inevitable to control the groundwater use to solve the problem (Aihara et al., 1969). Consequently, systematic observations of pumping up volumes, groundwater level changes and land subsidence records for the effective groundwater use have been carried out for preventing the land subsidence from a geological point of view (Research Committee on Land subsidence Prevention in Southern Kanto District, 1974).

2   Kanto groundwater basin and underground fluid resources

The Kanto Paleo-submarine basin can be called the Kanto groundwater basin from the standpoint of the distribution and the fluidity of the underground fluid resources (Nirei and Furuno, 1986). The bottom of the groundwater basin corresponds to the base of Kanto tectonic basin, which is situated at 3000–2500 m below the surface. The sediments of the Kazusa sub-groundwater basin in the basin contains brine groundwater (fossil sea water) including natural gas and iodine (Figs. 1, 2). The equivalent layer of the Kazusa Group that extended to the western and northern part of the Kanto plain is composed of alternating beds of coarse- to fine-grained deposits is suitable for groundwater. The lower part of the Shimosa sub-groundwater basin contains groundwater colored with humid material from the northern to central part of the Boso peninsula, and the colored groundwater is unsuitable for drinking. In the upper part of the Shimosa sub-groundwater basin, most of the aquifer layers contain fresh water; this basin had the largest pumpage volume of water in the Kanto groundwater basin. However, the more pumping up volume increased, the lower the groundwater level became in the Shimosa upper sub-groundwater basin. Which is also resulted in lowering of groundwater level even in the alluvial deposits as well as subsiding of ground surface. It is now widely confirmed that the amount of ground subsidence is affected by the thickness of alluvial deposits (Research Institute of Environmental Geology, 1979).

3   Status of underground fluid resource use

Natural gas and groundwater have been produced as the underground fluid resources in the Kanto groundwater basin. The former is mainly derived from the Kazusa sub-groundwater basin. Annual production of natural gas was 520 million Nm^3 in 1980, 460 million Nm^3 in 2003 and 440 million Nm^3 in 2011. The natural gas produced from the Kazusa sub-groundwater basin was contained in fossil sea water of the Kazusa group. Accordingly, it was inevitable to pump up the fossil sea water when releasing natural gas, which frequently caused a land subsidence problem. Therefore, the pumping is now limited to the area of hills and the Pacific coast in the southern Kanto, a considerable distance from the metropolitan area. Recently, the groundwater is mainly pumped up at the marginal area of the Shimosa lower sub-groundwater basin and the upper Shimosa sub-groundwater basin. Annual pumped volume was 3.4 million $m^3 day^{-1}$ in 1978 and 3.1 million $m^3 day^{-1}$ in 2003 (Prefectural Governors' Committee for Land subsidence Prevention in Kanto District, 1983, 2008) and 2.8 million $m^3 day^{-1}$ in 2015 (Ministry of Environment, 2015). These data were obtained in the limited area where reporting of pumped volume is required of groundwater users by regulation for preventing land subsi-

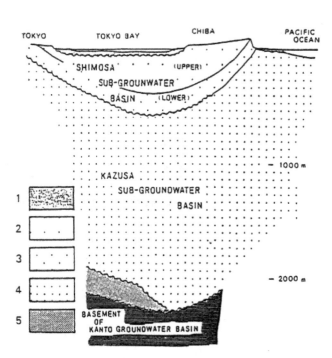

Figure 1. Schematic section of the Kanto groundwater basin: 1: alluvial deposits, 2: upper part of the Shimosa Group, 3: lower part of the Shimosa Group, 4: Kazusa Group, 5: Miura Group.

Figure 2. Kanto groundwater basin and the distribution of monitoring wells and benchmarks. Large dots: monitoring wells, small dots: benchmarks. A–E: monitoring wells which data were used in Fig. 3.

dence. This area covers about half of the Kanto groundwater basin. Consequently, the actual pumping volume is estimated to have been twice as much as the values mentioned above. The groundwater has been used for (1) drinking water supply, (2) industries, (3) agriculture (4) buildings and (5) other uses. The used volume increases generally according as drinking water supply, industries and so on increase in scale. However the used volume shows a little difference based on each autonomy on the Kanto groundwater basin and the scale of the volume used by the agriculture and the building is different at each prefectural government. The volume of other uses is the smallest amount of the volume which is the fifth order. The fact that the drinking water supply make up the largest amount of the ground water use may be the consequence of the concept that necessary of groundwater use is for the benefit of inhabitants (Shibasaki, 1976; Research Group for Water Balance, 1976).

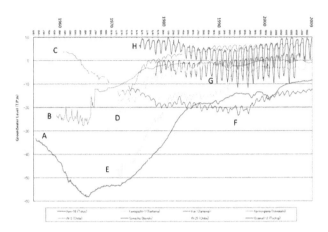

Figure 3. Historical change of the groundwater level in the Kanto groundwater basin. A: Ken-15 (Tokyo), B: Kannongawa (Kawasaki), C: W-2 (Chiba), D: W-25 (Chiba), E: Kawaguchi-1 (Saitama), F: Kuki (Saitama), G: Sowachu (Ibaraki), H: Oyama1–2 (Tochigi).

4   Control of pumping volume and change of ground level and land subsidence

Groundwater and brine groundwater were pumped up disorderly by companies, farmers and local governments for industry, agriculture, drinking water supply and production of natural gas and iodine before 1960s. However, the regulation by national government began in the middle 1960s and local governments began from the 1970s as progress of the subsidence became remarkable in order to prevent land subsidence.

Each local government has attempted to regulate the pumping volume of the groundwater in order to prevent land subsidence from 1970s. For example, users are required to report the pumping volume compulsorily. Criteria of depth and diameter of discharge pipes were set up (Prefectural Governors' Committee for Land subsidence Prevention in Kanto District, 1983). An agreement to control the pumping of fossil sea water containing natural gas was also made between the Chiba prefecture administration and each gas production company. The condition of groundwater levels in the whole Kanto groundwater basin has recently been clarified. The number of observation wells was 375 in total as of 1982 (Furuno et al., 1983), 459 as of 1991 and about 500 as of 2010s. The amount of land subsidence has been measured by yearly precise leveling. The number of benchmarks attained 4880 and the total length of the leveling, 7363.6 km as of 1983, 4945 benchmarks and 7387.6 km as of 1986, about 5000 benchmarks and 7500 km as of 2010s, The groundwater management in the shimosa sub-groundwater basin is described as below.

5   Control of groundwater over-pumping

In the Keihin industrial zone (coastal zone of Tokyo, Yokohama and Kawasaki) located in the south western part of the Kanto groundwater basin, the land subsidence caused by over-pumping of groundwater were recognized in early 1940s. As a result, industrial aqueducts were established to replace the groundwater pumping from the coastal zone of Kawasaki and Yokohama. In the industrial zone, the pumping was controlled in the 1950s by regulations on the depth of wells and the diameter of discharge pipes. In the industrial zone of Tokyo and its suburbs, the pumping volume was controlled in the 1960s by two laws concerning industrial water and pumping at building sites. However, the countermeasures for land subsidence were often loosely applied by groundwater users, especially the industrial users. In the 1970s, with the rapid growth of the Japanese economy, land subsidence have been recognized as the most important problems. In such a social background, not only the national government enacted the strict criteria for the regulation concerning the industrial water and the pumping up law for building use, but also the prefectural governments in the southern part of the Kanto groundwater basin strictly applied these regulations. As a result, the groundwater level and the land subsidence stopped gradually. It was even recognized in certain regions that the groundwater level rose. However, excessive regulations produce another problem in some areas. Tokyo Station and Ueno Station in Tokyo set up deep underground platforms before early 1960s when groundwater level was low. Groundwater level rose from late 1960s and buoyancy acted on the building. Those stations had to set up anchors and weights as a countermeasure of buildings rising by buoyancy. Rising of groundwater level in man-made strata and Holocene deposit increase the risk of liquefaction-fluidization. We need to keep proper groundwater level.

6 Change of groundwater level

It was in the early 1960s that the groundwater level fell to 60 m below sea level in the coastal industrial zone of Tokyo metropolis. After that, the recovery of the groundwater level was monitored according to the pumping up control. In the 1970s the regulations were strictly adhered to in the southern Kanto groundwater basin. Accordingly, in the years from 1975 to 1980, the groundwater level recovered 30–40 m in the area where groundwater level lowered. Thus the groundwater level showed a recovering tendency in the lowest groundwater level area. The area where groundwater level dropped 20–30 m below sea level has expanded. In other words, a center of groundwater depression moved to the northern part because the pumping control in the southern Kanto groundwater basin is applied strictly. Figure 3 shows historical groundwater change of the monitoring wells and Fig. 4 shows groundwater level of Kanto groundwater basin (Shimosa sub-groundwater) in 2011 based on the monitoring wells. As Fig. 4 shows, groundwater level is not so low as the former 60 but 20 m below sea level. The lowest groundwater level is stable around 10–20 m below sea level recently. We should think about the use of appropriate groundwater based on monitoring. Drinking water should have top priority.

7 Change of land subsidence

The land surface has been sinking to below sea level in the downtown Tokyo area for mixed industry and residential uses (Shitamachi) since 1910s. Coincidentally, the areas showing ground level lower than mean sea level and high water level of ordinary spring tide of Tokyo Bay were widely scattered. Recently, however, these phenomena have been recognized in alluvial area not only in Shitamachi of Tokyo, but also in suburbs of Tokyo such as Kawasaki and Yokohama in Kanagawa Prefecture, Funabashi, Gyotoku and Urayasu on western Chiba Prefecture. This land subsidence has been stabilized by the regulation of the pumping. The land subsidence was stopped in certain areas and the ground surface slightly uplifted by the recovery of the groundwater level. On the other hand, in the area from central to northern part of the Kanto groundwater basin, land subsidence has begun to sink as a result of over-pumping. That is, land subsidence area is moving northward. Land subsidence before 1970s spread lowland area around Tokyo Bay and Kujyukuri in Boso Peninsula which faced the Pacific Ocean (Fig. 5).

8 Conclusion

It is necessary to establish a more effective monitoring system for continuous observation of the groundwater pumping, groundwater level and land subsidence, and to decide the pumping volume in consideration of the change of the groundwater level and the ground movement with the law

July 2011

Figure 4. Confined groundwater level in the Kanto groundwater basin (especially the Shiosa sub-groundwater basin). Lowest contour line shows 10 m below sea level.

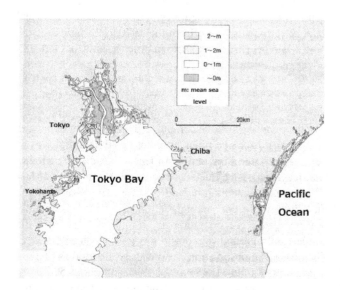

Figure 5. Lowland area around Tokyo Bay and Kujyukuri which faced the Pacific Ocean.

of dynamic equilibrium between man and nature due to the relation among human groups. In other words, the groundwater resources can be used while the health condition of the groundwater basin is examined by periodical health checks. The long-term monitoring of land subsidence and groundwater is important for the sustainable use of the groundwater in the groundwater basin.

Acknowledgements. We sincerely appreciate members of Research Institute of Environmental Geology, Chiba and members of the local governments of Kanto district who have developed a monitoring system and kept on recording land subsidence and groundwater level.

References

Aihara, S., Ugata, H., Miyazawa, K., and Tanaka, U.: Problems on groundwater control in Japan, in: Land subsidence, IAHS-UNESCO, 2, 635–644, 1969.

Furuno, K., Nemoto, K., Takanashi, Y., Yada, T., Takizawa, H., Oshima, K., Saotome, W., and Nirei, H.: The Monitoring of land subsidence and groundwater head in the Kanto groundwater basin, Bulletin of Chiba Prefectural Research Institute for Environmental Pollution, 15, 99–108, 1983.

Ministry of Environment: Review of land subsidence, available at: http://www.env.go.jp/water/jiban/chinka.html (last access: 20 September 2015), 2015 (in Japanese).

Nirei, H. and Furuno, K.: Development of Quaternary Resources and Environmental Protection: Status of Underground fluid resources use in the Kanto groundwater basin, in: Recent Progress of Quaternary Research in Japan, edited by: Kaizuka, S., Aso, M., Endo, K., et al., National Committee for Quaternary Research in Japan, Tokyo, 11, 71–80, 1986.

Nirei, H., Kusuda, T., Suzuki, K., Kamura, K., Furuno, K., Hara, Y., Satoh, K., and Kazaoka, O.: The 1987 East Off Chiba Prefecture Earthquake and It's Hazards, Memoirs of the Geological Society of Japan, 35, 31–46, 1990.

Prefectural Governors' Committee for land subsidence Prevention Kanto District: Report of groundwater volume pumping up, 228 pp., 1983 (in Japanese).

Prefectural Governors' Committee for land subsidence Prevention Kanto District: Report of groundwater volume pumping up, 124 pp., 2008 (in Japanese).

Research Committee on Land subsidence in Southern Kanto District: Report of research on land subsidence in Southern Kanto District, 297 pp., 1974 (in Japanese).

Research Group for Water Balance: Groundwater basin management, Tokai University Press, Tokyo, 242 pp., 1976 (in Japanese).

Research Institute of Environmental Geology: The Data Analysis on the Urayasu Land Subsidence Observation Well System, Research Institute of Environmental Geology, Chiba, Project Research Report No. 4, 86 pp., 1979. (in Japanese)

Shibasaki, T.: Carrying capacity of groundwater development and conservation, Groundwater basin management, Tokai University Press, Tokyo, 47–75, 1976 (in Japanese).

11

Spatial-temporal variation of groundwater and land subsidence evolution in Beijing area

K. Lei[1], Y. Luo[1], B. Chen[2], M. Guo[3], G. Guo[1], Y. Yang[1], and R. Wang[1]

[1]Beijing Institute of Hydrogeology and Engineering Geology, Beijing 100195, China
[2]College of Resources Environment and Tourism, Capital Normal University, Beijing 100048, China
[3]Beijing Bureau of Geology and Mineral Exploration and Development, Beijing 100195, China

Correspondence to: K. Lei (279459260@qq.com)

Abstract. Precipitation is the main recharge source of groundwater in the plain of Beijing, China. Rapid expansion of urbanization has resulted in increased built-up area and decreased amount of effective recharge of precipitation to groundwater, indirectly leading to the long-term over-exploitation of groundwater, and induced regional land subsidence. Based on the combination of meteorological data, groundwater level data, interferometric synthetic aperture radar (InSAR; specifically persistent scatterer interferometry, PSI), geographic information system (GIS) spatial analysis method and rainfall recharge theory, this paper presents a systematic analysis of spatial-temporal variation of groundwater level and land subsidence evolution. Results show that rainfall has been decreasing annually, while the exploitation of groundwater is increasing and the groundwater level is declining, which is has caused the formation and evolution of land subsidence. Seasonal and interannual variations exist in the evolution of land subsidence; the subsidence is uneven in both spatial and temporal distribution. In 2011, at the center of mapped subsidence the subsidence rate was greater than $120 \, \text{mm} \, \text{a}^{-1}$. The results revealed good correlation between the spatial distribution of groundwater level declines and subsidence. The research results show that it is beneficial to measure the evolution of land subsidence to dynamic variations of groundwater levels by combining InSAR or PSI, groundwater-level data, and GIS. This apprpach provides improved information for environmental and hydrogeologic research and a scientific basis for regional land subsidence control.

1 Introduction

Land subsidence caused by groundwater extraction can be a severe geologic hazard, causing structural damages to infrastructure and environmental damages to land an water resources and can lead to a series of other hazards related to the formation of ground fissures (Galloway et al., 1998; Galloway and Hoffmann, 2007). Some studies attribute the leading cause of regional land subsidence in Beijing China to the overexploitation of groundwater (Chen, 2000). Large-scale and long-term overexploitation of groundwater leads to declining groundwater levels, the deformation(compaction) of clay sediments in the aquifer system, and land subsidence. As the capital of China, Beijing is a metropolis with serious water shortage problems (Lei et al., 2014). More than two-thirds of the water supply comes from groundwater. Long term overexploitation of groundwater has led to continuously declining water levels and plays an important role in the formation and development of regional land subsidence (Jia et al., 2007). In the center of mapped subsidence, the cumulative settlement is as much as to 1302 mm; and some areas are still subsiding at a high rate of $100–120 \, \text{mm} \, \text{a}^{-1}$.

This study combines groundwater level data, the PSI technique, GIS spatial analysis methods and rainfall recharge theory to analyze spatial-temporal variations of the groundwater system and evolution of land subsidence in the Beijing plain area, China. The study also analyzes the spatial and temporal variability of rainfall groundwater recharge derived from precipitation. The relation between groundwater-level variations and the evolution of land subsidence is demonstrated.

2 Study area

Beijing ($39°28'$–$41°05'$ N, $115°25'$–$117°35'$ E), located in the north of China, is divided into three parts: northern mountains, western mountains and south-eastern plains. The climate in Beijing is characterized by uneven seasonal precipitation, with over 70 % of the annual precipitation distributed between June and August. During recent decades, rainfall has been declining continuously attributed in part to the impact of global climate change; the average annual rainfall was 479.49 mm, equivalent to 84.2 % of the average annual rainfall for the period of record.

3 Methodologies

3.1 Estimation of rainfall recharge to ground water

Kriging was employed to spatially interpolate observed precipitation data from 20 national weather stations in Beijing. The principle of Kriging interpolation is based on:

$$Z_0^* = \sum_{i=1}^{n} \lambda_0^i Z_i, \tag{1}$$

where, Z_0^* is Kriging interpolation values; Z_i is observations of the position $\chi_i (i = 1, 2, \ldots, n)$. Groundwater recharge from rainfall was estimated by using

$$Q = 10^3 a \cdot X \cdot F, \tag{2}$$

where Q is rainfall infiltration recharge (m^3); a is the infiltration coefficient; X is average annual rainfall (mm); and F is infiltration area.

Landsat TM images of three periods (2000, 2006, 2009) covering the Beijing area were used to extract spatial-temporal information on the expansion of built-up areas by the supervised classification method. Then, loss in rainfall recharge(precipitation minus groundwater recharge derived from infiltration of precipitation) was calculated by GIS raster calculation using groundwater recharge from rainfall and the built-up area.

3.2 Deformation monitoring using PSI

Persistent scatterers (PSI) can effectively reduce the spatial and temporal decorrelation effects and reduce the error component caused by atmospheric delay versus traditional In-SAR. Through error corrections on atmospheric, orbit and digital elevation model (DEM) phase components, accurate deformation rates can be extracted (Ferretti et al., 2001; Hooper, et al. 2004). The PSI method was used in this study:

$$\phi_{x,i} = \phi_{\text{def},x,i} + \phi_{\varepsilon,x,i} + \phi_{\text{atm},x,i} + \phi_{\text{orb},x,i} + \phi_{n,x,i}, \tag{3}$$

where, $\Phi_{\text{def},x,j}$ is the deformation phase of viewing direction; $\Phi_{\varepsilon,x,j}$ is the residual topographic phase introduced by

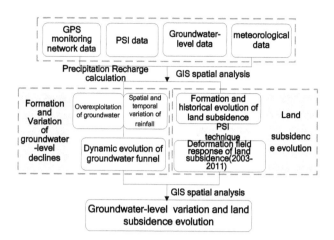

Figure 1. Framework diagram of the comprehensive analysis method.

errors in the external DEM; $\Phi_{\text{atm},x,j}$ is the phase difference of atmospheric delays between two satellite transits; $\Phi_{\text{orb},x,j}$ is the phase component of orbit error; $\Phi_{n,x,j}$ is the noise component; x, j is row number, line number for pixels.

3.3 Comprehensive analysis method

Combined with groundwater level data, meteorological data, PSI data and GIS were used to analyze groundwater level variations and the evolution land subsidence. The spatial and temporal variability of rainfall and its recharge to groundwater, is described in terms of the spatial and temporal correlations of groundwater level declines and land subsidence (Fig. 1).

4 Results and analysis

4.1 Spatial and temporal variability of precipitation and recharge to groundwater

4.1.1 Spatial and temporal variability of rainfall

Meteorological data was compiled from 20 weather stations for six typical periods (1997, 2000, 2003, 2006, 2009 and 2011) to develop time series contour maps of rainfall using the Kriging method (geostatistical tool of ArcGIS) (Fig. 2). The results show the spatial and temporal variation of rainfall in Beijing. In 1997, the rainfall was mainly concentrated in the northeast region and the average precipitation was 438 mm. In 2003, the rainfall was mainly concentrated in Yan-Qing and Fang-Shan where the average precipitation was 457 mm. In 2006, the rainfall was mainly concentrated in Ping-Gu and the average precipitation was 491 mm. In 2009–2011, there was a significant difference of precipitation between eastern areas and western areas. The precipitation in the central and eastern area was greater than 572 mm and distributed evenly.

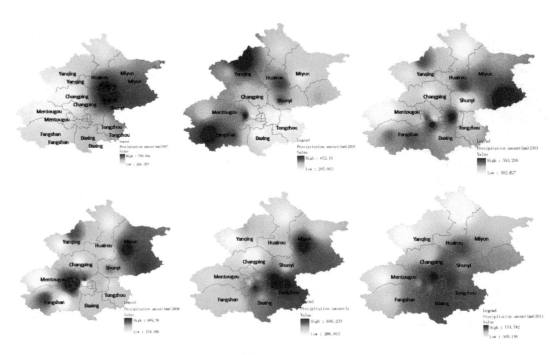

Figure 2. Spatial and temporal variability of rainfall (1997–2011).

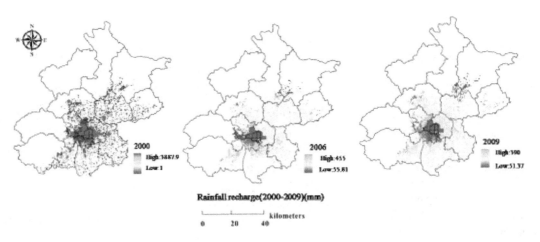

Figure 3. Spatial-temporal distribution of effective recharge to groundwater from rainfall (2000–2009).

4.1.2 Groundwater recharge from rainfall

With the decline of precipitation, groundwater recharge is decreasing considerably due to the growth of impervious areas, which result from the urbanization of Beijing.

Land use derived from Landsat TM images of three periods (2000, 2006, 2009) showed in Fig. 3 that the built-up area was 2246.94 km^2 in 2000, growing to 2344.97 km^2 in 2006, and growing further to 2767.22 km^2 in 2009 – a more than 20 % increase since 2000. With the expansion of impervious area (built-up area), and reduced groundwater recharge from rainfall led to the over-exploitation of groundwater.

Recharge from rainfall to the groundwater system was obtained by using GIS raster calculations and statistical analysis of groundwater recharge from rainfall accounting for the built-up area. In 2000, effective rainfall recharge was 18.81 × 10^8 m^3, accounting for 48 % of the total annual rainfall (39.41 × 10^8 m^3), residual rainfall not contributing to groundwater recharge (loss) was 52 % in 2000, 61 % in 2006, and 62 % in 2009. The results suggest that with the increasing built-up area, the loss amount of rainfall recharge to groundwater was increasing. Two-thirds of the Beijing water supply is from groundwater. With urban area expansion, water requirements and groundwater demand and exploitation are increasing annually. Thus, groundwater levels have been declining, which leads to formation and evolution of subsidence.

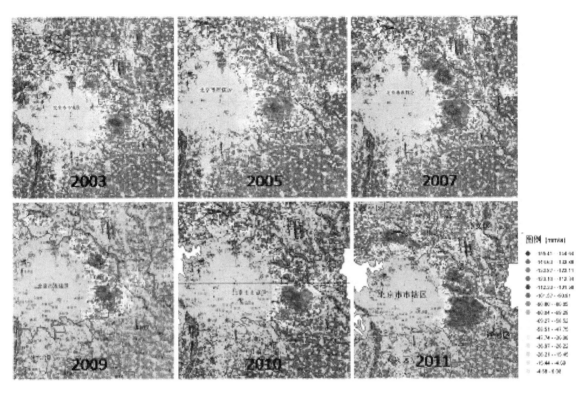

Figure 4. Land subsidence mapped using PSI technique (2003–2011).

4.2 Land subsidence evolution to resulting from variation of groundwater levels

4.2.1 Spatial-temporal distributions of land subsidence (2003–2011)

The time series land subsidence information was acquired using the InSAR PSI technique and the result was validated by leveling survey measurements.

According to the time-series PSI results, the subsidence expanded steadily during 2003–2011. In 2011, the cenrer of subsidence was subsiding 120 mm a^{-1}, and about 1087 km^2, mainly in the Changping, Shunyi, Chaoyang and Tongzhou Districts was subsiding more than 30 mm a^{-1} (Fig. 4).

4.2.2 Relationship between groundwater level variation and land subsidence evolution

The correlation between groundwater level variation and land subsidence evolution was analyzed by comparing groundwater level contoursfor select years during 2003–2011, with the average settlement rate obtained by the PSI technique during 2003–2011 (Fig. 5). Greater settlement rates are fairly well correlated with areas of greater groundwater-level declines-as indicated by the depth to water-level contours, mainly distributed in Tianzhu of Shunyi District, Chaoyang District and north-western Tongzhou District. The groundwater level was declining continuously with an average rate of 2.18 m a^{-1} and a maximum rate of 3.82 m a^{-1} in the area of maximum regional groundwater-level declines.

5 Conclusions

1. The paper analyzed the spatial and temporal variability of rainfall and its recharge to groundwater during 2000–2009. With declining precipitation, groundwater recharge was decreasing heavily due to reduced precipitation and to the growth of impervious areas. Meanwhile, water requirements and groundwater exploitation were increasing annually. Groundwater level has been declining, which has led to formation and evolution of a regional cone of depression in the groundwater system and land subsidence.

2. The land subsidence (during 2003–2011) was measured and mapped using PSI. In 2011, the settlement rate at the center of subsidence was larger than 120 mm a^{-1}, 1087 km^2 was subsiding at a rate larger than 30 mm a^{-1}.

3. From the comparison between variation of groundwater levels and subsidence, the results suggested that the spatial distribution of groundwater-level declines was fairly well correlated with the mapped subsidence, but not entirely.

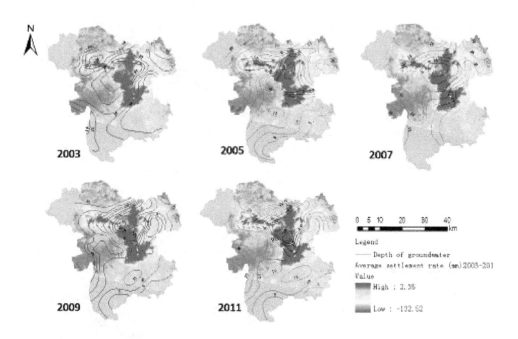

Figure 5. Groundwater contours (depth to water) and land subsidence.

Acknowledgements. This research was funded by the Beijing Municipal Science and Technology Project (Z131100005613022).

References

Chen, C. X.: Thinking on Land Subsidence Caused By Groundwater Exploitation, Hydrogeology and Engineering Geology, 27, 45–60, 2000.

Ferretti, A., Prati, C., and Rocca, F.: Permanent scatterers in SAR interferometry, IEEE T. Geosci. Remote, 39, 8–20, 2001.

Galloway, D. L. and Hoffmann, J.: The application of satellite differential SAR interferometry-derived ground displacements in hydrogeology, Hydrogeol. J., 15, 133–154, 2007.

Galloway, D. L., Hudnut, K. W., Ingebritsen, S. E., Phillips, S. P., Peltzer, G., Rogez, F., and Rosen, P. A.: Detection of aquifer-system compaction and land subsidence using interferometric synthetic aperture radar, Antelope Valley, Mojave Desert, California, Water Resour. Res., 34, 2573–2585, 1998.

Hooper, A., Zebker, H., Segall, P., and Kampes, B.: A New Method for Measuring Information on Volcanoes and Other Natural Terrains Using InSAR Persistent Scatterers, Geophys. Res. Lett., 31, 1–5, 2004.

Jia, S. M., Wang, H. G., Zhao, S. S., and Luo, Y.: A Tentative Study of the Mechanism of Land Subsidence in Beijing, City Geology, 2, 20–26, 2007.

Lei, K. C., Chen, B. B., Jia, S. M., Wang, S. F., and Luo, Y.: Formation and Genetic Mechanism of Land Subsidence Based on PS-InSAR Technology in Beijing, Spectrosc. Spect. Anal., 34, 2185–2189, 2014.

Production induced subsidence and seismicity in the Groningen gas field – can it be managed?

J. A. de Waal, A. G. Muntendam-Bos, and J. P. A. Roest

State Supervision of Mines, The Hague, the Netherlands

Correspondence to: J. A. de Waal (j.a.dewaal@minez.nl)

Abstract. Reliable prediction of the induced subsidence resulting from gas production is important for a near sea level country like the Netherlands. Without the protection of dunes, dikes and pumping, large parts of the country would be flooded. The predicted sea-level rise from global warming increases the challenge to design proper mitigation measures. Water management problems from gas production induced subsidence can be prevented if measures to counter its adverse effects are taken timely. This requires reliable subsidence predictions, which is a major challenge. Since the 1960's a number of large, multi-decade gas production projects were started in the Netherlands. Extensive, well-documented subsidence prediction and monitoring technologies were applied. Nevertheless predicted subsidence at the end of the Groningen field production period (for the centre of the bowl) went from 100 cm in 1971 to 77 cm in 1973 and then to 30 cm in 1977. In 1984 the prediction went up again to 65 cm, down to 36 cm in 1990 and then via 38 cm (1995) and 42 cm (2005) to 47 cm in 2010 and 49 cm in 2013. Such changes can have large implications for the planning of water management measures.

Until 1991, when the first event was registered, production induced seismicity was not observed nor expected for the Groningen field. Thereafter the number of observed events rose from 5 to 10 per year during the 1990's to well over a hundred in 2013. The anticipated maximum likely magnitude rose from an initial value of less than 3.0 to a value of 3.3 in 1993 and then to 3.9 in 2006. The strongest tremor to date occurred near the village of Huizinge in August 2012. It had a magnitude of 3.6, caused significant damage and triggered the regulator into an independent investigation. Late 2012 it became clear that significantly larger magnitudes cannot be excluded and that values up to magnitude 5.0 cannot be ruled out. As a consequence the regulator advised early 2013 to lower Groningen gas production by as much and as fast as realistically possible. Before taking such a decision, the Minister of Economic Affairs requested further studies. The results became available early 2014 and led to the government decision to lower gas production in the earthquake prone central area of the field by 80 % for the next three years. In addition further investigations and a program to strengthen houses and infrastructure were started.

Important lessons have been learned from the studies carried out to date. It is now realised that uncertainties in predicted subsidence and seismicity are much larger than previously recognised. Compaction, subsidence and seismicity are strongly interlinked and relate in a non-linear way to production and pressure drop. The latest studies by the operator suggest that seismic hazard in Groningen is largely determined by tremors with magnitudes between 4.5 and 5.0 even at an annual probability of occurrence of less than 1 %. And that subsidence in 2080 in the centre of the bowl could be anywhere between 50 and 70 cm. Initial evaluations by the regulator indicate similar numbers and suggest that the present seismic risk is comparable to Dutch flooding risks.

Different models and parameters can be used to describe the subsidence and seismicity observed so far. The choice of compaction and seismicity models and their parameters has a large impact on the calculated future subsidence (rates), seismic activity and on the predicted response to changes in gas production. In addition there are considerable uncertainties in the ground motions resulting from an earthquake of a given magnitude and in the expected response of buildings and infrastructure. As a result uncertainties in subsidence and seismicity become very large for periods more than three to five years into the future. To counter this a control loop based on interactive modelling, measurements and repeated calibration will be used. Over the coming years, the effect of the production reduction in the centre of the field on subsidence and seismicity will be studied in detail in an effort to improve understanding and thereby reduce prediction uncertainties. First indications are that the reduction has led to a drop in subsidence rate and seismicity within a period of a few months. This suggests that the system can be controlled and regulated. If this is the case, the integrated loop of predicting, monitoring and updating in combination with mitigation measures can be applied to keep subsidence (rate) and induced seismicity within limits. To be able to do so, the operator has extended the field-monitoring network. It now includes PS-InSAR and GPS stations for semi-permanent subsidence monitoring in addition to a traditional network of levelling benchmarks. For the seismic monitoring 60 shallow (200 m) borehole seismometers, 60 + accelerometers and two permanent downhole seismic arrays at reservoir level will be added. Scenario's spanning the range of parameter and model uncertainties will be generated to calculate possible subsidence and seismicity outcomes. The probability of each scenario will be updated over time through confrontation with the measurements as they become available. At regular intervals the subsidence prediction and the seismic risk will be re-evaluated. Further mitigation measures, possibly including further production measures will need to be taken if probabilities indicate unacceptable risks.

1   Introduction

By end 2013 production induced subsidence above the centre of the Groningen gas field in the Netherlands amounted to some 33 cm with subsidence increasing at a rate of some 7–8 mm per year (van Thienen-Visser and Breunese, 2015a; Pijpers, 2014a). Such an amount of subsidence requires the design and implementation of proper mitigation measures in a near sea-level country like the Netherlands (de Waal et al., 2012). The country would largely flood without the presence of dykes and active pumping and the predicted sea-level rise from global warming increases the challenge.

As reported for a number of other oil and gas fields (Lee, 1979; Merle et al., 1976; de Waal and Smits, 1986; Hettema et al., 2002; Mallman and Zoback, 2007; Kosloff and Scott, 1980) subsidence rate in Groningen was initially significantly lower than predicted on the basis of laboratory rock-mechanical measurements carried out on samples taken from the reservoir. At a later stage the subsidence rate accelerated, coming closer to expected values and in line with the notion of "delayed subsidence" in (Hettema et al., 2002). While the subsidence accelerated, earthquakes started to occur. Over time these have increased in numbers and strength (Fig. 1). The strongest tremor so far with a magnitude of 3.6 occurred near the village of Huizinge in 2012, causing damage to thousands of homes and raising anxiety among those living in the area (Dost and Kraaijpoel, 2012). The mechanism behind the earthquakes is understood to be (differential) compaction at reservoir level reactivating existing faults that have become critically stressed as a result of production induced stress changes. Better understanding of the physics of reservoir compaction and fault reactivation is important to predict the further evolution of the Groningen subsidence and seismicity and to assess if and how the resulting seismic hazards and risks can be managed, e.g. through production measures such as changes in production rates, off-take pattern, nitrogen injection etc. and/or by executing a strengthening program for houses and infrastructure.

After discussing some general aspects of the Groningen field and its production history, the evolution of subsidence and seismicity since the start of production will be discussed in more detail with emphasis on the lessons that have been learned so far. Thereafter the recent developments will be discussed that followed the 2012 insight that risks from reservoir compaction induced seismicity in Groningen might be significantly higher then previously realised (Muntendam-Bos and de Waal, 2013).

2   Field properties and production history

The Groningen gas field was discovered in 1959 in the Lower Permian Rotliegend at a depth of some 3000 m by exploration well Slochteren-1 near the Dutch village of Kolham (van Hulten, 2009). The structure containing the Groningen reservoir is a NNW-SSE trending high formed by normal faulting during the Late Jurassic to early Cretaceous (Stäuble and Milius, 1970; Whaley, 2009). The high has an overall horst geometry, lying between the Eems Graben to the east and the Lauwerszee Trough to the west. The reservoir is covered by Late Permian Zechstein carbonate, anhydrite and halite evaporates which provide an excellent seal. The field covers an area of some 900 km^2 and is located below a relatively densely populated area with some 250 000 houses including several urban centres. The average thickness of the gas-bearing sandstone interval is some 100 m. The field has excellent reservoir properties with porosity averaging around 17 % and permeabilities in the hundreds of mD (Mijnlieff and Geluk, 2011; van Ojik et al., 2011). Some 1800 larger and smaller faults have been mapped based on the available 3-D seismic using ant tracking on the 3-D seismic cube (NAM, 2013).

History matching and pressure measurements in observation wells indicate all of the faults to be non-sealing with the exception of a number of larger faults at the field boundaries. The original volume of recoverable Groningen gas reserves is estimated at some 2800 BCM (100 Tcf), which makes it

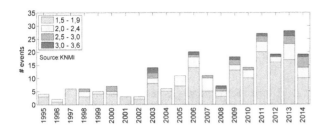

Figure 1. Groningen seismicity ($M > 1.5$) vs. time.

Figure 2. Groningen gas production (BCM/year) vs. time.

one of the larger gas fields worldwide. The gas is produced through more than 300 wells distributed over 22 production clusters. It contains slightly more then 15 % nitrogen. Gas appliances in the Netherlands and surrounding countries are optimised for this. As a result Groningen gas cannot be replaced by gas with a different composition without enough time to replace these appliances economically. An option, which also works short-term, is to mix higher methane content gas from other sources with nitrogen. For a substantial replacement volume this is not a small task given the amounts of nitrogen required. Groningen production levels over time are shown in Fig. 2.

During the early seventies of the previous century the field was produced at high rates as it was anticipated that the value of the gas would be eroded away by cheap nuclear power. When it was realised that this would not happen, the production philosophy was changed drastically and Groningen became the regional swing producer. Preference was given to production from smaller onshore and offshore fields, the development of which is stimulated by tax measures (Small Fields Policy). The policy was so successful that the annual Groningen production dropped from values above 90 BCM in 1976 to as low as 20 BCM by 2000. Thereafter production from the "Small Fields" started to decline with Groningen reaching a new annual high of 54 BCM in 2013. Presently more then two-third of the recoverable gas has been produced and first stage compression has been introduced. Over time the capacity of the Groningen field to act as the regional swing producer will decline and this role will be taken over gradually by large underground gas storage facilities.

3 Subsidence

3.1 Early days

Already during the early stages of field development it was realised that the gas production from the Groningen field could lead to a significant amount of surface subsidence. Detailed rock-mechanical measurement and modelling studies were executed (NAM, 1973) leading in 1973 to a predicted maximum subsidence of 100 cm at the end of field life, then expected around 2020. The prediction was based on an analytical linear elastic nucleus of strain model, supported by calculations with a linear elastic finite element numerical

model. The use of a linear elastic model seemed justified by the results of a large number of zero pore pressure laboratory compaction measurements carried out on core samples taken from the reservoir. An extensive monitoring network was installed around the same time. It involved a significant extension of the existing local geodetic surface network, monitoring of near surface layer compaction (unrelated to gas production) and a number of dedicated down-hole reservoir compaction monitoring wells with radioactive bullets shot into the formation (NAM, 1973).

By 1973 it had become clear that both reservoir compaction and surface subsidence were occurring at a much lower rate then had been predicted. This remained the case also after applying a Biot correction factor to account for the effect of the field pore pressure, bringing the 1971 prediction down to 77 cm in 1973. In 1976 the predicted maximum subsidence at the end of field life was brought down to 30 cm by calibrating a linear elastic model against the 1977 levelling data (Schoonbeek, 1976). The reason for the large difference between the laboratory measured and the field observed rock compressibility was not clear. It was speculated to be caused by core damage from drilling, transport and laboratory procedures applied to the core samples prior to testing. A number of studies were started to investigate this and other potential explanations.

3.2 The rate type compaction model

The additional studies looked at the relation between reservoir compaction and surface subsidence, core disturbance, the validity of the effective stress concept, the in-situ stress state, possible previous deeper burial, pressure lag (between different permeability layers) and loading rate (de Waal, 1986). The study included an extensive re-analysis of the available data from Groningen and from other fields worldwide with significant subsidence, as well as rock-mechanical laboratory experiments on the effects of stress state, pore pressure, previous deeper burial, loading path, core disturbance and (changes in) loading rate. Based on the outcomes, the cause for the discrepancy in Groningen was postulated to be the large increase in loading rate that occurs in the reservoir at the start of depletion. Such an increase in loading rate cannot be easily simulated in laboratory experiments and even if this could be done, the rates at which laboratory experiments can be carried out remain much higher than those occurring in the field over geological times. Validating this explanation for field application is therefore problematic. The newly developed rate-type compaction model explained the difference between the initial field observed and the lab-derived compressibility for Groningen and explained the observed reduced and/or delayed subsidence above a number of other depletion-drive oil and gas reservoirs (de Waal, 1986; de Waal and Smits, 1988).

The physical explanation for the laboratory observed rate sensitivity was postulated to be time-dependent friction at the sliding mineral contact points (Dieterich, 1978, 1994) between (assemblies of cemented) grains or on microscopic fractures responsible for the non-elastic component in the compaction. Using a simple model incorporating the time dependent friction equation proposed by Dieterich, results in a differential equation popular in soil mechanics to describe the known rate dependent compaction behaviour of sands and soils at lower stress levels (Bjerrum, 1967; Kolymbas, 1977; Vermeer and Neher, 1999).

The use of the new model for predicting subsidence in Groningen was accepted in 1986. Subsidence at the end of field life was predicted to be around 65 cm with a one-sigma uncertainty of 10 cm.

3.3   Back to linear models

Based on differences between the observed development of the subsidence above the Groningen field and predictions based on the rate type compaction model, the Groningen field operator in 1990 requested a review of the various subsidence predictions by two independent MIT advisors of high international standing (Toksöz and Walsh, 1990). The reviewers advised NAM to discount the prediction of Groningen subsidence on the basis of the rate type compaction model as they did not expect that loading or strain rate could have a significant effect on the deformation of the consolidated reservoir rock in the Groningen field. They advised the operator to return to the use of a linear elastic model calibrated against the field observed subsidence, leading to a new prediction for the ultimate maximum subsidence of 36 cm. Further adjustments increased the prediction to 38 cm in 1995 and to 42 cm in 2005 to honour results of new surface levelling data. By 2005 it had become clear again that a linear model could not describe the field behaviour. The operator introduced a bi-linear compaction model in which the uniaxial compressibility increases after a given amount of pressure drop after which it stays at that higher value. The pivot point and the compressibility before and after the pivot point were fitted to honour new levelling survey measurements as they came in over time. Using this approach, the predicted maximum subsidence at the end of field life was increased to 47 cm in

2010, including 2 cm of additional subsidence from the use of a lower gas abandonment pressure.

3.4   Back to non-linear models

By 2012 other Rotliegend gas reservoirs in the Netherlands showed continuing subsidence at a more or less constant gas pressure at the end of the production period. A bi-linear compaction model cannot explain such behaviour. For the time being NAM decided to switch to a time decay model, similar to a model proposed earlier by (Houtenbos, 2007) to describe the subsidence behaviour of compacting reservoirs. Mossop (Mossop, 2012) postulated a non-uniform heavy tailed permeability distribution resulting in a non-uniform pressure distribution within the bulk reservoir rock to be the explanation.

At the request of the regulator a new investigation into the physical background of the observed subsidence delays was commenced taking the permeability distribution explanation as only one of a number of possibilities. In addition salt flow, bottom and lateral aquifer inflow, intrinsic sandstone rate sensitivity, core disturbance, stress state and stress path are again being investigated.

One of the alternatives presently under consideration is a modified version of the rate-type compaction model abandoned for consolidated sandstones in 1990. The new formulation (Pruiksma et al., 2015) combines the features of the original rate type model with the soil mechanics isotach concept that enables transitions between different loading rates. The new isotach formulation combines a direct (linear elastic) strain component and a creep component, which elegantly explains the moderate loading rate dependence of the isotach compressibility. This way a number of limitations of the original rate type model have been resolved (de Waal et al., 2015), be it at the cost of an additional material property representing the linear elastic component of the deformation. The fit between predictions based on the modified isotach model and the measured subsidence in Groningen is better than that obtained with the time-decay model or the bi-linear model (van Thienen-Visser and Breunese, 2015).

The predicted impact of drastic changes in production rate during later stages of the production period is very different for the different compaction models. For the bi-linear model the predicted effect on compaction and subsidence is instantaneous, while for the isotach model the response is more gradual over a period of up to a year. The time-decay model predicts that it will take 5 to 10 years for the rock to respond to significant changes in production rate.

4   Induced seismicity

4.1   History

Until 1991 no induced seismicity was registered in the Groningen field, nor was it expected. The independent study by MIT experts (Toksöz and Walsh, 1990) estimated the probability of an induced seismic event with a magnitude above 3.0 at less then 10 % for the next 50 years. If such an event would occur at all, they expected it to happen on the sealing faults at the field edges, not in the field centre. History proved different. For a number of other fields in the Netherlands induced seismicity had been linked to gas production a few years earlier (BOA, 1993). While subsidence in Groningen was accelerating, induced seismicity started to occur. The first production-induced earthquake with a local magnitude M_L of 2.4 was recorded at Middelstum in 1991. The frequency and magnitude of the events thereafter continued to increase over time as shown in Fig. 1. The local seismic network was upgraded a number or times and is complete for magnitudes above 1.5 since 1995. To date, close to a thousand gas production induced earthquakes have been registered in Groningen. Most events have been of a small magnitude ($M_L < 1.5$), while by mid 2015 some 250 earthquakes had magnitudes $M_L \geq 1.5$. In contrast to early expectation, the tremors do not occur on the sealing faults at the field boundaries. They occur mainly in more central areas where cumulative compaction and the density of faults with unfavourable offsets are highest. The mechanism behind the earthquakes is now generally considered to be (differential) compaction at reservoir level reactivating offset faults as originally proposed by (Roest and Kuilman, 1994).

Initially doubts on the relation between gas production and observed seismicity continued, even after an official investigation in 1993 (known as the BOA study) concluded that the two were linked. Based on an analysis of the seismicity, initial estimates were made of the maximum probable magnitude that could be expected to occur. The 1993 BOA study came to an initial estimated value of 3.3. KNMI later increased this estimate to 3.3–3.5 in 1995 (de Crook et al., 1995), to 3.7 in 2006 (de Crook et al., 1998) and then to 3.9 in 2006 (van Eck et al., 2006). In (de Lange et al., 2011) it is estimated that induced seismic events with magnitudes below 3.9 would not result in severe structural damage to buildings and hence the associated risk was deemed acceptable until things changed by end 2012. As of 2008 State Supervision of Mines (the Dutch regulator) became more and more concerned about the increasing frequency of the induced seismicity and the increasing magnitudes. TNO was asked to carry out a study into possible explanations (Muntendam-Bos, 2009). The conclusions and recommendations of this study led to a request to the operator by end 2009 for further research into possible mechanisms and feasible remedies, the results of which needed to be included in the required 2013 update of the Groningen winningsplan.

4.2   New insights

By mid 2012 the larger than predicted annual frequency of earthquakes with a magnitude equal or above 3.0 further increased the concerns of State Supervision of Mines (SSM)

and intensified discussions with the operator. On 16 August 2012 an induced seismic event with a magnitude of 3.6 occurred near the village of Huizinge. It was the highest magnitude event registered in Groningen to that date, causing damage to thousands of homes and raising anxiety among those living on or near the field (Dost and Kraaijpoel, 2012). It triggered SSM into commencing its own investigation, including a new statistical analysis of all the available Groningen seismicity data (Muntendam-Bos and de Waal, 2013). The results were worrying. The SSM analysis indicated a continuous and systematic increase in the number of events and their magnitude (Fig. 3). The same figure suggests that smaller magnitude events might well have started earlier, without being detected given the limitations of the local seismic network at the time. The SSM analysis also showed that an upper magnitude bound – generally deemed to be 3.9 – could not be derived from a statistical analysis of the Groningen seismicity while other methodologies to arrive at such an upper estimate could not exclude values up to at least a magnitude 5 (Muntendam-Bos and de Waal, 2013). The results were confirmed by studies carried out by the operator (NAM, 2013, Bourne et al., 2014). In addition the SSM analysis suggested that increases and decreases in the annual number of seismic events could be linked to increases and decreases in the annual gas production rate with a delay of 6 to 12 months (Fig. 4). Based on the results of its investigation, SSM judged the seismic risk in Groningen as "high" and advised early January 2013 to reduce gas production by as much and as fast as realistically possible.

The SSM advice was followed by a year of concerted data acquisition and further studies, while production continued and reached its highest annual level in nearly 20 years (54 BCM). The results of the studies became available by end 2013 and suggested that the local seismic activity rate in Groningen increases exponentially with cumulative reservoir compaction (NAM, 2013). If correct it implies that production reductions in areas with large compaction are most effective to reduce future seismicity. Based on this insight the earlier SSM advice became more focussed: stop gas production in the central area of the Groningen field - where the largest amount of reservoir compaction takes place - for at least three years (Staatstoezicht, 2014). The ministerial decision, based on this advice, was to reduce the production in the centre of the field by 80 %. It was implemented on 17 January 2014. Initially the operator was allowed to increase production in areas outside the centre of the field to partly compensate for the production reduction in the centre. These production increases were capped in December 2014 after early observations suggested they resulted in increases in seismicity in these areas outside the field centre. It was realised from the start that the implemented production measures would probably only work temporarily (if at all) as pressure depletion in the centre of the field will resume after a few years given the continued production elsewhere. The large-scale data acquisition and study program was therefore

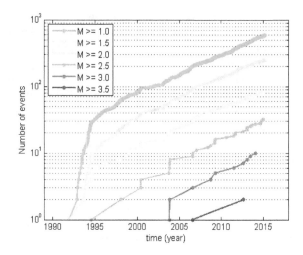

Figure 3. SSM analysis of the Groningen induced seismicity.

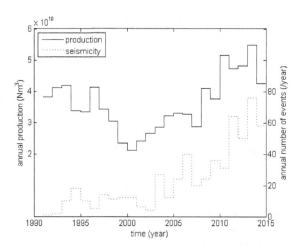

Figure 4. Groningen annual production and annual induced seismicity ($M > 1.0$).

continued and extended and a program to strengthen houses and infrastructure was started.

5 Recent developments

5.1 Field response

The main question related to the January 2014 production measures were how effective they would be to reduce the seismicity and on what time scale. Time is required for the pressure signal resulting from the shut-in at the central clusters to spread through the reservoir. Given the Groningen reservoir properties it takes some three months for that signal to reach a distance of 3 km (van Thienen-Visser and Breunese, 2015). As (differential) compaction is the engine behind the induced seismicity, the time scale on which compaction responds to stress changes is a second major factor. The time decay model predicts a typical response time of five to seven years. Linear and bi-linear compaction models predict an instant response. The rate type isotach model predicts part of the response to be immediate and part of it to happen over the next 6 to 12 months. In addition there is a delay from the time it takes for the pressure effect from the production reduction to diffuse from the wells into the reservoir.

Another factor influencing the response is the seismological model linking seismicity to compaction. On the one hand it can cause additional delays as a result of rate and state friction on the fault surfaces where the seismicity is generated. On the other hand the postulated exponential rise of seismic activity rate in combination with compaction creep would dampen the response significantly. Given the epistemic uncertainties around the compaction model and the seismological model in combination with the absence of empirical data, it was very uncertain how the seismicity would respond and a "wait and see" approach was taken. To be able to do so, the operator has extended the field monitor-

ing to include PS-InSAR and a network of differential GPS stations for semi-permanent subsidence monitoring. For the seismic monitoring 60 shallow (200 m) borehole seismometers, 60 + accelerometers and two temporally installed downhole seismic arrays at reservoir level in the centre of the field are now in place. The latter will be replaced by permanent installations shortly.

Conclusions on the seismic response to the reduced production in the central area of the field within a period of a year are a challenge given the statistical character of the seismicity and the challenge to accurately measure compaction rate changes over relatively short periods. Nevertheless results to date look promising. A double-double difference analysis of the GPS stations in the centre and those at the edge of the field carried out by Statistics Netherlands (Pijpers, 2014a) shows a clear break in subsidence rate around April 2014 as shown in Fig. 5. Seismic event density maps over the Groningen field for three consecutive periods of a year are discussed in (van Thienen-Visser, 2015b). A clear change in the pattern of seismic events is visible during the third period with seismicity reducing significantly in the central area of the field after April 2014 (Fig. 6). An independent second study by Statistics Netherlands (Pijpers, 2014b) confirms the statistical significance of the changes in seismic activity. Although not yet scientifically conclusive at a 99 % confidence level these results support decision making on remediating measures to reduce future seismic risk as they suggest that the seismicity can be controlled and regulated at relatively short notice.

5.2 Risk

Early 2013 SSM estimated the seismic risk level in Groningen as "high", based on the realisation that events with a magnitude well above 3.9 could not be excluded. Based on SSM's analysis an assumed upper magnitude limit of 5.0 re-

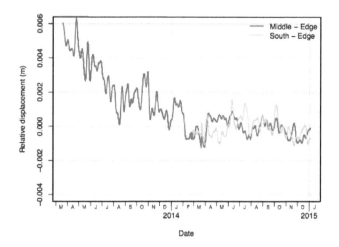

Figure 5. Analyses of GPS data showing break in subsidence rate around April 2014.

sulted in an annual probability of 3 % for an earthquake with a magnitude 4.5 or higher, a level at which serious damage cannot be excluded as houses in the Netherlands are not built to sustain seismic ground motions. By end 2013 a more detailed risk analysis was made for the central area of the field (Staatstoezicht, 2013; Muntendam et al., 2013). Based on a probabilistic analysis of the ground motions that can occur and taking into account the fragility of the local housing stock, data made available by the operator was translated into annual LPR (local personal risk) numbers and a Group risk F-N (frequency-number) plot. Results show that the seismic risk levels in Groningen are considerable and comparable to the highest flooding risk levels in the Netherlands (Fig. 7). Subsequent work by a dedicated impact assessment expert committee during 2014 estimated the number of houses in Groningen exceeding an LPR of 10^{-4} to be between 40 000 and 90 000 (Stuurgroep NPR, 2015). Supporting a widely accepted minimum LPR level of 10^{-5}, the committee recommends strengthening the housing stock to 10^{-4}, assuming actual LPR levels will come down when more refined and less conservative calculations become available. A final norm needs to be developed taking into account acceptable seismic risk, security of supply issues and financial consequences.

5.3 Measurement and control loop

Target is the cyclic development of a well organised risk management plan with clear objectives, well defined risk norms and control measures with proven effectiveness. To start, scenario's spanning the range of parameter and model uncertainties will be generated to calculate possible subsidence and seismicity outcomes. The probability of each scenario will be updated over time through confrontation with measurements as they become available. At regular intervals the subsidence prediction and the seismic risk will be re-evaluated. Further mitigation measures, possibly including

Figure 6. Increases and decreases in seismic event density after production reductions (from van Thienen-Visser et al, 2015b).

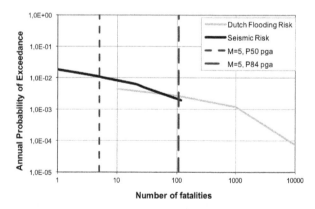

Figure 7. Calculated seismic Group Risk Loppersum area (from Muntendam et al., 2015).

further production measures and further building strengthening measures will need to be taken if probabilities indicate unacceptable risks. A possible framework for a measurement and control loop enabling adjustments when seismicity metrics indicate a risk of exceeding acceptable risk is shown in Fig. 8. To apply the concept it needs to be part of a "Seismic Risk Management Protocol" that should contain the following elements:

1. agreed metrics to express the seismic risk;

2. a norm for the amount of seismic risk considered acceptable considering societal issues, security of supply and financial consequences, translated into an enforceable, operational criterion;

3. a probabilistic assessment of the seismic risk (taking into account all uncertainties):

a. as a function of location and time;

b. under different production scenarios;

c. taking into account the effect of strengthening programs;

4. an agreed procedure, including regular analysis of proper signal parameters like GPS subsidence rate and seismic event density.to update the predicted probabilistic seismic risk as new data is gathered;

5. a measurement plan;

6. agreed 'SMART" measures (production/pressure-maintenance/strengthening) to adjust the seismic risk using a multi-objective optimisation (minimising seismic risk, while maximising security of supply and state income);

7. an independent audit system to ensure compliance and to verify the technical integrity of the underlying work.

6 On-going and future work

The on-going large study and data acquisition effort by the operator is continued and extended to arrive at a better understanding of the gas production induced reservoir compaction, surface subsidence and induced seismicity in Groningen. Aim is to reduce the large epistemic and aleatory uncertainties, to arrive at better monitoring and prediction of subsidence and seismicity, to better estimate seismic risk as a function of location and time and to assess the possibility to implement a measurement and control loop to contain seismic risk.

Regular surface levelling campaigns continue, complemented by semi-continuous PS-InSAR measurements and a network of permanent GPS stations distributed over the field. The existing network of shallow seismic monitoring wells is being upgraded to some 60 deep (100–200 m) boreholes spread over the field. The expectation is that this results in a completeness level of 0.5 by end 2015, leading to a tenfold

increase in the detectable number or seismic events and a threefold reduction of the aleatory uncertainty in the seismicity activity level. In addition some 60 + accelerometers and two permanent downhole seismic arrays at reservoir level are being installed. The accelerometers will help to reduce the uncertainties in the GMPE (ground motion prediction equation) relating earthquake magnitude to surface acceleration.

Extensive statistical (trend) analyses will be carried out to investigate the effects of compaction (rate) and compaction distribution on the spatial and temporal changes in seismic activity rate and b value (determining the ratio between weak and strong events). A similar analysis will be carried out to investigate the effect of production rate and production distribution. The improved detection limit of the extended seismic network is expected to be very beneficial for this type of work. The outcomes of the various analyses will be used to validate the present (activity rate) seismological model developed by the operator and to introduce model improvements if necessary. Potential examples are the incorporation of (nonlinear) compaction rate dependence and the incorporation of a critical stressing rate above which the seismic activity rate increases rapidly (Llenos et al., 2009; Toda et al., 2002).

Better static and dynamic subsurface models are being constructed to model reservoir property distributions (e.g. porosity, permeability, faults, compressibility). An example is the latest structural model for which more than 1800 faults have been mapped at reservoir level using ant tracking on the 3-D seismic cube (NAM, 2013). Another example is the detailed characterisation of the near surface layers to a depth of several hundred meters, important for seismic wave attenuation calculations. The subsurface models will be updated as new data becomes available while the operator and TNO have started on inversion of the surface subsidence data to better determine the amount and distribution of the downhole reservoir compaction.

The operator is combining the static, dynamic and geomechanical reservoir models with a seismological model, a ground motion predication equation, a database of building types and locations and with fragility curves for the various building types to arrive at probabilistic (Monte Carlo) predictions of seismic hazard and seismic risk under a number of different production scenarios and housing reinforcement programs (Bourne et al., 2014).

7 Lessons

A number of important lessons can be learnt from the history of subsidence and induced seismicity in Groningen. First it is clear that epistemic uncertainties were largely underestimated from the start and during a long period thereafter. This is illustrated by the large changes over time in the predicted maximum subsidence at the end of field life in the centre of the Groningen field. It illustrates that initial uncertainties, prior to the start of production, are perhaps as large as a fac-

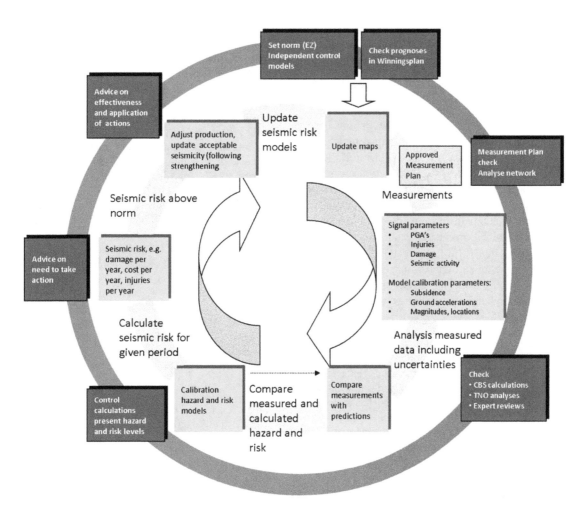

Figure 8. A possible seismic risk "measurement and control" loop.

tor of two (up or down) relative to the predicted expectation case. This has meanwhile been confirmed for a number of other fields in the Netherlands (de Waal et al., 2012). Major potential contributors are uncertainties in subsurface models, unknown influx of bottom and lateral aquifers, core damage, non-linear rock compressibility, the difference in loading rate between laboratory and field, the very large change in loading rate at the start of production, the in-situ stress state, reservoir burial history, salt flow, the relation between subsidence and compaction etc. Rather than predicting a single number it is therefore much better to provide a range of possible outcomes, only reducing the range when certain scenarios become too unlikely given field measurements (Nepveu et al., 2010). And to keep the range of possible outcomes, as it narrows over time, in line with the range that the area can sustain e.g. by timely taking appropriate mitigation measures where and when required.

Secondly it seems that non-linear compaction is the norm rather than the exception with the rate type isotach compaction model as a plausible explanation for the transition zone that follows the large increase in loading rate at the start

of the production period. The effects are well known from Soil Mechanics where a long time geological loading manifests itself as an apparent over consolidation from previous deeper burial (which it is not). This also implies that a certain degree of additional subsidence can be expected as a result of creep after the production period has ended. Subsidence and seismicity response time following the significant production adjustment in January 2014 seem in the order of a couple of months. Although not rigidly proven at a 95 % confidence level, it does seem to favour the rate type isotach model over the time decay model which uses a single 5–7 year response time to explain the initial subsidence delay after the start of production.

Also for induced seismicity, history proved very different from the original expectations. At the start of the production period induced seismicity was not expected and when it happened it took many years before its linkage to the gas production was widely accepted, an good illustration of the strength of paradigms. Contrary too initial expectation the seismicity has focussed in the central area of the field where the faults are not sealing. Also initial expectations on the ability to predict an upper bound on the likely magnitude to occur proved wrong and it is now realised that there is significant seismic risk that needs to be contained through a measurement and control loop including production measures and a housing and infra-structure strengthening programs.

The response time of a few months of subsidence rate and seismic activity rate to the production measures that have been taken suggests that the risk can be managed by adjusting production rates. To sustain a reduced seismicity level over longer periods of time, field-wide production measures will need to be taken as pressure decline in the centre of the field will re-start within a few years given the continued production from production clusters outside of the centre of the field. It is still unclear if production reductions will only smear the risk out over time (film-rate effect) or if real reductions in the total number of earthquakes can be expected. As a minimum, production reduction will buy time for further study and to strengthen buildings and infrastructure.

Acknowledgements. The authors want to thank the management of State Supervision of Mines for permission to publish this paper. We also thank staff at TNO, KNMI, CBS (Statistics Netherlands) and NAM for stimulating discussions and support although they may not agree with all of the interpretations and conclusions of this paper. We are particularly grateful for the contributions and support provided by Karin van Thienen-Visser and Jaap Breunese at TNO and by Jan van Herk at State Supervision of Mines.

References

Bjerrum, L.: Engineering geology of Norwegian normally consolidated marine clays as related to settlements of buildings, in: Geotechnique, Seventh Rankine Lecture 17, 81–118, 1967.

BOA (Begeleidingscommissie Onderzoek Aardbevingen): Eindrapport multidisciplinair onderzoek naar de relatie tussen gaswinning en aardbevingen in Noord-Nederland, Koninklijk Nederlands Meteorologisch Instituut, Afdeling Seismologie, De Bilt, 76 pp., 1993 (in Dutch).

Bourne, S. J., Oates, S. J., Elk, J. van, and Doornhof, D.: A seismological model for earthquakes induced by fluid extraction from a subsurface reservoir, J. Geophys. Res. Solid Earth, 119, 8991–9015, doi:10.1002/2014JB011663, 2014.

de Crook, Th., Dost, B., and Haak, H. W.: Analyse van het seismische risico in Noord Nederland, KNMI TR-168, de Bilt, 30 pp., 1995 (in Dutch).

de Crook, Th., Dost, B., and Haak, H. W.: Seismisch risico in Noord-Nederland, KNMI, 1998 (in Dutch).

de Lange, G., Oostrom, N. G. C. van, Dortland, S., Borsje, H., and de Richemont, S. A. J.: Gebouwschade Loppersum, Deltaris 1202097-000, 84 pp., 2011 (in Dutch).

de Waal, J. A.: On the rate type compaction behaviour of sandstone reservoir rock, PhD dissertation, Delft University of Technology, 166 pp., 1986.

de Waal, J. A. and Smits, R. M. M.: Prediction of reservoir compaction and surface subsidence: field application of a new model, SPE Formation Evaluation 3, 347–356, 1988.

de Waal, J. A., Roest, J. P. A., Fokker, P. A., Kroon, I. C., Breunese, J. N., Muntendam-Bos, A. G., Oost, A. P., and van Wirdum, G.: The effective subsidence capacity concept: How to assure that subsidence in the Wadden Sea remains within defined limits, in: Geologie en Mijnbouw Neth. J. of Geoscience, 91, 385–399, 2012.

de Waal, J. A., van Thienen-Visser, K., and Pruiksma, J. P.: Rate type isotach compaction of consolidated sandstone, 49th US Rock mechanics/Geomechanics symposium (ARMA), San Francisco, 28 June–1 July, ARMA 15-436, 2015.

Dieterich, J. H.: Time-dependent friction and the mechanics of stick-slip, Pageoph., 116, 790–806, 1978.

Dieterich, J. H.: A constitutive law for rate of earthquake production and its application to earthquake clustering, J. Geoph. Res. 99, 2601–2618, 1994.

Dost, B. and Kraaijpoel, D.: The August 16, 2012 earthquake near Huizinge (Groningen), KNMI publication TR 16/1/2013, 26, 2012.

Hettema, M., Papamichos, E., and Schutjens, P.: Subsidence delay: field observations and analysis, Oil and Gas Science and Technology, in: Revue dIFP, Energies Nouvelles, 57, 443–458, 2002.

Houtenbos, A. P. E. M.: Subsidence and gas production: an empirical relation, SubsvsProdFinal.doc, 1–8, available at: http://bodemdaling.houtenbos.org/resources (last access: 5 November 2015), 2007.

Kolymbas, D.: A rate-dependent constitutive equation for soils, Mech. R. Comms., 4, 367–372, 1977.

Kosloff, D. and Scott, R. F.: Finite element simulation of Wilmington oil field subsidence: II Nonlinear Modelling, Tectonoph., 70, 59–183, 1980.

Lee, L. L.: Subsidence earthquake at a California oil field, in: Evaluation and prediction of subsidence, edited by: Saxena, S. K., New York, ASCE, 549–564, 1979.

Llenos, A. L., McGuire, J. J., and Ogata, Y.: Modelling seismic swarms triggered by aseismic transients, Earth Planet. Sci. Lett., 281, 59–69, 2009.

Mallman, E. P. and Zoback, M. D.: Subsidence in the Louisiana coastal zone due to hydrocarbon production, J. of Coastal Res. SI, 50, 443–449, 2007.

Merle, H., Kentie, C., van Opstal, G., and Schneider, G.: The Bachaquero study – a composite analysis of the behavior of a compaction drive/solution gas drive reservoir, J. of Petr. Techn., 27, 1107–1115, 1976.

Mijnlieff, H. F. and Geluk, M.: Palaeotopography-governed sediment distribution-a new predictive model for the Permian Upper Rotliegend in the Dutch sector of the Southern Permian Basin, in: The Permian Rotliegend of the Netherlands, edited by: Grötsch, J. and Gaupp, R., SEPM Special Publication No. 98, 147–159, ISBN: 978-1-56576-300-5, 2011.

Mossop, A.: An explanation for anomalous time dependent subsidence, in: Proc. 46th US Rock Mech. Symp., 24-27 June, Chicago, 60–68, 2012.

Muntendam-Bos, A. G.: Geïnduceerde trillingen in of nabij het Groningen gasveld, een aanzet tot een geomechanische verklaring, available at: http://www.nlog.nl/nl/reserves/reserves/AGE09-10100_RapportAardtrillingenGroningenveld_PUBLIEK_NLOG.pdf (last access: 5 November 2015), 2009 (in Dutch).

Muntendam-Bos, A. G. and de Waal, J. A.: Reassessment of the probability of higher magnitude earthquakes in the Groningen gas field: technical report, available at: http://www.researchgate.net/publication/271764011_Reassessment_of_the_probability_of_higher_magnitude_earthquakes_in_the_Groningen_gas_field (last access: 5 November 2015), February 2013.

Muntendam-Bos, A. G., Roest, J. P. A., and de Waal, J. A.: A guideline for gas production induced seismic risk, The Leading Edge, special issue Injection Induced Seismicity, 672–677, 2015.

NAM: The analysis of surface subsidence resulting from gas production in the Groningen area, the Netherlands, Trans. Royal Dutch Soc. of Geol. and Mining Eng., edited by: Nederlandse Aardolie Maatschappij B. V., 28, 1973.

NAM: Wijziging winningsplan Groningen 2013, inclusief technische bijlage Groningen winningsplan 2013, Versie 29, November 2013 (partly in Dutch).

Nepveu, M., Kroon, I. C., and Fokker, P. A.: Hoisting a Red Flag: An Early Warning System for Exceeding Subsidence Limits, Math Geosci. 42, 187–198, doi:10.1007s11004-009-9252-2, 2010.

Pijpers, F.: Phase0 report1, Significance of trend changes in ground subsidence in Groningen, Statistics Netherlands Scientific paper, CBS, 14 pp., 2014a.

Pijpers, F.: Phase0 report2, Significance of trend changes in tremor rates in Groningen, Statistics Netherlands Scientific paper, CBS, 18 pp., 2014b.

Pruiksma, J. P., Breunese, J. N., van Thienen-Visser, K., and de Waal, J. A.: Isotach formulation of the rate type compaction model for sandstone, Int. J. Rock Mech. Min., 78, 127–132, doi:10.1016/j.ijrmms.2015.06.002, 2015.

Roest, J. P. A. and Kuilman, W.: Geomechanical Analysis of Small Earthquakes at the Eleveld gas Reservoir, in: Proc. Eurock '94, Balkema, Rotterdam, 573–580, 1994.

Schoonbeek, J. B.: Land subsidence as a result of natural gas extraction in the province of Groningen, SPE European Spring Meeting, 8–9 April, Amsterdam, the Netherlands, 20 pp., SPE 5751, 1976.

Staatstoezicht op de Mijnen: Risico Analyse Aardgasbevingen Groningen, available at: https://www.sodm.nl/sites/default/files/redactie/RisicoanalyseaardgasbevingenGroningen.pdf (last access: 5 November 2015), Den Haag, December 2013 (in Dutch).

Staatstoezicht op de Mijnen: Advies winningsplan 2013/Meet- en Monitoringplan NAM Groningen gasveld, available at: https://www.sodm.nl/sites/default/files/redactie/AdviesSodMwinningsplanGroningen2013.pdf (last access: 5 November 2015), 2014 (in Dutch).

Stäuble, A. J. and Milius, G.: Geology of the Groningen gas field, Netherlands Am. Ass. Petrol. Geol., Mem., 14, 359–369, 1970.

Stuurgroep NPR: Impact assessment Nederlandse Praktijk Richtlijn aardbevingsbestendig bouwen, 8 January 2015 (in Dutch).

TNO: Toetsing van de bodemdalingsprognoses en seismische hazard ten gevolge van gaswinning van het Groningen veld, R11953, Eindrapport, Utrecht, 218 pp., 2013 (in Dutch).

Toda, S., Stein, R. S., and Sagiya, T.: Evidence from the AD 2000 Izu islands swarm that stressing rate governs seismicity, Nature, 418, 58–61, 2002.

Toksöz, M. N. and Walsh, J. B.: An evaluation of the subsidence program directed by Nederlandse Aardolie Maatschappij, NAM EP90-20043, 51 pp., 1990.

van Eck, T., Goutbeek, F. H., Haak, H. W., and Dost, B.: Seismic hazard due to small-magnitude, shallow-source, induced earthquakes in the Netherlands, Eng. Geol., 87, 105–121, doi:10.1016/j.enggeo.2006.06.005, 2006.

van Hulten, F. F. N.: Brief history of petroleum exploration in the Netherlands, in: Symposium fifty years of petroleum exploration in the Netherlands after the Groningen discovery, EBN, TNO, Geo-Energy and PGK, 15–16 January 2009, Utrecht, 2009.

van Ojik, K., Böhm, A. R., Cremer, H., Geluk, M. C., De Jong, M. G. G., Mijnlieff, H. F., and Djin Nio, S.: The rationale for an integrated stratigraphic framework of the Upper Rotliegend depositional system, in: The Permian Rotliegend of the Netherlands, edited by: Grötsch, J. and Gaupp, R., SEPM Special Publication No. 98, 37–48, ISBN: 978-1-56576-300-5, 2011.

van Thienen-Visser, K. and Breunese, J.: Induced seismicity of the Groningen gas field: history and recent developments, The Leading Edge, special issue Injection Induced Seismicity, 665–671, 2015a.

van Thienen-Visser, K., Fokker, P., Nepveu, M., Sijacic, D., Hettelaar, J., and van Kempen, B.: Recent developments of the seismicity of the Groningen field in 2015, TNO R10755, 2015b.

Vermeer, P. and Neher, H.: A soft soil model that accounts for creep, in: Beyond 2000 in computational geotechnics-10 years of Plaxis international, Balkema, Rotterdam, ISBN: 90 5809 040 X, 1999.

Whaley, J.: The Groningen gasfield, GEO ExPro Magazine, Vol. 6–4, available at: http://www.geoexpro.com/articles/2009/04/the-groningen-gas-field (last access: 5 November 2015), 2009.

13

Modelling ground rupture due to groundwater withdrawal: applications to test cases in China and Mexico

A. Franceschini[1], **P. Teatini**[1], **C. Janna**[1], **M. Ferronato**[1], **G. Gambolati**[1], **S. Ye**[2], and **D. Carreón-Freyre**[3]

[1]Department ICEA, University of Padova, Padova, Italy
[2]School of Earth Sciences and Engineering, Nanjing University, Nanjing, China
[3]Laboratorio de Mecanica de Geosistemas, Mexican National University, Queretaro, Mexico

Correspondence to: A. Franceschini (franc90@dmsa.unipd.it)

Abstract. The stress variation induced by aquifer overdraft in sedimentary basins with shallow bedrock may cause rupture in the form of pre-existing fault activation or earth fissure generation. The process is causing major detrimental effects on a many areas in China and Mexico. Ruptures yield discontinuity in both displacement and stress field that classic continuous finite element (FE) models cannot address. Interface finite elements (IE), typically used in contact mechanics, may be of great help and are implemented herein to simulate the fault geomechanical behaviour. Two main approaches, i.e. Penalty and Lagrangian, are developed to enforce the contact condition on the element interface. The incorporation of IE incorporation into a three-dimensional (3-D) FE geomechanical simulator shows that the Lagrangian approach is numerically more robust and stable than the Penalty, thus providing more reliable solutions. Furthermore, the use of a Newton-Raphson scheme to deal with the non-linear elasto-plastic fault behaviour allows for quadratic convergence. The FE – IE model is applied to investigate the likely ground rupture in realistic 3-D geologic settings. The case studies are representative of the City of Wuxi in the Jiangsu Province (China), and of the City of Queretaro, Mexico, where significant land subsidence has been accompanied by the generation of several earth fissures jeopardizing the stability and integrity of the overland structures and infrastructure.

1 Introduction

The exploitation of subsurface resources, including freshwater aquifers, involves various environmental problems. One of these is land subsidence. Recently, attention has been directed to the activation of pre-existing regional faults, as well as the possible generation of new fractures, caused by groundwater withdrawals. Indeed, the activation of faults can trigger or induce a seismic activity (González et al., 2012). Another problem related to the fault activation is the creation of preferential pathways for fluid leakage. This problem can be also relevant in the case of storage of hydrocarbons or carbon dioxide in the subsoil. Activation of faults and the generation of fissures have a direct effect on the land surface and may cause unacceptable damage to engineered structures and disruption to infrastructures. This is especially problematic

in densely populated areas, such as in China (Wang et al., 2007), Mexico (Carreón-Freyre, 2010; Carreón-Freyre and Cerca, 2006), as well as in many semiarid sedimentary basins worldwide (e.g., Azat and Shahram, 2010). A different but somewhat related problem is the hydraulic fracturing or so-called "fracking" technique, which involves the intentional generation of underground fractures in the rock by injecting fluids at high pressure to create preferential pathways that facilitate the extraction of hydrocarbons. A reliable mathematical model to predict the consequences of subsurface fluid withdrawal must take into account the presence of discontinuities, if any, in the porous medium.

At present, the existing numerical-mathematical models to simulate the discontinuities in the porous medium are based on two techniques: (i) use of a continuous model where the fault is characterized by physical properties different from those of the hosting medium (Rutqvist et al., 2007); (ii) introduction of a discontinuity surface, which behaves as an internal boundary and can be present or absent depending on the stress state. At a numerical level, in the context of the Finite Element Method, the first approach implies a finer discretization of the neighbourhood of the fault, substantially to simulate a layer of different material, while the second requires the introduction of Interface Finite Elements. The first method is simpler, as it involves no real discontinuities, but the difficulties and limitations rest on the proper characterization of a continuous medium that simulates a discontinuity. With Interface Finite Elements (Goodman et al., 1968; Ferronato et al., 2008), there are two approaches to prescribe the necessary mechanical-geometric constraint conditions, that are the non-penetration of solid bodies and the stress state consistent with some failure criterion, e.g., Mohr-Coulomb (Labuz et al., 2012). The first method, referred to as penalty (Bathe, 2006), consists of the introduction of very stiff springs between the faces of the rupture, which impose the condition of non-penetration until they break from overcoming a stress limit. From a mathematical point of view, this method is not exact, because the elastic springs have displacements for any non-zero stress value, while, in nature, in case of a closed fracture the relative displacements are exactly zero. Furthermore, numerically, this method causes a strong ill-conditioning of the stiffness matrix (Ferronato et al., 2012), because of the introduction of the penalty coefficients. However, mainly for the ease of implementation, this approach is widely used. Alternatively the constraint conditions can be imposed by using Lagrange multipliers (Bathe, 2006; Simo et al., 2000), namely in an analytically exact way. The Lagrange multipliers, which physically represent the contact stresses, are additional unknowns, so that the problem is numerically more complex. The increase of the computational cost however, is offset by a more robust convergence in all non-linear steps and a more stable numerical behaviour.

With a Lagrange approach, it is of paramount importance, in case of a rupture sliding, to account for the relative orientation of the shear stress vector with respect to the relative displacement. As the Mohr-Coulomb failure criterion is planar, there is no information on three-dimensionality. A way to compute such an orientation is based on an application of the criterion of "Maximum Plastic Dissipation" (Wriggers, 2006), which uniquely provides the direction of the shear stress as a function of the relative displacement so as to maximize the friction work. By implementing this criterion in the derivation of the variational formulation, we get a numerical scheme that converges quadratically.

2 Numerical modelling of fault mechanics

A ground rupture is a discontinuity in a porous medium, whose behaviour is governed by the combination of stress

Table 1. Geomechanical properties used for the two test cases. E: Young modulus; ν: Poisson's ratio; e: void ratio.

City	Material	E (MPa)	ν	e
Queretaro City	Sand	15	0.25	0.78
	Clay	15	0.25	0.78
	Rock	4900	0.25	0.78
Wuxi	Sand	12	0.30	0.77
	Clay	8	0.30	0.77
	Rock	1200	0.30	0.77

and geometry of the close surroundings. To model such behaviour, we choose a failure criterion that prescribes the rupture activation. The shear stress limit τ_L is given by (Labuz et al., 2012):

$$\tau_L = c - \sigma_n \tan\phi \qquad (1)$$

where c is the cohesion and σ_n the normal stress acting on the fault, taken to be negative in compression. Equation (1) defines the modulus but not the direction of the limit shear stress. According to the criterion of "Maximum Plastic Dissipation", the maximization of the friction work $W_f = \tau_L^T u_r$, with u_r the relative tangential sliding and τ_L the limit shear stress, yields:

$$\tau_L = \tau_L \frac{u_r}{\|u_r\|_2} \qquad (2)$$

Notice that boldface characters denote vectors. In case of a fault opening, i.e. $\sigma_n > 0$, the two sides of the rupture become completely free to move with no loading.

To attain equilibrium, the minimization of the functional representing the total potential energy produces the principle of virtual work:

$$\delta\Pi = \delta W_i - \delta W_e \qquad (3)$$

where the internal virtual work reads $\delta W_i = \int_\Omega \delta\varepsilon^T \hat{\sigma} dV$, with $\hat{\sigma}$ the total stress, and the external virtual work is $\delta W_e = \int_\Omega \delta u^T b dV + \int_{\delta\Omega} \delta u^T t dS$, with b and t the volume and surface forces, respectively. The contribution provided by the work of the discontinuity must be added to Eq. (3). The total rupture area is denoted by Γ, while the sliding portion, where Eqs. (1) and (2) hold, is indicated with $\bar{\Gamma}$. If opened, the rupture is an unloaded boundary with no contributions to the virtual work. Otherwise, the virtual work δW_f reads:

$$\delta W_f = \int_{\Gamma\backslash\bar{\Gamma}} \delta u^T \lambda dS + \int_{\Gamma\backslash\bar{\Gamma}} \delta\lambda^T u dS + \int_{\bar{\Gamma}} \delta u_r^T \tau_L dS \qquad (4)$$

As long as the discontinuity is close, both stresses and displacements are independent variables and can have virtual

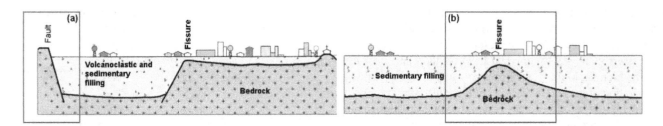

Figure 1. Sketches of the typical geological setting ground rupture in Queretaro City (a) and Wuxi, China (b). The simulated configurations are highlighted by the red boxes.

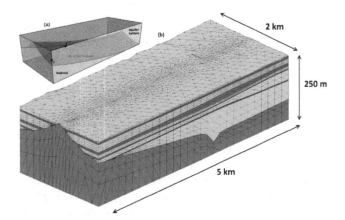

Figure 2. Test case 1: perspective views of the geological setting (a) and the 3-D FE-IE grid (b). The materials are highlighted in various colours: rock in red, aquifers in blue, and aquitards in green.

variations. Conversely, in case of sliding, only the displacement is an independent variable, as the shear stress is fully defined. In this work, we adopt a one-way coupled approach, so the pressure is a priori known and can be treated as an external force, separating it from the term of total stress, according to Terzaghi's principle (Bishop, 1959), that is $\hat{\sigma} = \sigma - \alpha p i$, where σ are the effective stress, α is the Biot coefficient, p is the pressure and i is the vector form of the Kronecker delta. So we may write the complete conditional formulation as:

$$\int_{\Omega} \delta \boldsymbol{\varepsilon}^T \boldsymbol{\sigma} dV + \int_{\Gamma \backslash \overline{\Gamma}} \delta \boldsymbol{u}^T \boldsymbol{\lambda} dS + \int_{\Gamma \backslash \overline{\Gamma}} \delta \boldsymbol{\lambda}^T \boldsymbol{u} dS + \int_{\overline{\Gamma}} \delta \boldsymbol{u}_r^T \boldsymbol{\tau}_L dS$$
$$= \int_{\Omega} \delta \boldsymbol{\varepsilon}^T \alpha p i dV + \int_{\Omega} \delta \boldsymbol{u}^T \boldsymbol{b} dV + \int_{\partial \Omega} \delta \boldsymbol{u}^T \boldsymbol{t} dS \qquad (5)$$

2.1 Numerical model

The displacements and stresses are approximated by the functions belonging to two different finite Hilbert spaces that, according to the Finite Elements Method, have a dimension equal to the nodal grid number of the entire domain and of the surface of the discontinuity, respectively. In general, we

can write:

$$\boldsymbol{u}^h(\boldsymbol{x}) = \mathbf{N}_u(\boldsymbol{x})\boldsymbol{u}, \quad \boldsymbol{\lambda}^h(\boldsymbol{x}) = \mathbf{N}_\lambda(\boldsymbol{x})\boldsymbol{\lambda} \qquad (6)$$

With the assumption of small displacements, deformations are calculated as $\boldsymbol{\varepsilon}^h = \mathbf{B}\boldsymbol{u}^h = \mathbf{LN}\boldsymbol{u}$, where $\mathbf{L}$ is the symbolic matrix of differential operators. Denoting by $\mathbf{D}$ the tangent stiffness matrix, the variation of the effective stress is $d\boldsymbol{\sigma}^h = \mathbf{D}d\boldsymbol{\varepsilon}^h$. Moreover, we have to introduce the relative displacements between the two rupture faces, $\boldsymbol{u}_r^h = \mathbf{S}\boldsymbol{u}$, where the matrix $\mathbf{S}$ properly maps the nodes of the global mesh. It is also necessary to express the stress on the discontinuity in a local reference system, using an appropriate rotation matrix $\mathbf{R}$, which binds the global system to the local one. Thus, the discretized form of the Eq. (2) becomes:

$$\boldsymbol{\tau}_{\mathbf{L}}^h = (c - \sigma_n \tan \phi) \frac{\mathbf{RN}_u \mathbf{S}\boldsymbol{u}}{\sqrt{(\mathbf{RN}_u \mathbf{S}\boldsymbol{u})^T \mathbf{RN}_u \mathbf{S}\boldsymbol{u}}} \qquad (7)$$

In Eq. (7), the normal stress to the rupture surface can be written as $\sigma_n = \boldsymbol{n}^T \boldsymbol{\lambda}$, where $\boldsymbol{n}$ is the unit vector normal to the discontinuity plane. The functional Eq. (5) is now written as:

$$\delta \boldsymbol{u}^T \int_{\Omega} \mathbf{B}\boldsymbol{\sigma}^h dV + \delta \boldsymbol{u}^T \int_{\Gamma \backslash \overline{\Gamma}} \mathbf{N}_u^T \mathbf{N}_\lambda \mathbf{R}^T \boldsymbol{\lambda} dS + \delta \boldsymbol{\lambda}^T$$
$$\int_{\Gamma \backslash \overline{\Gamma}} \mathbf{RN}_\lambda^T \mathbf{N}_u \boldsymbol{u} dS + \delta \boldsymbol{u}^T \int_{\overline{\Gamma}} (c - \boldsymbol{n}^T \boldsymbol{\lambda} \tan \phi) \frac{\mathbf{H}\boldsymbol{u}}{\sqrt{\boldsymbol{u}^T \mathbf{H}\boldsymbol{u}}} dS$$
$$= \delta \boldsymbol{u}^T \int_{\Omega} \mathbf{B}^T \alpha p i dV + \delta \boldsymbol{u}^T \int_{\Omega} \mathbf{N}_u^T \boldsymbol{b} dV \qquad (8)$$
$$+ \delta \boldsymbol{u}^T \int_{\partial \Omega} \mathbf{N}_u^T \boldsymbol{t} dS$$

Equation (8) must hold for every virtual displacement and stress. So, two non-linear equations are obtained. To solve this system, we use the Newton-Raphson scheme. Computing the Jacobian, we obtain a non-symmetric saddle point matrix. The non-symmetry is caused by the contribution due to the principle of "Maximum Plastic Dissipation" which introduces a component of virtual work where the displacements are independent but the stresses are dependent variables.

3   FE-IE model applications

The algorithm presented above has been implemented and tested. We use linear interpolation functions for displace-

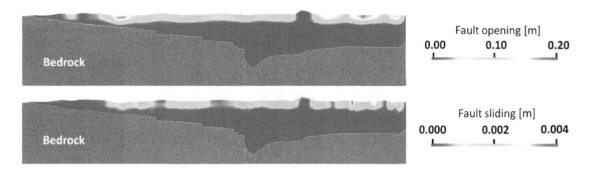

Figure 3. Test case 1: opening and sliding of the ground rupture after 10 years of water withdrawal.

Figure 4. Test case 1: land subsidence after 10 years of water withdrawal.

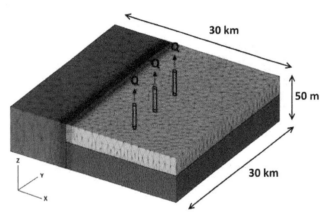

Figure 5. Test case 2: 3-D mesh with the materials highlighted in various colours. Rock is red, aquifer is blue, and clay is green. The fault divides rock from the sedimentary clay-aquifer system.

ments and constant piecewise functions for the Lagrangian multipliers. This choice was performed for compatibility reasons with the stresses of the 3-D FE grid. The discretization relies on the classic tetrahedral FE and the fault is a union of tetrahedra faces.

We address two applications: (i) the first concerns earth fissuring in Wuxi, near Shanghai, China, (Wang et al., 2007) caused by groundwater withdrawal from a shallow aquifer, with shallow confined bed; (ii) the second case concerns the City of Queretaro, Mexico (Carreón-Freyre and Cerca, 2006), placed above a graben structure with a volcanoclastic and sedimentary filling (stiff clay, silt and sandy materials) where several earth fissures have been generated by groundwater withdrawal; A sketch of the hydrogeological setting of the two test cases is shown in Fig. 1. For each case, we present the essential geometry and material properties, the history of pumping and the results, with special attention to the behaviour of both subsidence and rupture generation. Preliminary simulations with a simplified stress history are carried out.

3.1 Test case 1: Wuxi

In China, earth fissuring related to groundwater pumping has occurred since the 1970s. To cope with the rapid economic development large volumes of groundwater were withdrawal in the Wuxi area, Jiangsu Province, resulting in extensive land subsidence (Shi et al., 2007) and fissure development (Wang et al., 2007). Earth fissures cause serious damages, including cracking of buildings and failure in underground

pipelines, with huge economic losses. The geologic setting, which is characterized by an undulating shaped, relatively shallow rocky paleo-basement covered by Quaternary compressible sedimentary deposits from the Yangtze River, strongly enhances the risk of fissure development.

Geological information has been used to develop the static model. A rock ridge, characterized by a depth between few meters and about 100 m from the ground surface is buried below the Quaternary sedimentary sequence (Fig. 2a). The large land subsidence measured in the area between the 1980s and 2000s, has been accompanied by the development of a fissure in correspondence with the tip of the subsurface ridge. The geometry of the main sandy and clayey layers forming the multi-aquifer system has been derived from a number of boreholes drilled down to the bedrock. The target of the study is to simulate the development and propagation of the ground rupture, using a realistic history of groundwater extraction.

The domain extends 2 km × 5 km in the horizontal plane and from the land surface down to 250 m depth in the vertical direction. A refined mesh consisting of 23 303 nodes and 121 942 tetrahedral elements is used to accurately represent the actual lithostratigraphic configuration (Fig. 2b). A num-

Figure 6. Test case 2: opening and sliding on the fault after 10 years of water extraction.

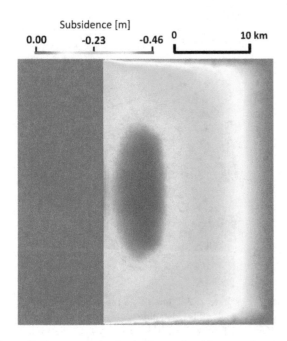

Figure 7. Test case 2: land subsidence after 10 years of water extraction.

ber of 1876 IE are introduced into the 3-D FE grid along the trace of the ridge tip, extending from the land surface to the bedrock top. The IE are characterized by a friction angle $\phi = 30°$ and zero cohesion. The geomechanical properties of the bedrock, aquifers, and aquitards are provided in Table 1.

A uniform pressure decline is prescribed in the sandy layers. The drawdown increases linearly from 0 to 20 m over 10 years. To account for a likely delay of pressure propagation in the clay layers, a half pressure change (i.e., 10 m at the end of the simulation period) is prescribed in the aquitards. No pressure change propagates in the bedrock.

The numerical results confirm the development of an earth fissure along the ridge tip. Due to the quasi-symmetric geological setting, the rupture is characterized by an opening and a negligible sliding. At the end of the simulation period the maximum opening amounts to a couple of decimeters (Fig. 3). Figure 4 provides the corresponding land subsidence distribution. The basement depth of burial significantly affects both the rupture opening and land subsidence.

3.2 Test case 2: Queretaro City

Subsidence related groundwater withdrawal and water-level decline has been documented in Queretaro City since the 1970s (Trejo-Moedano and Martinez-Baini, 1991). Local fractures in the fluvio-lacustrine sediments have been reported. The intensity of fracturing has increased and caused numerous problems to urban infrastructure. In the simplified case addressed in this study, an extensive shallow aquifer is covered with a undersaturated and stiff clayey sequence and confined on one side by a hard rock mass. The rock mass is separated from the aquifer by a regional pre-existing vertical fault. The clayey sediments have a thickness of 20 m, while the aquifer extends for 30 m in depth. The mesh covers a 30 km-side square area, one third of which is occupied by rock. In the z direction, the mesh extends from the free surface down to 50 m. Pumping takes place from the aquifer through three wells, as shown in Fig. 5, yielding a maximum pressure drawdown of about 20 m. In 10 years, a linearly increasing flow rate is implemented, up to the value of $300\,\mathrm{L\,s^{-1}}$ per well. The geomechanical parameters are provided in Table 1. Permeability of clayey sediments and rock is $K = 10^{-10}\,\mathrm{m\,s^{-1}}$, while permeability of aquifer is $K = 10^{-4}\,\mathrm{m\,s^{-1}}$. The fault has zero cohesion and friction angle $\phi = 30°$. The FE mesh consists of and 297 971 tetrahedra with 57 050 nodes. The fault is discretized by 8074 IE.

The model results provided by the IE and FE are shown in Figs. 6 and 7, respectively. In this geological setting the fault mainly slides. It can be noted that the fault opens negligibly (less than 3 cm), while the sliding in the fault plane approaches 40 cm, a value of the same order of magnitude as the maximum recorded subsidence, as can be seen in Fig. 7. This shows how important the presence of the fault is for the exact prediction of the land subsidence. The greatest calculated displacements along the fault are in the upper central part, i.e., in areas where the influence of the wells is largest and compaction of the underlying layers is the greatest. The displacements are quite regular and representative of the real order of magnitude.

4 Conclusions

The main subject of this investigation is how to simulate a geological discontinuity in a continuous medium. We used the Interface Finite Element method with a Lagrangian approach. At variance with existing algorithms, we introduce the principle of "Maximum Dissipation Plastic" in the formulation of the functional to be minimized. The solution is generally very smooth, as can be seen from the results discussed above.

The algorithm was implemented and preliminarily tested in two cases: (i) an aquifer with a shallow undulating bedrock representative of the geological setting in Wuxi, China. and (ii) a laterally rock confined aquifer covered with stiff clayey sediments, representative of the typical geological setting in Queretaro City In both cases the algorithm is robust and provides numerically stable and physically plausible solutions. Further improvements will be carried out for both the case studies, with the implementation of a more realistic evolution of the pressure head the calibration of FE-IE models on available measurements, and the development of future scenarios accounting for the expected management of the subsurface water resources.

Acknowledgements. The work has been developed within the IGCP Project 641 (M3EF3 – Deformation and fissuring caused by exploitation of subsurface fluids) funded by UNESCO.

References

Azat, E. and Shahram, S.: Land subsidence and fissuring due to ground water withdrawal in Yazd-Ardakan Basin, Central Iran, World Academy of Science, Engin. Technol., 71, 535–538, 2010.

Bathe, K.-J.: Finite Element Procedures. Prentice Hall, Upper Saddle River, New Jersey, 2006.

Bishop, A. W.: The principle of effective stress, Teknisk Ukeblad, 106, 859–863, 1959.

Carreón-Freyre, D.: Land subsidence processes and associated ground fracturing in Central Mexico, in: Land Subsidence, Associated Hazards and the Role of Natural Resources Development, edited by: Carreon-Freyre D., Cerca, M., and Galloway, D., IAHR Publ. no. 339, 149–157, 2010.

Carreón-Freyre, D. and Cerca, M.: Delineating the near-surface geometry of the fracture system affecting the valley of Queretaro, Mexico: Correlation of GPR signatures and physical properties of sediments, Near Surface Geophysics, EAGE (European Assoc. of Geoscientists and Engineers), 4, 49–55, 2006.

Ferronato, M., Gambolati, G., Janna, C., and Teatini P.: Numerical modelling of regional faults in land subsidence prediction above gas/oil reservoirs, Int. J. Numer. Anal. Meth. Geomech., 32, 633–657, 2008.

Ferronato, M., Janna, C., and Pini, G.: Parallel solution to ill-conditioned FE geomechanical problems, Int. J. Numer. Anal. Meth. Geomech., 36, 422–437, 2012.

González, P. J., Tiampo, K. F., Palano, M., Cannavó, F., and Fernández, J.: The 2011 Lorca earthquake slip distribution controlled by groundwater crustal unloading, Nature Geosci., 5, 821–825, 2012.

Goodman, R. E., Taylor, R. L., and Breeke T. L.: A model for the mechanics of jointed rock, J. Soil Mech. Foundation Div. (ASCE), 94, 637–659, 1968.

Labuz, J. F. and Zang, A.: Mohr–Coulomb Failure Criterion, Rock Mech. Rock Eng., 45, 975–979, 2012.

Rutqvist, J., Birkholzer, J. T., and Tsang, C.-F.: Coupled reservoir-geomechanical analysis of the potential for tensile and shear failure associated with CO_2 injection in multilayered reservoir-caprock systems, Int. J. Rock. Mech. Mining Sci., 45, 132–143, 2007.

Shi, X., Xue, Y., Ye, S., Wu, J.-C., Zhang, Y., and Jun, Y.: Characterization of land subsidence induced by groundwater withdrawals in Su-Xi-Chang area, China, Environ. Geol., 52, 27–40, 2007.

Simo, J. C. and Hughes, T. J. R.: Computational Inelasticity, Springer, New York, 2nd Edn., 2000.

Trejo-Moedano, A. and Martinez-Baini, A.: Soils cracking in the Querétaro zone, Proc. Symp. "Agrietamientos de suelos", Sociedad Mexicana de Mecánica de Suelos, México, 67–74, 1991 (in Spanish).

Wang, G. Y., You, G., Shi, B., Yu, J., Li, H. Y., and Zong, K. H.: Earth fissures triggered by groundwater withdrawal and coupled by geological structures in Jiangsu Province, China, Environ. Earth Sci., 57, 1047–1054, 2009.

Wriggers, P.: Computational Contact Mechanics. Springer, Heidelberg, 2nd Edn., 2006.

14

Land subsidence in the San Joaquin Valley, California, USA, 2007–2014

M. Sneed and J. T. Brandt

US Geological Survey, 6000 J Street, Placer Hall, Sacramento, CA 95819, USA

Correspondence to: M. Sneed (micsneed@usgs.gov)

Abstract. Rapid land subsidence was recently measured using multiple methods in two areas of the San Joaquin Valley (SJV): between Merced and Fresno (El Nido), and between Fresno and Bakersfield (Pixley). Recent land-use changes and diminished surface-water availability have led to increased groundwater pumping, groundwater-level declines, and land subsidence. Differential land subsidence has reduced the flow capacity of water-conveyance systems in these areas, exacerbating flood hazards and affecting the delivery of irrigation water.

Vertical land-surface changes during 2007–2014 were determined by using Interferometric Synthetic Aperture Radar (InSAR), Continuous Global Positioning System (CGPS), and extensometer data. Results of the InSAR analysis indicate that about 7600 km^2 subsided 50–540 mm during 2008–2010; CGPS and extensometer data indicate that these rates continued or accelerated through December 2014. The maximum InSAR-measured rate of 270 mm yr^{-1} occurred in the El Nido area, and is among the largest rates ever measured in the SJV. In the Pixley area, the maximum InSAR-measured rate during 2008–2010 was 90 mm yr^{-1}. Groundwater was an important part of the water supply in both areas, and pumping increased when land use changed or when surface water was less available. This increased pumping caused groundwater-level declines to near or below historical lows during the drought periods 2007–2009 and 2012–present.

Long-term groundwater-level and land-subsidence monitoring in the SJV is critical for understanding the interconnection of land use, groundwater levels, and subsidence, and evaluating management strategies that help mitigate subsidence hazards to infrastructure while optimizing water supplies.

1 Introduction and background

The extensive withdrawal of groundwater from the unconsolidated deposits of the San Joaquin Valley (SJV), California has caused widespread land subsidence–locally reaching 9 m by 1981 (Ireland, 1986). Long-term groundwater-level declines can result in a vast one-time release of "water of compaction" from compacting silt and clay layers in the aquifer system, which causes land subsidence (Galloway et al., 1999). Land subsidence in the SJV from groundwater pumping began in the mid-1920s (Poland et al., 1975; Bertoldi et al., 1991; Galloway et al., 1999), and by 1970, about half of the SJV, or about 13 500 km^2, had subsided more than 0.3 m (Poland et al., 1975).

Surface-water imports from the Delta-Mendota Canal (DMC) since the early 1950s and the California Aque-

duct since the early 1970s resulted in decreased groundwater pumping in some parts of the valley, which was accompanied by a steady recovery of water levels and a reduced rate of aquifer-system compaction in some areas (Ireland, 1986). During the droughts of 1976–1977 and 1987–1992, diminished availability of surface water prompted increased pumping of groundwater to meet irrigation demands. This increased groundwater pumping resulted in water-level declines and periods of renewed compaction. Following each of these droughts, recovery to pre-drought water levels was rapid and compaction virtually ceased (Swanson, 1998; Galloway et al., 1999). Similarly, during the more recent droughts of 2007–2009, and 2012–present, groundwater pumping has increased in some parts of the valley. Groundwater levels declined during these periods in response to increased pumping, approaching or surpassing historical low levels, which reinitiated compaction.

Groundwater pumping that resulted in renewed aquifer-system compaction and land subsidence caused serious operational, maintenance, and construction-design problems for the California Aqueduct, the DMC, and other water-delivery and flood-control canals in the SJV. Subsidence has reduced the flow capacity and freeboard of several canals that deliver irrigation water to farmers and transport floodwater out of the valley; structural damages have already required millions of dollars' worth of repairs, and more repairs are expected in the future (Bob Martin, San Luis and Delta-Mendota Water Authority, and Chris White, Central California Irrigation District, personal communication, 2010). Even small amounts of subsidence in critical locations, especially where canal gradients are small, can impact canal operations. On the DMC between the canal intakes and San Luis Reservoir, where less than 15 mm of subsidence was measured during 2007–2010 (Sneed et al., 2013), a 5-day window of opportunity to recharge the Reservoir in spring 2014 fell short because of reduced flow capacity (Bob Martin, San Luis and Delta-Mendota Water Authority, personal communication, 2014).

The objective of this paper is to describe the location, extent, and magnitude of land subsidence in the SJV during 2007–2014, which includes both drought and non-drought periods (http://www.ncdc.noaa.gov/cag/, assessed 18 April 2015). The SJV is a broad alluviated structural trough constituting the southern two-thirds of the Central Valley of California, that is a substantial source of the nut, fruit, and vegetable supply for the United States (Faunt, 2009).

2 Land subsidence and groundwater levels

Interferometric Synthetic Aperture Radar (InSAR), continuous Global Positioning System (CGPS), and extensometer data were used to determine the location, extent, and magnitude of aquifer-system compaction and resultant land subsidence. Analysis of interferograms generated from synthetic aperture radar images from the European Space Agency's ENVISAT satellite and the Japan Aerospace Exploration Agency's ALOS satellite acquired between 2008 and 2010 indicated 50–540 mm of subsidence in two large agriculturally-dominated areas in the SJV. One area is centered near the town of El Nido (2100 km^2) and the other near the town of Pixley (5500 km^2; Fig. 1). The period 2008–2010 is shown in Fig. 1 because interferograms covering the entire study area were generated for this period only. Because suitable InSAR data were not available for 2010–2014, CGPS data collected during 2007–2014 were used to generate land subsidence time series, which confirmed the InSAR-derived rates and generally indicated that these rates continued or accelerated through 2014 (Fig. 2). Extensometer data collected at four sites during 2012–2014 (data were

not available for 2007–2011) were used to generate aquifer-system compaction time series at locations along the major canals, and generally indicated higher compaction rates during the growing season of 2014, the third consecutive year of drought, than for the previous two growing seasons (Fig. 3). To help explain the variability in location and magnitude of land subsidence, computed subsidence and compaction were compared with water-level measurements retrieved from US Geological Survey and California Department of Water Resources databases (Figs. 2 and 3).

3 El Nido

The largest subsidence magnitude in the SJV during 2007–2014 was measured near El Nido. The interferograms are the only measurements that captured the maximum magnitudes because the CGPS stations and extensometers are located on the periphery of the most rapidly subsiding area (Fig. 1); however, InSAR data were only available for 2008–2010. The interferograms indicated a local maximum of about 540 mm during January 2008–January 2010, or 270 mm yr^{-1}, which is among the highest rates ever measured in the SJV. The maximum subsidence measured at nearby CGPS station P303 was about 50 mm during that same time period, indicating a large subsidence gradient between the two locations (Fig. 1). Assuming the same rate of subsidence occurred during 2007–2014 as occurred during 2008–2010 at the local subsidence maximum near El Nido, about 2 m of subsidence may have occurred during 2007–2014.

Data from the three CGPS stations and two extensometers near the periphery of the El Nido subsidence area show seasonally variable subsidence and compaction rates (including uplift as elastic rebound during the fall and winter), but different characters over longer periods of time. Vertical displacement at P307 and P303 indicated subsidence at fairly consistent rates during and between drought periods (Fig. 2a). These fairly consistent subsidence rates indicate that these areas continued to pump groundwater despite climatic variations (possibly due to lack of surface water availability); residual compaction also may be a factor. Vertical displacement at P304, however, indicated that most subsidence occurred during drought periods and very little occurred between drought periods (Fig. 2a). This suggests that this area received other sources of water, most likely surface water, when it was available between drought periods, and also that residual compaction was not very important in this area. Data from the extensometers 12S/12E-16H2 and 14S/13E-11D6 were available only during the recent drought period, so comparison to a non-drought period was not possible. CGPS and extensometer data indicated subsidence and compaction rate increases during 2014, the third year of drought (Figs. 2a and 3a). In parts of the El Nido subsidence area, where the planting of permanent crops (vineyards and

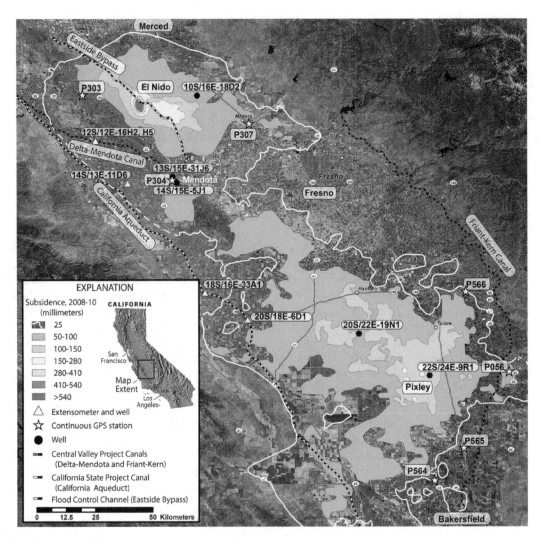

Figure 1. Map showing regions of subsidence derived from interferograms for 2008–2010, and locations of major canals, extensometers, Continuous GPS stations, and wells.

orchards) has increased, groundwater was either the primary source of water or groundwater pumping increased when surface-water availability was reduced, and groundwater levels declined to near or below historical lows during 2007–2010 and 2012–2014 (Figs. 2a and 3a). The correlation between high rates of compaction or land subsidence and water levels near or below historical lows indicates that the preconsolidation stress likely was exceeded; if so, the subsidence likely is mostly permanent.

4 Pixley

The Pixley subsidence area is larger than the El Nido subsidence area, but subsided at a lower rate during 2007–2014. Similar to the El Nido area, the interferograms provided the only measurements that captured the maximum magnitudes because the CGPS stations and extensometers are located on the periphery of the most rapidly subsiding

area (Fig. 1); however, InSAR data were only available for 2007–2010. The interferograms indicated a maximum subsidence of about 180 mm during January 2008–January 2010, whereas the maximum measured subsidence at nearby CGPS station P056 (40 km distant) was about 65 mm during that same time period (Figs. 1 and 2b). If the same rate of subsidence occurred during 2007–2014 as occurred during 2008–2010 at the local maximum near Pixley, then about 0.7 m of subsidence may have occurred during 2007–2014. Data from the four CGPS stations and two extensometers near the periphery of the Pixley subsidence area show seasonally variable subsidence and compaction rates (including uplift as elastic rebound during the fall and winter), but different characters over longer periods of time. Vertical displacement at P564 and P565 indicated that most subsidence occurred during drought periods and very little occurred between drought periods (Fig. 2b). This suggests that this area received other sources of water, most likely surface water, when it was avail-

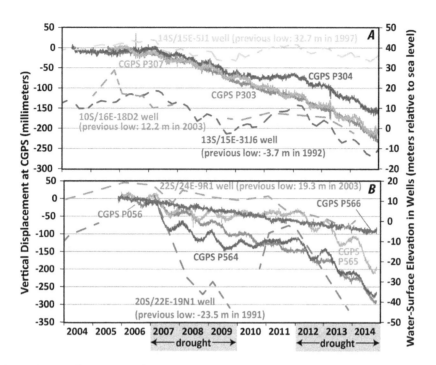

Figure 2. Graphs showing vertical displacement at selected CGPS stations and water-surface elevation in selected wells for 2004–2014 near (**a**). the El Nido subsidence area and (**b**). the Pixley subsidence area. (See Fig. 1 for CGPS and well locations.)

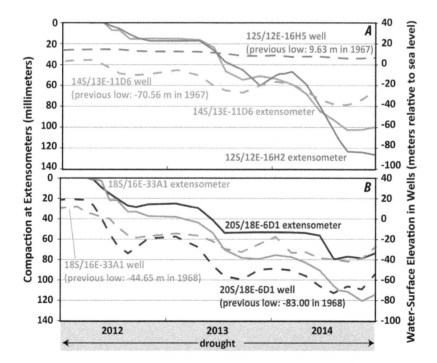

Figure 3. Graphs showing compaction and water-surface elevation at selected extensometers and associated wells for 2012–2014 near (**a**). the El Nido subsidence area and (**b**). the Pixley subsidence area. (See Fig. 1 for extensometer and well locations.)

able between drought periods, and also that residual compaction was not very important in this area. Vertical displacement at P056 and P566 indicated subsidence at fairly consistent rates during and between drought periods (Fig. 2b).

These fairly consistent subsidence rates indicate that these areas continued to use groundwater despite climatic variations (possibly due to limited surface water availability); residual compaction also may be a factor. Data from extensometers

18S/16E-33A1 and 20S/18E-6D1 were available only during the recent drought period, so comparison to a non-drought period was not possible. CGPS and extensometer data indicated subsidence and compaction rate increases during 2014, the third year of drought (Figs. 2b and 3b). In the Pixley area, groundwater pumping continued or increased when surface-water availability was reduced, and groundwater levels declined to near or below historical lows during 2007–2010 and 2012–2014 (Figs. 2b and 3b). The correlation between high rates of compaction or land subsidence and water levels near or below historical lows indicates that the preconsolidation stress likely was exceeded; if so, the subsidence likely is mostly permanent.

5 Summary and conclusions

Groundwater and surface water are generally used conjunctively in the SJV (Faunt, 2009). During recent drought periods (2007–2009 and 2012–present), groundwater pumping increased in areas where surface-water deliveries were curtailed; in response, groundwater levels declined. However, in areas where surface-water deliveries were normally an absent or minor component of the water supply, pumping was fairly steady during drought and non-drought periods; accordingly, groundwater levels declined at fairly consistent rates regardless of climatic conditions. Groundwater levels in both water-supply-source scenarios declined to levels approaching or surpassing historical low levels, which caused aquifer-system compaction and land subsidence that likely is mostly permanent.

Land use in some parts of the SJV has trended toward the planting of permanent crops (vineyards and orchards) at the expense of non-permanent land uses such as rangeland or row crops. This may have the effect of "demand hardening", which refers to the need for stable water supplies to irrigate crops that cannot be fallowed. As land use and surface-water availability continue to vary in the SJV, long-term groundwater-level and land-subsidence monitoring is critical because continued groundwater use in excess of recharge, which the historical record indicates is likely, would result in additional water-level declines and associated subsidence. Such long-term data can be used to better understand the interconnection of land use, groundwater levels, and subsidence, and to enable the evaluation of management strategies to mitigate subsidence hazards to infrastructure while optimizing water supplies. This knowledge will be critical for successful implementation of California's recent legislation aimed toward sustainable groundwater use without damage from land subsidence.

References

Bertoldi, G. L., Johnston, R. H., and Evenson, K. D.: Ground water in the Central Valley, California–A summary report, US Geological Survey Professional Paper 1401-A, 1991.

Faunt, C. C. (Ed.): Groundwater availability of the central valley aquifer, California, US Geological Survey Professional Paper, 1766, 2009.

Galloway, D. L., Jones, D. R., and Ingebritsen, S. E.: Land subsidence in the United States, US Geological Survey Circular 1182, 1999.

Ireland, R. L.: Land subsidence in the San Joaquin Valley, California, as of 1983, US Geological Survey Water-Resources Investigations Report 85–4196, 1986.

Poland, J. F., Lofgren, B. E., Ireland, R. L., and Pugh, A. G.: Land subsidence in the San Joaquin Valley, California, as of 1972, US Geological Survey Professional Paper 437-H, 1975.

Sneed, M., Brandt, J., and Solt, M.: Land subsidence along the Delta-Mendota Canal in the northern part of the San Joaquin Valley, California, 2003–2010, US Geological Survey Scientific Investigations Report 2013–5142, 2013.

Swanson, A. A.: Land subsidence in the San Joaquin Valley, updated to 1995, in: Borchers, J.W., ed., Land subsidence case studies and current research, Proceedings of the Joseph F. Poland Symposium on Land Subsidence, Sacramento, Calif., 4–5 October 1995, Association of Engineering Geologists, Special Publication no. 8, 75–79, 1998.

15

Factor analysis on land subsidence in the Nobi Plain, southwest Japan

A. Kouda[1], K. Nagata[2], and T. Sato[3]

[1]Teikoku International Corporation, 2-8 Hashimoto-cho, Gifu, Japan
[2]Geospatial Information Authority of Japan, 2-5-1 Sannomaru, Naka-ku, Nagoya, Japan
[3]Dept. of Civil Engineering, Gifu University, 1-1 Yanagido, Gifu, Japan

Correspondence to: A. Kouda (kouda@teikoku-eng.co.jp)

Abstract. The land subsidence of the Nobi Plain largely ceased with the commencement of pumping regulations beginning in 1975. However, a small amount of land subsidence, less than one centimeter, is still observed at the Delta zone in the southwest part of the Plain. The authors attempted to investigate the cause of the small amount of existing subsidence. The alluvial clay layer deposits are more than 15–20 m thick and the withdrawal is small. The decrease of the yearly average groundwater level has not been confirmed.

On the other hand, the seasonal change in the groundwater level is clearly observed. This investigation focuses on the seasonal change of groundwater level each year and its relation to the thickness of alluvial clay layer. The Delta zone was divided into several cells and a multiple regression analysis was applied to the seasonal change of the groundwater level and the thickness of alluvial clay layer of the cell. The study concluded that the small amount of land subsidence was caused of drawdown of piezometric head of groundwater every year during the summer.

1 Introduction

Three large rivers (Fig. 1), the Kiso R. to the east, the Nagara R. in the middle and the Ibi R. in the west, flow it into the Ise Bay which is located at the bottom of Fig. 1.

The area of the plain is almost 1300 km^2 and the population reached almost 6 million in 2010 which is a 16 % increases from 2005. Average precipitation has been recorded as 1543 mm year−1 since 1961. Groundwater withdrawal was recorded as 1 251 million m^3 year−1 in 1976, which breaks down to 743 million m^3 for industry, 228 million m^3 for drinking and 156 million m^3 for agricultural use.

Large amounts of land subsidence have occurred due to excess groundwater use (Fig. 2). Groundwater use has been regulated since 1975. As a result the piezometric heads have gradually increased and the rate of land subsidence has gradually decreased. However, small amounts of land subsidence have continued in the Delta zone of the southwest part since the 1980s (Fig. 3). Groundwater use in the Delta zone is small. However, 15–20 m of soft alluvial clay occur in this area.

In this paper, we conducted a factor analysis using the potential factors and the induction factor for the Delta zone to elucidate the cause of the small amount of ongoing land subsidence.

2 Multiple regression analysis of land subsidence

2.1 Features of land subsidence

Since 1989 the land subsidence area shows good agreement with the area of the alluvial clay layer with more than 15 m thickness (Fig. 4). Piezometric groundwater levels severely drop more than 2 m in the summer in the northern part of the area, A decrease of piezometric head in the Delta zone and is estimated to be about 0.2–2.0 m each year during the summer (Fig. 5). Figure 5 indicates that the area of greatest cumulative land subsidence does not necessarily align with that of groundwater level declines each year during the summer. Special attention is placed on the relationship between piezometric head decline and thickness of the alluvial clay layer and cumulative land subsidence amount.

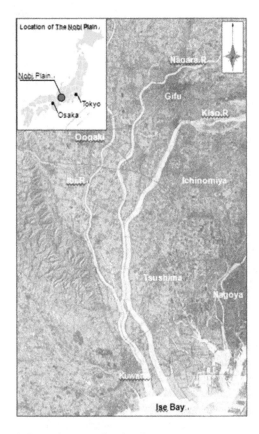

Figure 1. Reserch area and major rivers.

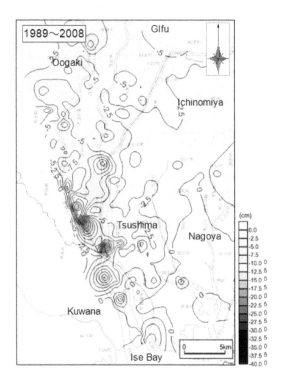

Figure 3. Cumulative land subsidence amount in the Nobi Plain (1989–2008).

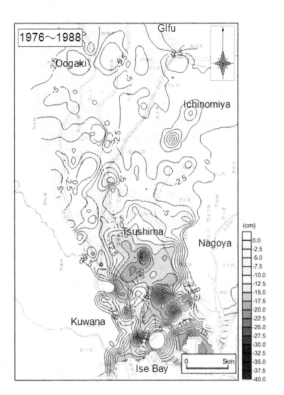

Figure 2. Cumulative land subsidence amount in the Nobi Plain (1976–1988).

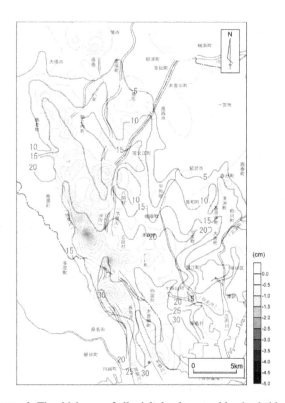

Figure 4. The thickness of alluvial clay layer and land subsidence area (m).

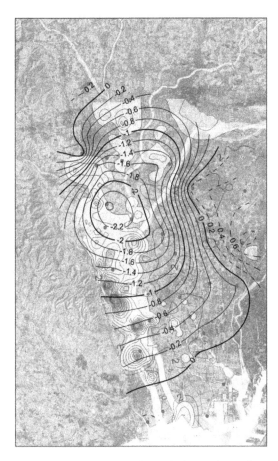

Figure 5. Decrease of piezometric head of groundwater in summer and land subsidence area (2007).

2.2 Model of the regression analysis

The research area is divided into 32 cells each with a length of 5.7 km in the east-west direction and 4.6 km in the north-south direction (Fig. 6). The measured values at the 307 benchmarks are applied to the calculation of the cumulative land subsidence of the cells. The results during 17 years of analysis, from 1989 to 2006, are shown in Fig. 7. The thickness of the alluvial clay layer is shown in Fig. 8. Seasonal variation of piezometric head of groundwater was calculated from differences in the yearly average groundwater level from 1989 to 2006 from the 28 observation wells. The amount of seasonal variation in each cell, ΔP is determined from the isolines of the seasonal variation, which is derived from the difference of the yearly average value of the piezometric head of groundwater during the summer (from June to September) and the winter (from October to May). The amount of ΔP is obtained from the difference between the yearly average value of piezometric head of groundwater in the winter and the yearly average value of piezometric head of groundwater in the summer. Positive values represent the decline in the piezometric head during the summer season (Fig. 9).

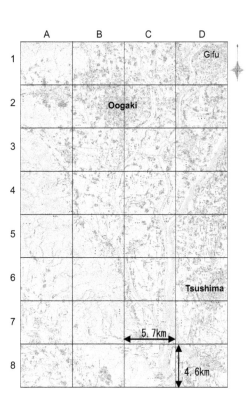

Figure 6. Analysis area.

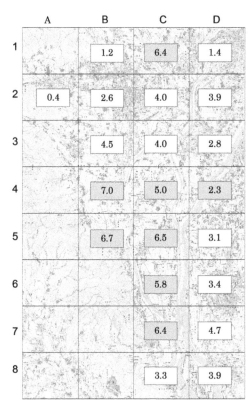

Figure 7. Average Cumulative value of land subsidence (1989–2006).

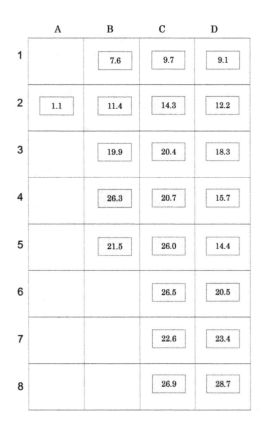

	A	B	C	D
1		7.6	9.7	9.1
2	1.1	11.4	14.3	12.2
3		19.9	20.4	18.3
4		26.3	20.7	15.7
5		21.5	26.0	14.4
6			26.5	20.5
7			22.6	23.4
8			26.9	28.7

Figure 8. Average thickness of alluvial clay layer.

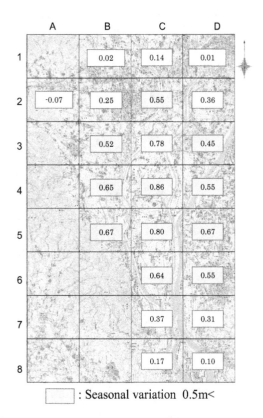

	A	B	C	D
1		0.02	0.14	0.01
2	-0.07	0.25	0.55	0.36
3		0.52	0.78	0.45
4		0.65	0.86	0.55
5		0.67	0.80	0.67
6			0.64	0.55
7			0.37	0.31
8			0.17	0.10

☐ : Seasonal variation 0.5m<

Figure 9. Seasonal variation of average value of groundwater level (1989–2006).

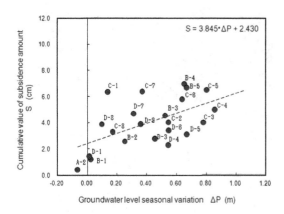

Figure 10. Cumulative value of land subsidence and seasonal variation groundwater level.

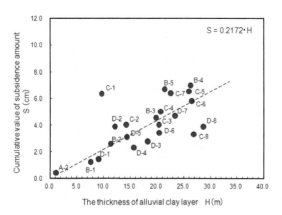

Figure 11. Cumulative value of land subsidence and thickness of the alluvial clay layer.

2.3 Multiple regression analysis results

The relationship between the amount of seasonal variation of groundwater level (ΔP) and the cumulative land subsidence (S) is shown in Fig. 10. Calculated values vary widely and the correlation coefficient is low, $r = 0.566$. One of the tendencies is that cumulative land subsidence is proportional to the increase of the amount of seasonal variation of groundwater level.

The relationship between the thickness of the alluvial clay layer (H) and the cumulative land subsidence (S) is shown in Fig. 11. The correlation coefficient is $r = 0.650$. Cumulative land subsidence is largely proportional to the increase of the thickness of the alluvial clay layer.

Multiple regression analysis provides the influence of (H) and (ΔP). The relationship between the measured and the calculated values of the cumulative land subsidence is shown in Fig. 12. The multiple correlation coefficient improves to $R = 0.704$.

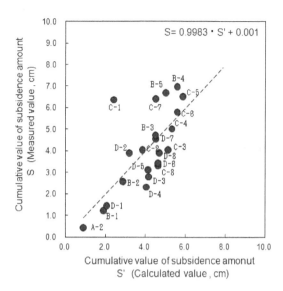

Figure 12. Comparison of Measured value and Calculated value of land subsidence.

3 Discussions and conclusions

Land subsidence occurs every year during the summer in the region where the thickness of soft alluvial clays deposits are greatest. The results of the multiple regression analysis suggests that the seasonal variation of the piezometric head is a suitable measure for describing the inducing factor of land subsidence,which is less than one centimeter per year in this area.

Consolidation curves under cyclical loading are below the normal consolidation line (NCL). It will takes a significant amount of time to reach the final land subsidence, which is greater than the NCL of the same level of an effective mean stress. Piezometric heads decline each year during the summer, which coincides with the cyclical loading curves in consolidation theory for alluvial clay layers.

This represents the main cause of continuous land subsidence over the long term at this site.

References

Gifu Prefecture: Report of Measures summary promote research for land subsidence revention, etc., 70–80, 2008.

Yasuhara, K. and Anderson, K. H.: Recompression of normally consolidated clay after cyclic loading, Soils and Foundations, 31, 83-94, 1991.

Land subsidence research meeting of tokai three Prefecture: Land subsidence and groundwater in Nobi Plain, 59–76, 1985.

16

Land subsidence caused by a single water extraction well and rapid water infiltration

I. Martinez-Noguez and R. Hinkelmann

Water Resources Management and Modeling of Hydrosystems, TU Berlin, Berlin, Germany

Correspondence to: I. Martinez-Noguez (isaac.martinez@wahyd.tu-berlin.de)

Abstract. Nowadays several parts of the world suffer from land subsidence. This setting of the earth surface occurs due to different factors such as earth quakes, mining activities, and gas, oil and water withdrawal. This research presents a numerical study of the influence of land subsidence caused by a single water extraction well and rapid water infiltration into structural soil discontinuities. The numerical simulation of the infiltration was based on a two-phase flow-model for porous media, and for the deformation a Mohr–Coulomb model was used.

A two-layered system with a fault zone is presented. First a single water extraction well is simulated producing a cone-shaped (conical) water level depletion, which can cause land subsidence. Land Subsidence can be further increased if a hydrological barrier as a result of a discontinuity, exists. After water extraction a water column is applied on the top boundary for one hours in order to represent a strong storm which produces rapid water infiltration through the discontinuity as well as soil deformation. Both events are analysed and compared in order to characterize deformation of both elements and to get a better understanding of the land subsidence and new fracture formations.

1 Introduction

Land subsidence processes can occur due to extraction of minerals in mining galleries, tunnels construction, fluids extraction (water, oil or gas) from natural reserves, decrease of groundwater level during prolonged extraction, natural land dissolution, compaction of soil materials or tectonic activity. In modern times, cities have grown rapidly and likewise have the agricultural and industrial activities. In order to satisfy the growing demands for the vital liquid, for which the superficial water bodies have not been sufficient, groundwater has been exploited to such an extent that its extraction has outpaced groundwater recharge. Especially in arid and semi-arid zones this has produced a deficit leading to a rapid decline of the groundwater table. In heterogeneous strata the fast decline of the groundwater table leads to a differential adjustment and in some cases to a formation of cracks and fractures on the surface of the ground. The valley of Queretaro city, Mexico (Fig. 1, right), is an example of active land subsidence. Since the 1970s, with the rapid growth of the city, the water demand increased rapidly and as a result the urban area is affected by differential compaction and formation of a reticular system of faults and fractures. Many of them appear on the surface and have caused economical damages in the last 40 years.

Groundwater overexploitation causing water table decline in unconfined aquifers leads to deformation of the porous media. This is mainly controlled by the increase in effective stress in the soil mass, reducing pore volume and, hence, the aquifer will compresse causing subsidence (Rojas et al., 2002; Martinez et al., 2013) and may cause the generation of earth fissures at land surface. Fault and fractures can influence groundwater flow in aquifer and aquitard layers (Carreón-Freyre et al., 2005). Structural discontinuities can perform as hydrological barriers. Such a barrier can result in different water levels on either side of the fault and as a result differential land subsidence occurs across the fault on the surface of the earth. Furthermore, it has been shown that fast water infiltration through a pre-existing fracture zone into a soil system could perform as a mechanism for fracturing and triggering of land subsidence (Martinez et al., 2013). Groundwater depletion occurs at scales ranging from a single well to regional aquifer systems. The extents of the resulting effects depend on several factors including pumpage and natural dis-

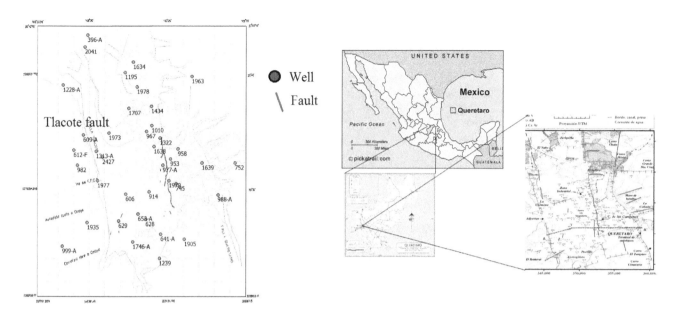

Figure 1. Queretaro, urban area, well distribution and faults and fractures location after Pacheco (2007); State of Querétaro within Mexico.

charge rates, physical properties of the aquifer, and natural and human-induced recharge rates.

Figure 1 left shows the urban area of Queretaro city (light green), faults and fractures (red line) and some of the active wells in the city (circles). In many of the cases the wells at both sides of the faults show a considerable difference of piezometric levels. An example of different water levels across a foult are given by, the static levels of the wells number 612-F at the left side of the fault Tlacote and the well number 1313-A of the year 2011 are 153.9 and 140 m respectively, which is a difference almost of 14 mm. The same behaviour exists also in wells number 1322 and 1638 located at the right and left sides of the fault Epigmenio Gonzales, respectively, where the water depths in 2011 were 116 and 129.5 m, respectively (Table 1).

The hydrogeological system is complex because of the heterogeneity of the soil layers, the basement, the fracture system and the well system. Beside the complexity of the system, another problem is the lack of data. In the literature there are several examples of analysis of subsidence produced by one single well, for example Ziaie and Rahnama (2007) presented in a study the relationship between classical soil parameters and subsidence driven for a pumping single well. Also Martinez et al. (2013) demostrated that fast water infiltration could be a mechanism for fracture generation and for triggering land subsidence. In this research both elements are analysed.

The aim of this research is to simulate the dynamic groundwater table of a homogeneous aquifer with a fault zone which has been caused by a single water extraction well and fast water infiltration after a strong rainfall. Although only idealized hydrogeological systems are modelled, the re-

Table 1. Water depths of some wells in Queretaro in 2011.

	Well no.	Water depth 2011 (m)
Fault Tlacote		
right side	2427	133.1
right side	1313-A	140
left side	612-F	153.9
Fault E. Gonzales		
right side	1322	116
left side	1638	129.5
right side	953	132.2

sults help us to understand the land subsidence intensity resulting from both events, extraction and rapid infiltration and the interrelation of them.

2 Model concepts

2.1 Two-phase flow in porous media

If two fluid phases are not or only slightly miscible into each other, a two-phase flow model concept for porous medium can be applied (Hinkelmann, 2005). The continuity equation must be fulfilled for each phase α, one for the liquid phase w (water) and one for the gas phase n (air):

$$\frac{\partial \left(S_\alpha \varphi_\alpha \rho_\alpha\right)}{\partial t} + \operatorname{div}\left(\rho_\alpha \underline{v}_\alpha\right) - \rho_\alpha q_\alpha = 0. \tag{1}$$

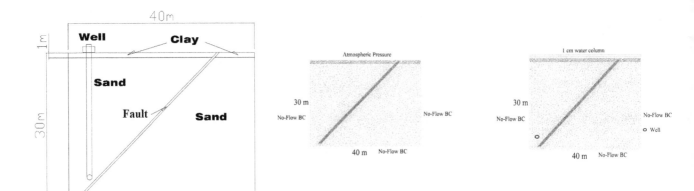

Figure 2. Idealised system (left). Unstructured grid of a 40 m × 30 m system with higher resolution around the fault zone (middle); Complete 1 cm water column on the top boundary (right).

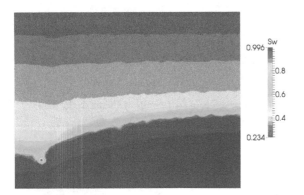

Figure 3. Water saturation after water extraction of a single system without fracture.

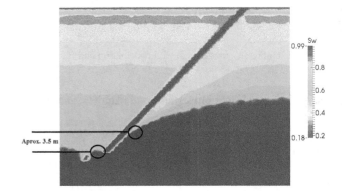

Figure 4. Water saturation in the case of a single well extraction with a fault zone.

Here, φ_α is the void space filled with phase α, S_α the saturation, ρ_α the density, $\underline{v}_\alpha$ the filter or Darcy velocity and q_α a sink or source of the phase.

To describe the Darcy velocity of each phase, the generalized form of the Darcy law can be used:

$$\underline{v}_\alpha = -\frac{k_{r\alpha}}{\mu_\alpha}\mathbf{K} \cdot (\operatorname{grad} p_\alpha - \rho_\alpha g), \tag{2}$$

where $k_{r\alpha}$ is the relative permeability, $\mathbf{K}$ the tensor of intrinsic permeability, μ_α the dynamic viscosity, p_α the pressure and g the gravity. $\frac{k_{r\alpha}}{\mu_\alpha} = \lambda_\alpha$ is referred to as the mobility of phase α.

The relative permeability for liquid phase w and gas n phase are computed with the Brooks–Corey (Brooks and Corey, 1964) relationship:

$$k_{rw} = S_e^{\frac{2+3\lambda}{\lambda}}, \tag{3}$$

$$k_{rn} = (1 - S_e)^2 \left(1 - S_e^{\frac{2+\lambda}{\lambda}}\right). \tag{4}$$

The parameter λ characterizes the grain-size distribution. A small value describes a single grain-size material, while a

larger value indicates highly non-uniform material. S_e stands for the effective (water) saturation defined as:

$$S_e = \frac{(S_w - S_{wr})}{(1 - S_{nr} - S_{wr})}. \tag{5}$$

The pore volume is completely filled with the wetting and the non-wetting phase saturation, S_w, S_n:

$$S_w + S_n = 1. \tag{6}$$

At the interface between both, the difference between the phase pressure of gas p_n and liquid phase p_w is called capillary pressure.

$$p_c = p_n - p_w \tag{7}$$

The authors would like to mention that a Richards model concept also would have been suitable for the simulations. Further information about the model is found e.g. in Hinkelmann (2005).

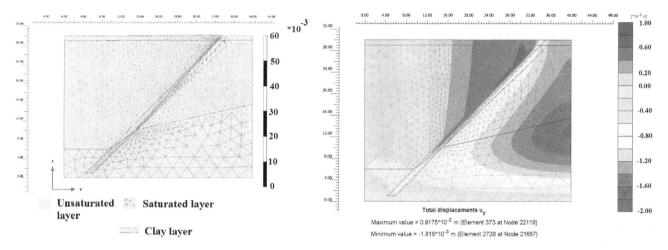

Figure 5. Deformed mesh after water extraction (left). Maximum total vertical displacement after water extraction (right).

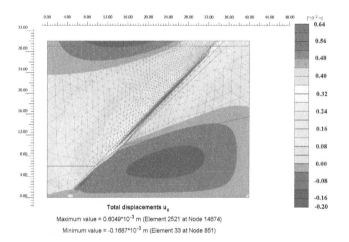

Figure 6. Maximum total horizontal displacement after water extraction.

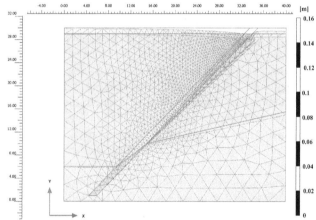

Figure 7. Deformed mesh after fast water infiltration.

2.2 Linear-elastic perfectly plastic model with Mohr–Coulomb failure criterion: deformation

Plasticity refers to irreversible strains. As soil behaviour is highly non-linear and irreversible, a linear-elastic perfectly plastic Mohr–Coulomb model has been applied (Waterman et al., 2004). This model involves five input parameters, i.e. Young's modulus E, Poisson's ratio υ, friction angle φ, dilatancy angle ψ and cohesion C. For simplification, in this investigation the dilatancy angle is taken equal to zero.

Material models for soil and rock are generally expressed as a relationship between infinitesimal increments of effective stress ("effective stress rates") and infinitesimal increments of strains ("strain rates"). This relationship is expressed in Hooke's law:

$$\underline{\dot{\sigma}}' = \underline{\underline{M}}\underline{\dot{\varepsilon}},$$

where $\underline{\underline{M}}$ is a material stiffness matrix.

This model decomposed strains and strain rates into elastic and plastic parts:

$$\underline{\varepsilon} = \underline{\varepsilon}^{e} + \underline{\varepsilon}^{p} \qquad \underline{\dot{\varepsilon}} = \underline{\dot{\varepsilon}}^{e} + \underline{\dot{\varepsilon}}^{p}. \tag{8}$$

If Eq. (8) is inserted into the Hook's law Eq. (8) leads to:

$$\underline{\dot{\sigma}}' = \underline{\underline{\dot{D}}}^{e}\dot{\varepsilon}^{e} = \underline{\underline{\dot{D}}}^{e}\left(\dot{\varepsilon} + \dot{\varepsilon}^{p}\right) \tag{9}$$

$\underline{\underline{D}}^{e}$ is the elastic material stiffness matrix $\underline{\dot{\sigma}}'$ effective stress rates.

A plastic potential function g is introduced. The plastic strain rates are written as:

$$\underline{\dot{\varepsilon}^{p}} = \lambda \frac{\partial g}{\partial \underline{\sigma}'}, \tag{10}$$

in which λ is the plastic multiplier. For elastic behaviour λ is zero and for plastic behaviour λ is positive.

These model concepts are implemented in the tool PLAXIS. Further information about the model is found in Waterman et al. (2004).

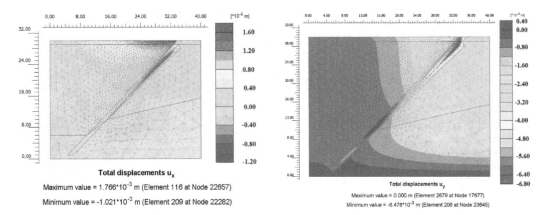

Figure 8. Maximum total vertical displacement after fast water infiltration (left). Maximal total horizontal displacement after fast water infiltration (right).

3   Idealized system and parameter

A rectangular system of 40 m length and 30 m depth was simulated. A 2-D steady-state model of a two-layered system with a clay layer on the top, a sand layer underneath and a 1 m fault zone was set up. A fracture zone was set up with a 45° inclination with respect to the horizontal (Fig. 2, left). In this study the fault zone was idealized as a porous medium damage band with higher permeability (blue in Fig. 2, middle) compared to the sand and the clay layers (green and red, respectively, in Fig. 2, middle). An unstructured grid was generated with 9798 cells with a higher resolution around the fault zone as shown in Fig. 2 middle. The boundary condition (BC) at the top of the system is a Dirichlet BC with a fixed atmospheric pressure. At the bottom, left and right of the domain a fixed Neumann no flow BC was chosen. For all simulations, as an initial condition, full water saturation was considered.

The flow parameters used in the numerical simulation for the sand, clay and fracture zone are described in Table 2.

4   Results

4.1   Reference case

Firstly a system without fracture was modelled as a reference to be compare later with the results of the other setups. Figure 3 shows the normal water depletion cone that appears around a well extraction.

4.2   Water withdrawal of a single well

With difference to the reference case, in the first step of simulation, the water extraction of a single well near a high permeable fault zone, a barrier effect is produced, i.e. the water level at both sides of the fault zone are different. The results of water saturation of the system with a 45° high permeability fault zone and a single well extraction (Fig. 4) show a

Table 2. Soil parameters for two-phase flow modeling.

		Sand	Clay	Fault zone
S_{wr}	[−]	0.15	0.15	0.05
S_{ar}	[−]	0.01	0.01	0.01
K	[−]	5×10^{-11}	3.5×10^{-13}	Variable
λ	[−]	2.3	3.5	2.2
P_d	[Pa]	700	2400	200
φ	[−]	0.37	0.43	0.32

difference of approximately 3.5 m on both sides of the fault zone.

This barrier produces an irregular soil deformation, especially on the top of the surface (Fig. 5, left). The results of the simulation show a maximum total vertical displacement of approximately 1.8×10^{-3} m (Fig. 5, right) and approximately a 0.6×10^{-3} m maximum total horizontal displacement (Fig. 6).

4.3   Fast water infiltration

To simulate an accumulation of water caused by a runoff after a strong rainfall a 1 cm water column on the complete surface was applied. The BC at the top of the system is a Dirichtlet BC with a fixed water pressure $p_n = p_{atm} + \rho g h$ with $h = 1$ cm simulating an accumulation of a 1 cm water column (Fig. 2, right).

As shown in Fig. 7, which highlights the deformed mesh after fast water infiltration, the maximal displacements take place in the upper part of the fault zone.

The results show that after 30 min of fast water infiltration through the high permeable fault zone a maximum horizontal displacement of approximately 1.8×10^{-3} m and a vertical displacement of 6.8×10^{-3} m take place (Fig. 8).

5 Conclusions

The present study is a conceptual model of land subsidence produced water extraction from a single well and fast water infiltration through a fault zone.

A two-layered system consisting of a sand layer and a small clay layer on the top with an inclined high permeable fault zone was investigated.

The differences of the water levels on both sides of the fault zone produced by the single well water extraction lead to different settlements on both sides. However, there is a greater impact on the vertical than on the horizontal displacement. The fast water infiltration impacts the displacements in both directions, vertical and horizontal, and the horizontal displacement can cause plastic deformation and also could produce new fracture formations.

If the fault zone is narrower, for example 40 cm, the infiltration through the fault zone is not significant and the water spreads out more into the partially saturated zone because of capillarity forces and gravity (Martinez et al., 2013). This phenomenon increased the deformation of the soil layers, specially the horizontal one. The results also show that a fault zone with high permeability could act as hydrological barrier. This phenomenon will be studied in more detail in future work.

References

Brooks, R. H. and Corey, A. T.: Hydraulic properties of porous media, in: Hydrology papers, Vol. 3, Colorado State University, Fort Collins, CO, 1964.

Carreón-Freyre, D., Cerca, M., Luna-González, L., and Gámez-González F. J.: Influencia de la estratigrafía y estructura geológica en el flujo de agua subterránea del Valle de Querétaro, Revista Mexicana de Ciencias Geológicas, 22, 1–18, 2005.

Hinkelmann, R.: Efficient numerical methods and information processing techniques for modeling hydro-and environmental systems, in: Lecture Notes in Applied and Computational Mechanics, Vol. 21, Springer, Berlin, Heidelberg, 2005.

Martinez, I., Hinkelmann, R., and Savidis, S.: Fast water infiltration: a mechanism for fracture formation during land subsidence, Hydrogeol. J., 21, 761–771, doi:10.1007/s10040-013-0971-6, 2013.

Pacheco, J.: Modelo de subsidencia del Valle de Querétaro y predicción de agrietamientos superficiales, PhD thesis, UNAM Mexico, 2007.

Rojas, E., Arzate, J., and Arroyo, M.: A method to predict the group fissuring and faulting caused by regional groundwater decline, Eng. Geol., 65, 245–260, 2002.

Waterman, D., Al-Khoury, R., Bakker, J. K., Bonnier, P. G., Burd, H. J., Soltys, G., Vermeer, P. A.: PLAXIS, 2D, Version 8, Manual, PLAXIS, Delft, The Netherlands, available at: http://www.plaxis.nl/shop/135/info/manuals, last access: 8 March 2014, 2004

Ziaie, A. and Rahnama, M. B.: Prediction of single well land subsidence due to ground water drainage, Int. J. Agric. Res., 2, 349–358, 2007.

17

Land subsidence risk assessment and protection in mined-out regions

A. Zhao[1,2] and A. Tang[2,1]

[1]School of Civil Engineering, Harbin Institute of Technology, Harbin, China
[2]School of Civil Engineering, Heilongjiang University, Harbin, China

Correspondence to: A. Tang (tangap@hit.edu.cn)

Abstract. Land subsidence due to underground mining is an important hazard that causes large damages and threatens to social and economic activities. The China government has started a national project to estimate the risk of land subsidence in the main coal production provinces, such as Heilongjiang, Anhui and Shanxi Provinces. Herein, the investigation methods for land subsidence identification were reported, some types of land settlement are summarized, and some successful engineering measures to mitigate the subsidence are discussed. A Geographical Information System (GIS) for land subsidence risk assessment is developed and is based on site investigations and numerrical simulation of the subsidence process. In this system, maps of mining intensity and risk ranks are developed.

1 Introduction

In China, land subsidence in mined-out regions is an important hazard for sustainable economic and social development, particularly in coal mine areas. By 2003, land subsidence attributed to mining production and groundwater pumping with a cumulative subsidence larger than 200 mm affected an area of more than $79\,000\,\text{km}^2$ (Xue et al., 2005; Wu et al., 2008; Zhang et al., 2014) in mainland China. The subsidence damaged mine, building and mining infrastructure,caused the deaths of many miners, and damaged the affected lands and their their land uses. In these affected areas, agricultural production and industry had to be curtailed, and a large number of the native inhabitants became homeless and had to move from the subsidence regions. Thus, the land subsidence in mined-out regions caused vast economic losses and posed great risks for local social, agricultural and industrial developments. Since 2000, the China government has implemented a national project to real-time monitor and assess the hazard and risk of the mined-out regions, and the core of this project is land subsidence risk assessment and prevention. After 15 years, the project has made great progress. Highly effective methods to mitigate subsidence have been implemented in the main coal producing regions, although the project plan has not completely solved these problems.

Many specialists have done a lot of research on the land subsidence mechanism and how to estimate the subsidence effects and asses the safety of structures built on the sinking areas. In the beginning of 19th century, the vertical-line theory (VLT) and Gnoto-normal line theory (GNLT) had been used to calculate land subsidence induced by minning in Belgium (Klatesch, 1983), VLT and GNLT proposed the equation to calculate the subsidence: $w = m\cos\alpha$, here, w is the subsidence, m is the thickness of mine strata, and α is the dip angle of mine strata. In the middle of the 19th century, many more theories were developed to predict land subsidence and explore its mechanism, such as bisector theory (Jlcinsky, 1876–1884), round arch theory (Fayol, 1885), strip-belt zone theory (House, 1897), cantilever beam theory (Halbnam, 1903), and influence line theory (Schimizx, 1932), among others (Liu, 1965). Many scholars have studied how land subsidence affects the safety of mining engineering, and a number of scientific publications on subsidence emanating from European countries and proposed several alternative methods (Tugrul et al., 2013). In the 1970s, Jones and Spencer (1977) discussed how mining subsidence damged highway structure in USA. Since the 1950s, Chi-

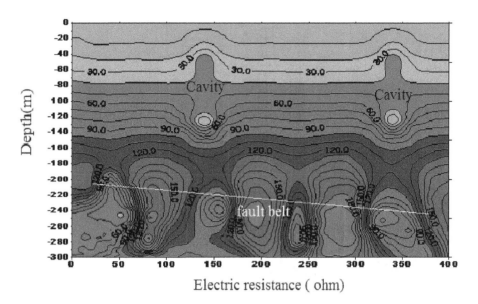

Figure 1. Electromagnetic exploring for cavity identification.

nese scholars have studied the land subsidence mechanism and protective methods, and many new theories and practical engineering measures have been referred (He, 1992).

In this paper, a comprehensive risk assessment framework and investigation methods for land subsidence identification are introduced, several typical subsidence mechanisms are summarized, and some successful engineering measures to mitigate the subsidence are discussed. A GIS for land subsidence risk assessment is developed and is based on site investigations and numerical simulation of subsidence process. In this system, maps of mining intensity and risk ranks are developed.

2 Identification and types of land subsidence in mined-out regions

In mining regions, because land subsidence is closely related to the underground excavation, it is commonly used to explore the underground cavity. In order to discover these cavities, geophysical exploration techniques have been mostly applied, such as induced polarization electrical method, 3-D seismic exploration, magnetic and electromagnetic exploration, ground penetrating radar, light detection and ranging (LIDAR), geophysical computerized tomography, electrical prospecting, and gravity methods. Generally, all of these methods have been successfully applied in identifying mined-out cavities within less than 300 m depth below land surface. In China, the first three methods have been most applied. Figure 1 shows a typical result for electromagnetic exploration of cavities.

The traditional theories divide subsiding materials above the mine into three belts or zones or areas: the bending belt, the fault and cracking belt, and the collapsing belt (from the

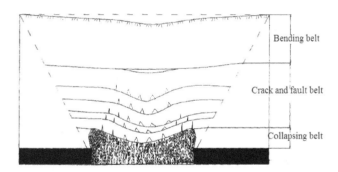

Figure 2. "Three belt" model of land subsidence types in mined-out region.

surface to the underground cavity). In the bending belt, the soil and rock are deformed but for the most parts are not mechanically failed, there are few fractures developed mainly in the vertical direction, the strata have little differential displacement and subsidence with no strata separation cracks. The thickness of the bending belt depends on the overlying rock characteristics and depth of mined-out cavity (Fig. 2). If the rock is soft and the cavity is deep, the bending deformation is more widely distributed. For the fault and crack belt, there are many cracks and faults of both horizontal and vertical directions in the rocks, and layer separation cracks are usually observed. The number and width of the rocks' cracks and faults are greater and inversely related to the distance between land surface and the cavity. In the top layer of this crack belt, the orientation of the cracks and faults is vertical toward the cavity, but in the bottom layer, the orientation is horizontal, and some strata are separated, and easily slip toward the cavity. In the collapsing belt, the rock is completely damaged, and partly or completely fill the mined-out cavities

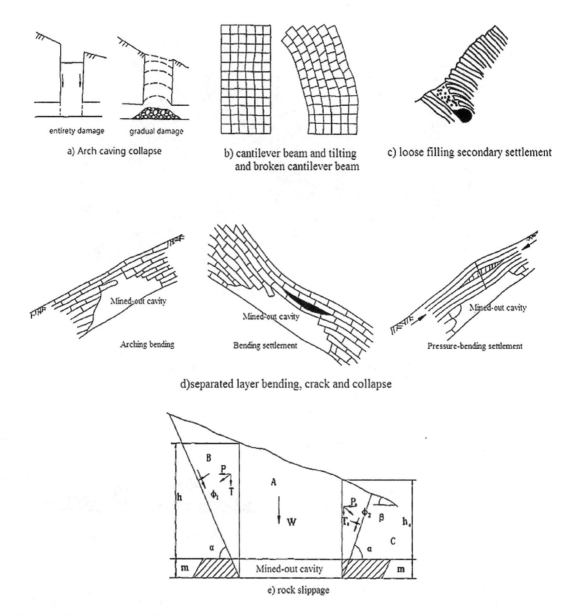

Figure 3. Land subsidence patterns in mined-out region.

depending on the rock types and thickness of mined strata. The height of the collapsing zone can be calculated as:

$$h = \frac{m}{(k-1)\cos a},$$ (1)

where h is the height of the collapsing zone, m is the thickness of the mining stratum, k is the hulking coefficient of the cracked rock (Table 1), and α is the inclination angle of the stratum. This traditional theory describes the rock deformation characteristics, but does not describe the land subsidence mechanism. Therefore, based on this theory alone, it is difficult or not suitable to design or plan reliable protective measures against land subsidence induced by different rock damage mechanisms. Since the 1950s, many important theories of subsidence processes in mining regions were developed by Chinese scientists, such as the rheological theory of the overburden strata in mining fields (Liu, 1995; Ran and Gu, 1998), nonlinear smoothing finite element theory and large deformation theory (He and Peng, 1994; He and Zhao, 1995), probability influence function theory (He, 1992), stochastic and fuzzy model theories for land subsidence prediction (Liu, 1965; Li et al., 2010), chaotic support vector machine theory (Li et al., 2008), and the thin-layer plate roof caving model theory (Zhang and Cao, 2015). According to the theoretical, numerical and in situ observations, mining-related land subsidence in China was classified into seven main types: (1) arch caving collapse (Fig. 3a), (2) cantilever beam caving of natural rock strata, including tilting and broken strata (Fig. 3b), loose filling caving secondary settlement (Fig. 3c), (4) separated layer bending, crack and

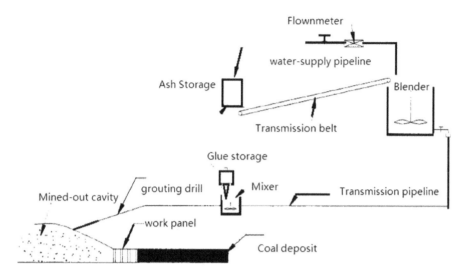

Figure 4. Simplified flowchart of a popular grouting filling method used in China.

Table 1. Hulking coefficient values of cracked rock (k).

Rock Type	Initial crack status	Residual status (compacted)
sand	1.05–1.15	1.01–1.03
Clay	< 1.2	1.03–1.07
Cracked coal	< 1.2	1.05
Clayed shale	1.40	1.10
Sandy shale	1.60–1.80	1.10–1.15
Hard sandstone	1.50–1.80	
Soft rock		1.02
Middle-hardness rock		1.025
Hard rock		1.03

collapse (Fig. 3d), (5) rock slippage (Fig. 3e), (6) rock softening and creep due to flooding and groundwater penetration, (7) collapsing due to overburden load, dynamic loads, and groundwater-level variation from natural or man-made causes (e.g. climatic variation, flooding and groundwater extraction).

3 Main engineering treatment methods for subsidence protection

Reclamation and protection of the areas affected by land subsidence are the important tasks for achieving sustainable development in mining regions. The Chinese government has paid much more attention to this since the 1950s, and during the recent past two decades, vast investments have been made to treat land subsidence. Based on the reclamation and protection practices, several principal engineering treatment methods have shown some success.

3.1 Grouting filling method

For a jointed rock mass and cavity, the grouting filling method (GFM) has been a highly effective method to control land subsidence. In our practice, the principle GFMs have included goaf-stowing with cement materials (GSC), ash grouting (AG), and low density slurry (LDS) methods. The following parameters must be considered to obtain satisfactory results with GFMs: grouting pressure, injection velocity, fluid type and density, crack density of rock mass, buried depth of cavity, volume of the collapsing belt, hydrology, construction technique and equipment. The most popular GFM in China is illustrated in Fig. 4. Here, the main filling material is ash from a power plant. The glue is a mixture of aluminium-alumine, inorganic materials and additives. Its relative density is 3.0 and its density is $1.2\,\mathrm{t\,m^{-3}}$, and it has a volumetric water content of 90 %. After mixing with water, the glue mixture can have a bearing capacity of 0.5–1.0 MPa in one hour. Typically, the mass fraction of the component of grouting materials are ash (50–70 %), glue (5 %), and water (25–45 %).

3.2 Anchor and anti-slide pile supporting system

If the cavity is buried at less than 100 m depth and is overlain by hard rock mass, the anchor supporting system (ASS) or anti-slide pile supporting system (APSS) can be used to mitigate land subsidence. However, ASS and APSS are more effective in preventing the lateral deformation of the cavity. In China, ASS and APSS are usually applied to treat the subsidence in metal (gold, iron and cobber etc.) mining regions. For ASS and APSS, the important parameters include the anchorage depths of piles and anchors, pile and anchor length and spacing, pile cross section, and ratio of height and width, and stability and strength of the supporting systems. From 1995, new ASS and APSS technique have been

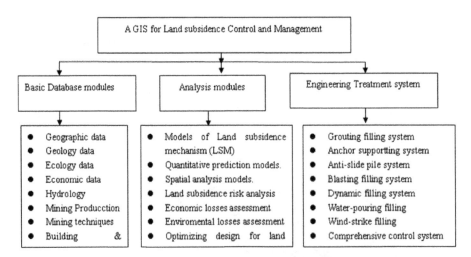

Figure 5. A GIS focussed land subsidence management system for in mined-out regions.

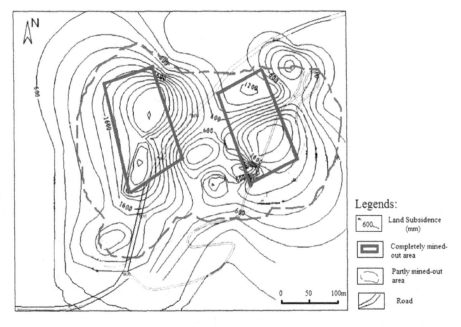

Figure 6. One of the typical maps of mining intensity and risk ranks based on GIS.

continually applied in coal mining regions, like Heilongjiang province, Shanxi province, Inner-Mongolia province and An-hui province.

3.3 Comprehensive treatment method

Dynamic compaction, blasting and filling, water-pouring fill-ing (WPF) and wind-strike filling methods are used to control the subsidence in mined-out regions in China. For a large and deep subsidence area with little population, the blasting and filling method is often used, and is capable of rapidly con-trolling subsidence for relatively short durations. Where ac-ceptable environmental conditions permit, the reclaimed land use is typically recreation and agriculture. Dynamic com-

paction is applied to treat the shallow land subsidence, par-ticular surface-based transportation system. WPF is often ap-plied to prevent the deep mined-out cavity from subsiding, in those cases, it is necessary where there is rich and easily ex-tracted groundwater. However, subsidence control with WPF method often takes a long time before it is optimally effec-tive due to long-term creep deformation and softening with soil-rock-water interaction. Wind-strike-filling is also used to treat shallow mined-out cavities where fine sand and clay are easily available.

4 A GIS for land subsidence management in mined-out regions

Managing land subsidence in mined-out regions is a complex. It entails multi-disciplinary knowledge of the geology, hydrology, geography, ecology, soil and rock mechanics, engineering, management, and risk assessment. A powerful data management and spatial analysis platform, GIS has been applied as a development tool for land subsidence management systems in China since 2000. Presently, almost all of land subsidence management systems for mining areas are based on GIS. The general management framework, flowchart and function for these systems are described in Fig. 5. One of the typical maps of mining intensity and risk ranks based on GIS is illustrated in Fig. 6. For a GIS system based on land subsidence risk assessment, prediction and prevention, these sub-systems are often at least included inside them: (1) subsidence mechanism analysis system; (2) subsidence risk assessment,mitigation and warning system; (3) subsidence spatial distribution estimation system; (4) engineering techniques to controlling and treatment system; (5) daily management system; and (6) emergency response system.

5 Summary and conclusions

Land subsidence in mining regions is an important manmad hazard in China and throughout the world that poses large risks for land use and environmental protection. Despite the successful application of engineering treatment methods based in part on traditional theories of subsidence processes related to underground mining, more research on subsidence processes in mined-out regions is needed particularly with respect to application of modern monitoring techniques, and the development of numerical simulation and computational methods based on refined process models. As the result of new advance in mining engineering and the growing demand for energy and mineral products, mining depths will increase. Land subsidence induced by underground mining will face new challenges, related to the increased mining depths and the potential reactivation of previously treated land subsidence. Therefore, reliable new methodologies to forecast and prevent land subsidence, for example by reducing the disturbing effect on the rock mass, must be developed. Additionally, laws and ordinances related to restricting subsidence induced by mining must been made and strictly executed, otherwise, it is possible that land subsidence risk will rapidly increase for the need to high-speed economic development of China.

References

He, G.: Mining Land subsidence, China University of Mining and Technology Press, 1992.

He, M. and Peng, T.: The problem of large deformation in soft-rock engineering and practical analysis of floor-heaving, J. China Univ. Mining Technol., 4, 86–91, 1994.

He, M. and Zhao, J.: Mining theory and practice under building, China University of Mining and Technology Press, 1995.

Jones, C. J and Spencer, W. J.: The implication of mining subsidence for modem highway structure, UK, Wales: Proceedings of Large Ground Tunnel Movement and Structure, University of Wales, Cardiff, UK, Halstead Press, New York, 515–526, 1977.

Karmis, M., Agioutantis, Z., and Jarosz, A.: Subsidence prediction techniques in the United States: a state-of-the-art review, Min. Res. Eng., 3, 197–210, 1990.

Klatesch, H.: Mining subsidence engineering, Springer Verlag Press, Berlin, Germany, 1983.

Li, W. X., Liu, L., and Dai, L. F.: Fuzzy probability measure (FPM) based non-symmetric membership function: engineering examples of ground subsidence due to underground mining, Eng. Appl. Artif. Intel., 23, 420–431, 2010.

Li, W. X., Gao, C. Y., Yin, X., Li, J.-F., Qi, D.-L., and Ren, J.-C.: A visco-elastic theoretical model for analysis of dynamic ground subsidence due to deep underground mining, Appl. Math. Model., 39, 5495–5506, doi:10.1016/j.apm.2015.01.003, 2015.

Li, X., Wang, Q., Yao, J., and Zhao, G.: Chaotic time series prediction for surrounding rock's deformation of deep mine lanes in soft rock, J. Centre South University of Technology, 15, 224–229, 2008.

Liu, B. C.: The basic laws of land surface movement in coal mining, China Industry Press, Beijing, China, 1965.

Liu, T.: Influence of mining activities on mine rockmass and control engineering, J. China Coal Soc., 20, 1–5, 1995.

Ran, Q. and Gu, X.: A coupled model for land subsidence computation with consideration of rheological property, Chinese J. Geol. Hazard Control, 9, 99–103, 1998.

Tugrul, U., Hakan, A., and Ozgur, Y.: An integrated approach for the prediction of subsidence for coal mining basins, Eng. Geol., 166, 186–203, 2013.

Wu, J. C., Shi, X. Q., Xue, Y. Q., Zhang, Y., Wei, Z. X., and Yu, J.: The development and control of the land subsidence in the Yangtze Delta China, Environ. Geol., 55, 1725–1735, 2008.

Xue, Y., Zhang, Y., Ye, S. J., Wu, J., and Li, Q.: Land subsidence in China, Environ. Geol., 48, 713–720, 2005.

Zhang, B. and Cao, Sh.: Study on first caving fracture mechanism of overlying roof rock in steep thick coal seam, Int. J. Mining Sci. Technol., 25, 133–138, 2015.

Zhang, Y. Q., Gong, H. L., Gu, Z. Q., Wang, R., Li, X. J., and Zhao, W.: Characterization of land subsidence induced by groundwater withdrawals in the plain of Beijing city, China, Hydrogeology, 22, 397–409, 2014.

18

Mass movement processes triggered by land subsidence in Iztapalapa, the eastern part of Mexico City

M. González-Hernández[1], D. Carreón-Freyre[2], R. Gutierrez-Calderon[1], M. Cerca[2], and W. Flores-Garcia[1]

[1]Centro de Evaluación de Riesgos Geológicos CERG, Iztapalapa, Mexico City, Mexico
[2]Centro de Geociencias de la UNAM, Juriquilla, Querétaro, Mexico

Correspondence to: D. Carreón-Freyre (freyre@geociencias.unam.mx)

Abstract. Geological and structural conditions in the Basin of Mexico coupled with natural and anthropogenic factors, such as groundwater exploitation, provokes land subsidence and differential deformation. The study area is located in to the north of Iztapalapa, a municipality within Mexico City, in a site called "El Eden" with irregular topography. Where volcanic sequences overlie the lacustrine deposits of clays and silts and show displacements by the action of gravity. The displacement zone was delimited at the top of the slope by the formation of circular tensile fractures with stair -shaped geometries. At the base of the slope, compressive processes damaged housing, sidewalks and inclined light poles and trees. A NW-SE system of fractures was identified in which displacement velocities vary from a few millimeters to several centimeters per year. Which affects urban facilities. In this work a conceptual model of deformation is presented that integrates the geological and mechanical factors leading to landslide and land subsidence. A geophysical survey leads to evidence of how land subsidence processes increase the sliding slope.

1 Introduction

Mexico City (MC) is located within the Basin of Mexico, an endorheic basin whose morphology is the result of the interaction of volcanic and sedimentary processes. Nowadays the highest records of subsidence and ground fracturing are located at the northeast side of the city. More than 30 % of the total surface has been damaged by processes related to the subsoil deformation within the Iztapalapa area. From a geomorphologic point of view the study area is comprised of three units (Fig. 1): (1) the highland, formed by volcanic deposits, (2) the slope zone, consisting of alluvial sequences and reworked volcanic deposits and (3) the lacustrine plain, formed by silty clayey sequences interbedded with fluvial and volcanic materials. In the zones 2 and 3, the geological conditions coupled with groundwater extraction, seismicity, static and dynamic loads, facilitate the development of land subsidence and ground fracturing, causing significant damage to urban infrastructure.

In Iztapalapa subsidence ranges from 10 to 40 cm year^{-1} (López-Quiroz, 2009) and three systems of fracturing affect

the 10 % of its total surface (Carreón-Freyre, 2010). The study area called "El Eden", of about 0.15 km^2, is located at the-northeast sector of Iztapalapa became highly damaged by localized deformation and fracturing. The subsoil is composed of sequences of volcanic materials from the Santa Catarina volcanic range, interbedded with pyroclastics, fluvial and lacustrine deposits (that correspond to the former Texcoco Lake). The topography is irregular and the volcanic sequences that overlie the lacustrine deposits of clays and silts displaced by the action of gravity. The movement zone was identified at the top of the slope by the formation of circular tensile fractures with stair shaped geometries.

At the base of the slope, compressive processes damaged housing, sidewalks and inclined light poles and trees. In this work a conceptual model of deformation is presented that integrates the geological and mechanical factors leading to landslide and land subsidence. The average depth of the contact zone and areas of fracturing and compression and a NW-SE system of fractures are identified.

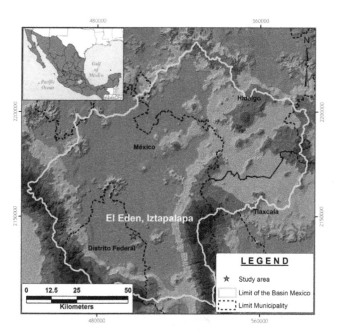

Figure 1. Location of the Mexico basin (in green) in the central part of Mexico. Whithin the Basin of Mexico is located the Mexico City (blackk dotted line, also called Distrito Federal) and the Iztapalapa municipality (red line). The location of "El Eden" Study Site is shown with a magenta star.

2 Geological setting

The morphological features of Mexico City are due to the interaction of faulting systems and volcanic activity from the Oligioceno to Miocene (Mooser at al., 1974). The subsoil is mainly composed of sedimentary rocks of the Upper Cretaceous, Tertiary and Quaternary volcanic and recent fluvio-lacustrine sequences. In the Iztapalapa zone higher structures correspond to volcanic buildings composed by andesites, basalts and pyroclastic rocks of the Pliocene-Pleistocene *Santa Catarina* Range, *Cerro de la Estrella* and *Peñón del Marques*. Along these structures the former Texcoco Lake covered the volcanic plain that was filled by lava flows, andesitic-basaltic pyroclastics (tuffs and ashes) as well as alluvial and lacustrine deposits (sand, silt and clays). From the mapping of subsidence and ground fracturing three main facture systems were identified in Iztapalapa: (a) the NE-SW system, aligned with the *Santa Catarina* Range (b) the WNW-ESE system aligned with the edge of *Texcoco* Lake and, (c) the system of tensile fractures surrounding the *Peñón del Marqués* (Fig. 2).

3 Ground fracturing in slope areas of Iztapalapa

Fractures mapped in Iztapalapa are concentrated around the border of the Santa Catarina Range and the Peñón del Marques volcano. These systems exhibit preferential directions of fracturing NE-SW and NW-SE, changes in the direction

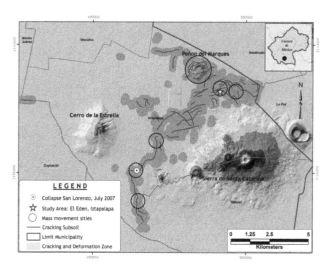

Figure 2. Detailed mapping of fracturing in Iztapalapa (red lines). The *Santa Catarina* Range is located to the south-eastern sector of Iztapalapa and the *Peñon del Marques* volcano is located to the north-east. Black circles correspond to the six indentified mass movements.

suggest that they are associated with the shoreline of the former Texcoco Lake and follow contacts between lake sediments (clays and silts) and buried lavas or pyroclastics (flow, surges and fall deposits) (Fig. 2). After mapping-in slope borders of the Santa Catarina range, six mass movement areas were identified. These areas are showing gravitational displacements that can be associated with the paleomorphology of the volcanic structures formed at the same period as Texcoco Lake, and that may be covered by volcanic-sedimentary deposits (Fig. 2).

In these areas slopes vary from 3 to 10° and the volcanic deposits are often interbedded with fluvio-lacustrine sequences (Fig. 3). The lithological contacts become sliding planes that determine a lateral spreading and fracturing of the surface materials associated with vertical and/or horizontal displacements. The contrasting physical and mechanical properties of lacustrine sequences (low shear strength and high compressibility) and pyroclasts materials (non-plastic, porous and non-cohesive), the irregular topography and coupled with groundwater exploitation facilitates the development of land subsidence. This phenomenon causes gradual changes in the thicknesses of the lacustrine sequences in turn modifying the original slope, this condition trigger a gravity process of "mass movement" with orders of magnitude ranging from some mm to many cm per year. These "masses" displace overlying irregular surfaces that are lithological contacts and generates shear fissures between both geological materials. The morphological features of this phenomenon consist of tensile fractures in the upper zone, with variable length and irregular geometry, often concave, with significant vertical displacements (0.3 to 1 m in average). The stress fractures have variable length, opening and direction. While

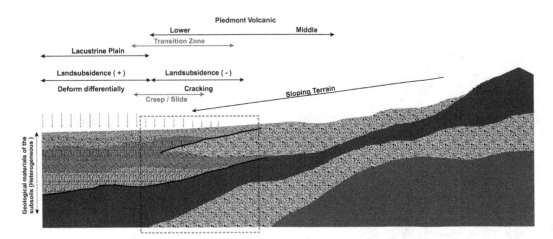

Figure 3. Schematic geological section of the piedmont zones in the *Santa Catarina* Range. Phenomena of land subsidence in the lacustrine plain (left) associated with mass movements processes (right) over a detachment surface are shown.

in the front area a compression process has been mapped by damage on sidewalks and walls, as well as distortion structural elements of houses. The depth of the lithological contact acting as a detachment surface depends on the fabric and morphology of the deposit (Fig. 3).

4 Field mapping at "El Eden" study site

From the six sites with identified mass movement processes, the site with the highest reported damage is called "El Eden", located at the northeast sector of Iztapalapa (Fig. 4). The subsoil sequence is composed by silty clays and volcanic deposits, lava and fall, identified through information collected from previous geotechnical boreholes and outcrops nearby the study area. The zones identified within "El Eden" mass movement are shown in the map of Fig. 4. The orange area, of about 38 000 m^2, corresponds to a tensile zone with NW-SE stepped fractures spaced every 45 m. The yellow area, of about 35 000 m^2, indicates a high deformed compressive zone. The levelling survey in the study site shows an irregular topography that can be associated with the paleo-morphology of a buried volcanic structure, part of the same Santa Catarina Range. Altitude variations ranges from 0.5 to 8 m and slopes obtained by the interpolation of levelling points vary from 0 to 12°.

A geophysical survey was performed to characterize the subsurface structure. The 2-D MASW method was applied for obtaining S wave patterns, with the analysis variation of velocities of shear waves in depth, it was possible to identify the depth of lithological contacts and the thicknesses of the prospected layers. A 40 m length profile with an estimated penetration depth of 15 m is shown in Fig. 5. A sequence of 3 layers was characterized: (1) A high plasticity lacustrine clay, at a depth from 0 to 2 m, with an S wave velocity less than 180 m s^{-1}; (2) Black sands interbedded with silt and pyroclastic fall deposits, at depths from 2 to 6 m depth, with

Figure 4. Location of "El Eden" Study Site in Iztapalapa. Orange colour indicate the tension area and the yellow colour the compressive area. Fractures are shown as red lines.

a s wave velocity varying than 180 to 300 m s^{-1}; and, (3) fractured lava and pyroclastic flows with S wave velocities greater than 300 m s^{-1}. The detachment surface of the mass movement was located a depth 3 m.

5 Conclusions

Land Subsidence in the Iztapalapa area of Mexico City is associated with mass movement processes in the slope-area of the two main volcanic structures, the *Santa Catarina* Range and the *Peñon del Marques* volcano. Slope changes due to land subsidence in the lacustrine plains causes instability in areas where the subsoil is composed of high plasticity clays from Texcoco Lake, along with interbedded sandy and silty ashes and fractured lava flows.

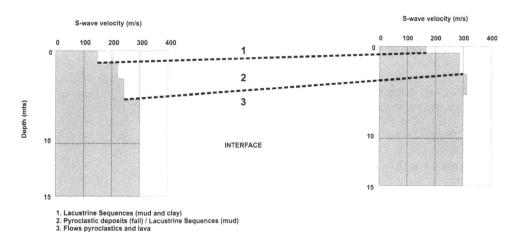

Figure 5. 1-D S-Wave velocities model from the 2-D MASW Method showing a 15 m depth of investigation. Three layer according to the shear velocities were identified (1) lacustrine clays, (2) sand, volcanic ashes; (3) volcanic flows and fractured lavas.

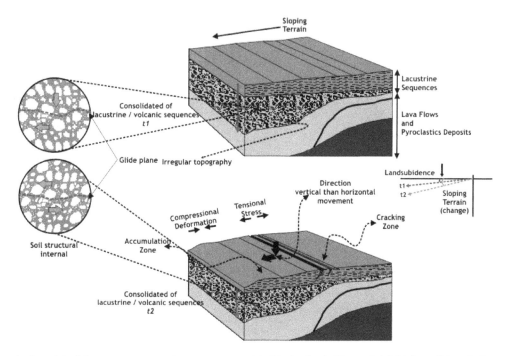

Figure 6. Conceptual model of the mass movement processes associated to land subsidence. (**a**) Initial conditions before consolidation of lacustrine sediments; (**b**) Mass movement showing the tension and compression zones in the upper and lower part of the slope respectively. Black arrows indicate the movement direction.

A conceptual model of the mass movement in the "El Eden" study site is presented (Fig. 6). A sliding slope was delimited into two strips, the upper zone and the front area. The upper zone is under tensional strees because of lateral spreading with stepped fracturing systems and parallel geometries. The front zone is under compressional stress causing significant damage to urban structures. The clayey layers behave as detachment surfaces that could be identified by the 2-D MASW method because of its low S wave velocities.

The mass movements associated with land subsidence processes are of greatest importance in Iztapalapa because the damage caused to housing and urban facilities (roads and pipes), where the demaged area covers approximately $74\,000\,\mathrm{m}^2$.

Acknowledgements. The authors acknowledge the support of the DGAPA-UNAM research project No. IN114714 and, the field support of the staff from *Centro de Evaluación del Riesgo Geológico* (CERG): Cesar M. Tovar Medina, Enrique Luna Sánchez, Felix A. Centeno-Salas and Said R. Zacarías Ramírez.

References

Carreón-Freyre, D.: Land subsidence processes and associated ground fracturing in Central Mexico, in: Land Subsidence, Associated Hazards and the Role of Natural Resources Development (Proceedings of EISOLS 2010, Querétaro, México), edited by: Carreón-Freyre, D., Cerca, M., and Galloway, D. L., Red Book Series Publication 339, IAHS Press, CEH Wallingford, UK, ISBN: 978-1-907161-12-4, ISSN: 0144-7815, 149–157, 2010.

López-Quiroz, P., Marie-Pierre Doin, M.-O., Tupin, F., Briole, P., and Nicolas, J.-M.: Time series analysis of Mexico City subsidence constrained by radar interferometry, J. Appl. Geophys., 69, 1–15, doi:10.1016/j.jappgeo.2009.02.006, 2009.

Mooser, F., Nairn, A. E. M., and Negendank, J. F. W.: Paleomagnetic investigations of the Tertiary and Quaternary igneous rocks: VIII A paleomagnetic and petrologic study of volcanics of the Valley of Mexico, Geologischen Rundschau, 63, 451–483, 1974.

19

Compaction parameter estimation using surface movement data in Southern Flevoland

P. A. Fokker[1], J. Gunnink[1], G. de Lange[2], O. Leeuwenburgh[1], and E. F. van der Veer[1]

[1]TNO, Utrecht, the Netherlands
[2]Deltares, Utrecht, the Netherlands

Correspondence to: P. A. Fokker (peter.fokker@tno.nl)

Abstract. The Southern part of the Flevopolder has shown considerable subsidence since its reclamation in 1967. We have set up an integrated method to use subsidence data, water level data and forward models for compaction, oxidation and the resulting subsidence to estimate the driving parameters. Our procedure, an Ensemble Smoother with Multiple Data Assimilation, is very fast and gives insight into the variability of the estimated parameters and the correlations between them. We used two forward models: the Koppejan model and the Bjerrum model. In first instance, the Bjerrum model seems to perform better than the Koppejan model. This must, however, be corroborated with more elaborate parameter estimation exercises in which in particular the water level development is taken into account.

1 Introduction

The Southern part of the Flevopolder (the Netherlands) was reclaimed in 1967–1968. After lowering the groundwater level, the sediments compacted, resulting in surface subsidence. The compaction is a combination of elastic and viscoplastic compaction of clay and peat, and of peat oxidation. A good prediction of future compaction and an extrapolation to other areas naturally requires knowledge of the subsurface lithology, the compaction parameters and the phreatic water level development. The present paper describes an exercise of parameter estimation using a limited number of borehole data in the Southern Flevopolder area. An Ensemble Smoother with Multiple Data Assimilation was employed to this end.

2 Available data

The Flevopolder is a large polder in the Netherlands, which was reclaimed from the IJsselmeer in 1967 (Fig. 1). The surface has an elevation of about 3 m below sea level. The composition of the subsurface had been mapped on a number of places prior to the reclamation. It is a variable layered structure which we categorized into four generic types: clay, humic clay, peat, and sand. For the present study, 10 measure-ment locations were selected, at which the surface level had been monitored yearly between 1967 and 1993 by geodetic levelling.

Data about the phreatic groundwater level are scarce. One of the few measurements covering the complete period is represented in Fig. 2. It shows the fast lowering of the water level at reclamation of the polder, followed by a gradual decrease during the next 7 years. This can be explained by the gradual adjustment of the phreatic level at the measurement location to the drainage level in the ditches.

3 Forward model

For the subsidence prediction we employed a combination of compaction in the clay, the humic clay and the peat, and oxidation in the humic clay and peat layers above the phreatic level. We used models for primary and secondary settlement for stresses larger than the pre-consolidation stress: the Koppejan model and the Bjerrum model, both with an instantaneous lowering of the groundwater level to a fixed value below the original surface level (Koppejan, 1948; Bjerrum, 1967). In combination with peat oxidation, the two models then predict the following compaction strain and associated

Table 1. Assimilation results.

	Koppejan Case 1 Prior	Koppejan Case 1 Estimate	Koppejan Case 2 Prior	Koppejan Case 2 Estimate	Bjerrum Prior	Bjerrum Estimate
Compaction parameters						
C_p (clay)	30 ± 5	33 ± 4	30 ± 20	50 ± 10		
CR (clay)					0.15 ± 0.08	-0.03 ± 0.06
C_p (humic clay)	30 ± 5	46 ± 4	30 ± 20	65 ± 10		
CR (humic clay)					0.15 ± 0.08	-0.12 ± 0.05
C_p (peat)	20 ± 4	30 ± 3	25 ± 20	57 ± 12		
CR (peat)					0.15 ± 0.08	-0.06 ± 0.04
C_s (clay)	320 ± 20	330 ± 17	320 ± 80	380 ± 55		
C_α (clay)					0.030 ± 0.015	0.010 ± 0.005
C_s (humic clay)	120 ± 12	140 ± 11	120 ± 80	270 ± 55		
C_α (humic clay)					$0.030 \pm .015$	$0.022 \pm .009$
C_s (peat)	80 ± 8	90 ± 6	100 ± 50	180 ± 34		
C_α (peat)					0.030 ± 0.015	0.015 ± 0.008
Oxidation parameters						
v_{ox} (humic clay)	0.010 ± 0.005	0.030 ± 0.002	0.020 ± 0.010	0.041 ± 0.003	0.010 ± 0.005	0.032 ± 0.002
v_{ox} (peat)	0.015 ± 0.005	0.021 ± 0.004	0.025 ± 0.010	0.027 ± 0.007	0.015 ± 0.005	0.025 ± 0.003
Phreatic levels (m below surface in 1968):						
ZF 3	2.0 ± 0.5	2.35 ± 0.36	2.0 ± 0.5	2.24 ± 0.33	2.0 ± 0.5	1.99 ± 0.35
ZF 6	2.0 ± 0.5	2.10 ± 0.25	2.0 ± 0.5	1.84 ± 0.33	2.0 ± 0.5	2.18 ± 0.25
ZF 11	2.0 ± 0.5	1.98 ± 0.09	2.0 ± 0.5	1.82 ± 0.17	2.0 ± 0.5	2.30 ± 0.19
ZF 15	2.0 ± 0.5	2.27 ± 0.08	2.0 ± 0.5	2.17 ± 0.08	2.0 ± 0.5	2.71 ± 0.11
ZF 19	2.0 ± 0.5	1.90 ± 0.06	2.0 ± 0.5	1.92 ± 0.08	2.0 ± 0.5	2.27 ± 0.09
ZF 20	2.0 ± 0.5	2.51 ± 0.15	2.0 ± 0.5	2.54 ± 0.06	2.0 ± 0.5	2.98 ± 0.19
ZF 26	2.0 ± 0.5	1.96 ± 0.07	2.0 ± 0.5	1.96 ± 0.20	2.0 ± 0.5	2.37 ± 0.10
ZF 30	2.0 ± 0.5	2.26 ± 0.07	2.0 ± 0.5	2.27 ± 0.07	2.0 ± 0.5	2.52 ± 0.11
ZF 33	2.0 ± 0.5	1.92 ± 0.08	2.0 ± 0.5	1.80 ± 0.08	2.0 ± 0.5	2.14 ± 0.10
ZF 36	2.0 ± 0.5	1.40 ± 0.06	2.0 ± 0.5	1.39 ± 0.05	2.0 ± 0.5	1.45 ± 0.12

subsidence as a function of time:

$$\varepsilon_{v,Koppejan} = \left[\frac{1}{C_p} + \frac{1}{C_s} \log \frac{t}{t_{ref}} \right] \times \ln \frac{\sigma'}{\sigma'_0} + \left[1 - \exp(-v_{ox}t) \right]$$

$$\varepsilon_{v,Bjerrum} = CR \log \frac{\sigma'}{\sigma'_0} + C_\alpha \log \frac{t}{t_{ref}} + \left[1 - \exp(-v_{ox}t) \right]$$

Here, σ' and σ'_0 are the actual and original effective vertical stresses; C_p and C_s are the primary and secondary compression coefficients above pre-consolidation pressure in the Koppejan model; v_{ox} is the oxidation rate; CR and C_α are the virgin compressibility (above pre-consolidation pressure) and the creep parameter or coefficient of secondary compression in the Bjerrum model.

4   Inverse model

The present study aimed at the assessment of the uncertainty of the model parameters in order to improve the his-

tory match of the ground level displacement and the reliability of the associated predictions. We considered as uncertain parameters the primary and secondary compression coefficients, the oxidation rates and the phreatic water levels. We assumed the measured lithology at every location as fixed. Of the four lithologies identified, we assumed the sand to be incompressible. Both models were thus left with 18 uncertain parameters: primary and secondary creep parameters for clay, for humic clay, and for peat (6 parameters); oxidation rates for humic clay and for peat (2 parameters); and groundwater levels for each location (10 parameters).

We define the vector m as the collection of adjustable model parameters; the vector d as the collection of data (surface level measurements vs. time) and the functional $G(m)$, working on the model parameters, as the forward model. The inverse problem is then formulated as the task of estimating the vector m for which $G(m)$ approaches the data vector d best. With additional information present in the form of a

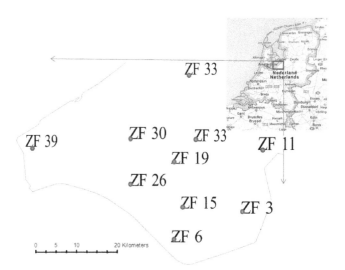

Figure 1. Map of the SE Flevopolder with the selected locations.

prior model (m_0) and covariance matrices of the measurements ($\mathbf{C}_d$) and of the prior model ($\mathbf{C}_m$), the conventional least-squares solution is obtained by maximizing the objective function J given by Tarantola (2005) (or by minimizing the exponent in the expression, $-\log(J)$):

$$J = \exp\left[-\frac{1}{2}(m - m_0)^T \mathbf{C}_m^{-1}(m - m_0) - \frac{1}{2}(d - G(m))^T \right.$$
$$\left. \mathbf{C}_d^{-1}(d - G(m))\right].$$

We used an ensemble approach in which the mean and the covariance of the model vector m_0 are mapped on an ensemble of N_e vectors. Different approaches exist to obtain an estimate of the model vector. A global update of the model using all available data can be achieved in a single step; this procedure is called an Ensemble Smoother (Emerick and Reynolds, 2012) – it is, however, suboptimal for non-linear problems as ours. When both data and parameters are assimilated to newly incoming data in subsequent time steps, the procedure is called a filter. EnKF is an example of a filter, but we did not use EnKF because all data were available at all times. Emerick and Reynolds (2013) and Tavakoli et al. (2013) state that the best approach for non-linear systems such as ours is to use an Ensemble Smoother with Multiple Data Assimilation (ES-MDA). In ES-MDA, the ensemble smoother is applied iteratively multiple times. In this way, correlations between parameters that result from the smoother are retained in subsequent steps. The advantage of ES-MDA above EnKF is that the data are used as given and that the procedure is computationally less demanding.

5 Results

The results of the estimation exercise with both models are presented in Table 1. For the Koppejan model we report re-

sults that were achieved after starting from two initial distributions – the first one with closer bounds than the second.

The results show that the main adjustments for the Koppejan model are in the oxidation rates and in the phreatic water levels. These parameters have the largest impact on the surface movement rates. The adjustments of the oxidation rates are larger for the less constrained case, as should be expected. For the phreatic water levels, the adjustments are similar. There is quite a large variability in the resulting estimate of the phreatic water level among the different locations – this is related to the types of lithology present around the phreatic water level: peat oxidation has the largest influence on subsidence and boreholes with peat will result in a better constrained water level because the sensitivity to it is larger.

For the Bjerrum model, the largest effect is also related to the peat oxidation rates and the water levels, but the compaction parameters are also significantly constrained. For these parameters, the values of CR and C_α are anti-correlated – an increase in one of them can be partially compensated by a decrease in the other.

Figure 3 visualizes the data, the priors and the inversion results. While a large degree of scatter is present for the prior distribution of parameters, the smoother succeeds in constraining this uncertainty and obtaining a reasonable fit to the data. The assimilation is better when using the Bjerrum model.

6 Discussion

We wish to address some issues here that are striking from the results as presented above. The first is the relative insensitivity to the compaction parameters in the Koppejan model; the second is the appearance of unphysical negative numbers for the immediate compaction parameter CR in the Bjerrum model; the third is the high oxidation rates resulting for the humic clay and for the peat – the former even being the largest. In our opinion, these issues are related to a fourth issue: the fact that the match of the curves using estimated parameters is suboptimal in the first 6 or 7 years. We feel there is a strong relation with the fact that the parameter estimation with the present dataset was accomplished under the assumption of an immediate drop of the water level to a further constant value. This is a simplification that must be addressed in a follow-up study. Although only limited data on the water level are available, Fig. 2 shows that this assumption is not generally applicable. A first-order approximation could be to distribute a 1 m level drop to the final value over the first 7 years after reclamation. We are currently building such an extension into our models.

7 Conclusions

We estimated compaction parameters for clay, humic clay, and peat, oxidation rates for peat and for humic clay,

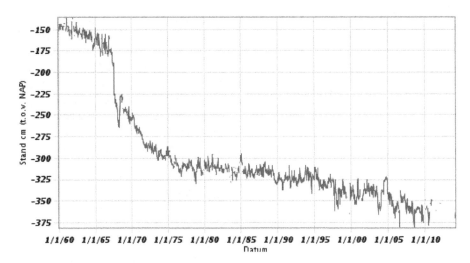

Figure 2. Hydraulic head in well B26E00030001, located on the boundary between the previously reclaimed Eastern Flevoland polder and the subsequently reclaimed Southern Flevoland polder. The reclamation of the Southern Flevoland polder in 1967 is followed by a gradual decrease of the level during approximately 7 years; after which the level remains virtually constant until 1995.

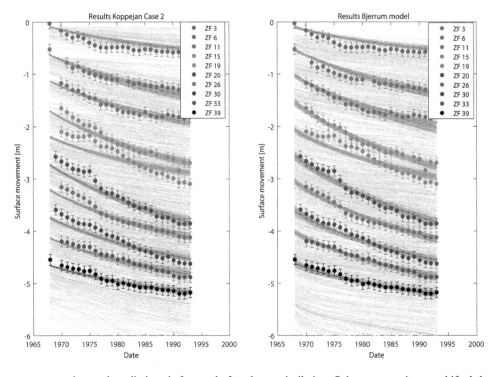

Figure 3. Surface movement data and predictions before and after data assimilation. Subsequent series are shifted downward by 0.5 m. Colored circles: data; colored dashed lines: average prior estimate; colored solid lines: average assimilated estimate. Light grey lines: prior ensemble predictions. Dark grey lines: posterior ensemble predictions. Case 2, with less constrained prior gives larger prior spread and better estimates than Case 1.

and phreatic water levels for 10 locations in the Southern Flevopolder, using surface movement parameters between the moment of reclamation in 1967 and 1993. The procedure, using an Ensemble Smoother with Multiple Data Assimilation, is very fast and gives insight into the variability of the estimated parameters and the correlations between them.

In first instance, the Bjerrum model seems to perform better than the Koppejan model. This must, however, be corroborated with more elaborate parameter estimation exercises in which in particular the water level development is taken into account.

References

Bjerrum, L.: Engineering geology of Norwegian normally consolidated marine clays as related to settlements of buildings, Géotechnique, 17, 81–118, 1967.

Emerick, A. and Reynolds, A. C.: Ensemble smoother with multiple data assimilation, Comput. Geosci., 55, 3–15, 2012.

Emerick, A. and Reynolds, A. C.: Investigation of the sampling performance of ensemble-based methods with a simple reservoir model, Comput. Geosci. 17, 325–350, 2013.

Koppejan, A. W.: A Formula Combining the Terzaghi Load Compression Relationship and the Buisman Secular Time Effect, Proc. Int. Conf. Soil Mech. Found. Eng., Rotterdam, 3, 32–38, 1948.

Tarantola, A.: Inverse Problem Theory and Methods for Model Parameter Estimation, SIAM, Paris, France, 2005

Tavakoli, R., Yoon, H., Delshad, M., ElSheikh, A. H., Wheeler, M. F., and Arnold, B. W.: Comparison of ensemble filtering alrorithms and null-space Monte Carlo for parameter estimation and uncertainty quantification using CO_2 sequestration data, Water Resour. Res., 49, 8108–8127, 2013.

20

Land subsidence of clay deposits after the Tohoku-Pacific Ocean Earthquake

K. Yasuhara[1] and M. Kazama[2]

[1]Institute for Global Change Adaptation Science, Ibaraki University, 2-1-1 Bunkyo, Mito, Ibaraki, 310-8512, Japan
[2]Graduate School of Engineering, Tohoku University, 2-1-1 Katahira Aoba-ku Sendai-shi, Miyagi, 980-8577, Japan

Correspondence to: K. Yasuhara (kazuya.yasuhara.0927@vc.ibaraki.ac.jp)

Abstract. Extensive infrastructure collapse resulted from the cataclysmic earthquake that struck off the eastern coast of Japan on 11 March 2011 and from its consequent gigantic tsunami, affecting not only the Tohoku region but also the Kanto region. Among the geological and geotechnical processes observed, land subsidence occurring in both coastal and inland areas and from Tohoku to Kanto is an extremely important issue that must be examined carefully. This land subsidence is classifiable into three categories: (i) land sinking along the coastal areas because of tectonic movements, (ii) settlement of sandy deposits followed by liquefaction, and (iii) long-term post-earthquake recompression settlement in soft clay caused by dissipation of excess pore pressure. This paper describes two case histories of post-earthquake settlement of clay deposits from among the three categories of ground sinking and land subsidence because such settlement has been frequently overlooked in numerous earlier earthquakes. Particularly, an attempt is made to propose a methodology for predicting such settlement and for formulating remedial or responsive measures to mitigate damage from such settlement.

1 Introduction

Land subsidence and ground sinking are as remarkable and extremely severe phenomena as the gigantic tsunami that struck Tohoku and Kanto in Japan after the 2011 Off the Pacific Coast of Tohoku Earthquake (hereinafter the "Tohoku Earthquake"). Moreover, the mechanisms of both events are extremely complex because several factors induced land subsidence and ground sinking from tectonic, geologic, and geotechnical aspects, exerting overlapping influences that persist even now, four years after the earthquake.

Roughly speaking, this land subsidence and ground sinking is classifiable into three categories: (i) land sinking along coastal areas because of tectonic movements (Imakiire and Koarai, 2012, Yasuhara et al., 2012;, (ii) settlement of sandy deposits followed by liquefaction, and (iii) long-term post-earthquake recompression settlement in soft clays caused by the dissipation of excess pore pressure. Land sinking described in category (i) induces inundation of wide coastal areas, particularly during severe storm-surges and strong ty-

phoons. However, land subsidence in the latter two categories (ii) and (iii) occurs in inland areas, strongly influencing local residents. A difference is apparent between the mechanisms of settlement of both types in that settlement included in category (ii) is related to saturated sandy deposits, whereas settlement in category (iii) is associated with saturated clayey deposits, which cause them to take a long time to cease. Based on differences among the causes and mechanisms related to the three kinds of land subsidence described above, suitable countermeasures against subsidence should be taken, corresponding to their different characteristics.

This paper presents two case histories of post-earthquake settlement of clay deposits from among the three categories of ground sinking and land subsidence described above because normally such settlement has been overlooked in numerous historical earthquakes.

2 Post-earthquake settlement of cohesive soil deposits in Miyagi

Generally speaking, liquefaction-induced settlement described in the previous section ceases in a very short time. As described later, post-earthquake long-term settlement is caused mainly by consolidation of clayey soils situated under sand deposits simply because the hydraulic conductivity of clay layers is much lower than that of sand deposits. For that reason, a long time is necessary for the dissipation of excess pore pressure generated during earthquakes (Yasuhara et al., 2001; Yasuhara and Matsuda, 2002; Matsuda et al., 2014).

Figure 1 depicts the variations of settlement at bench marks after 1974, when the total amount of settlement was greater than 200 mm up to 2010. Settlement measuring locations are benchmarks located in the eastern part of Sendai on the Quaternary alluvial lower plain of unconsolidated layers of around 50 m. It is apparent from Fig. 1 that a sudden increase of settlement occurred after the 1978 Miyagioki Earthquake. It then progressed at a higher rate of settlement than before the earthquake. The sudden increase of settlement in 1978 resulted from the collapse of clay soil particle structures: an immediate effect of the earthquake action.

Acceleration of post-earthquake settlement as shown in Fig. 1 originated from the superimposition of land subsidence before the earthquake and consolidation settlement caused by the dissipation of excess pore pressure generated during the earthquake.

Figure 2 depicts an example of variations of settlement over time before and after the Tohoku Earthquake in 2011. The tendency presented in Fig. 2 is almost identical as that presented previously in Fig. 1.

Kazama (2014) reported that post-earthquake settlement of clay deposits are caused by collapse of clay soil particle structures immediately after the earthquake and the recovery of collapsed structures to the new state of clay structures This mechanism for post-earthquake settlement of clay soils is understood from Fig. 3, which portrays the progress of settlement in the form of void ratio, e vs. logarithmic of vertical stress, log p. To validate the proposed concept for estimating post-earthquake settlement, undrained cyclic triaxial tests followed by drainage were conducted by Kazama using undisturbed and remolded clay specimens taken from the site where settlement had taken place. Subsequently, the results from undrained cyclic triaxial tests followed by drainage under the conditions of 0.4 for cyclic stress ratio and 0.2 Hz. for cyclic frequency are presented for comparison in Fig. 3 along with results from odometer tests on the undisturbed and remolded specimens of the same clay as that used in cyclic triaxial tests.

Void ratios after termination of undrained cyclic triaxial tests followed by drainage are shown on the vertical line corresponding to the consolidation pressure with $e - \log p'$ curves in Fig. 3. Three dotted circles are shown after each

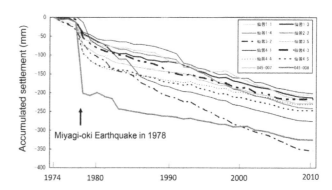

Figure 1. Examples of land subsidence variation in Sendai before and after the earthquake.

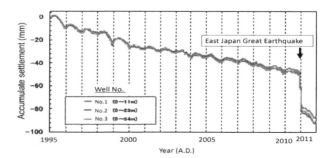

Figure 2. Variation of settlement before and after the Tohoku Earthquake (at Kamomachi of Sendai).

undrained loading followed by drainage, which simulates the circumstances under which three earthquakes struck separately after some intervals.

If initial void ratios under a current consolidation pressure p'_i for undisturbed and remolded conditions corresponds to cases before and after the earthquakes are assumed as $e_{i,u}$ and $e_{f,r}$, respectively, then post-earthquake settlement S_{post} is given as

$$S_{\text{post}} = \frac{e_{i,u} - e_{f,r}}{1 + e_{i,u}} H \tag{1}$$

where H stands for the clay layer height. If one assumes 2.15 and 1.85 for $e_{i,u}$ and $e_{f,r}$, respectively, and 10 m for the clay layer height, then one obtains 47.6 cm for post-earthquake settlement S_{post}.

Another methodology can be used to estimate the post-earthquake settlement of clay layers proposed by Yasuhara et al. (1991, 2001). It is given as

$$S_{\text{post}} = 0.225 C_c H \log(\frac{1}{1 - \frac{u_{\text{cy}}}{p'_c}}), \tag{2}$$

where C_c stands for the compression index of clay soil, H signifies the clay layer height, u_{cy} represents excess pore pressure generated by the earthquake, and p'_c denotes the consolidation or overburden pressure. When we take values

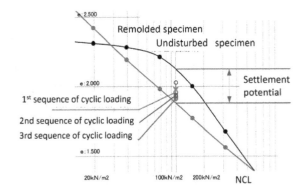

Figure 3. Void ratio vs. consolidation pressure relations for prediction of post-earthquake settlement of clay deposits (from Kazama, 2014).

Table 1. Index properties of clay beneath river dykes in Ibaraki.

Index	unit	Kujigawa clay	Jyouban clay
Void ratio e		2.17	2.40
Density of soil particle ρ_s	g/cm^3	2.67	2.68
Wet density ρ_t	g/cm^3	1.52	-
Natural water content w_i	%	81.0	93.7
Liquid limit w_L	%	73.0	70.7
Plasticity limit w_L	%	37.0	28.8
Plasticity index I_p		36.0	41.9
Liquidity index I_L		1.20	1.20
Sensitivity ratio S_t		17.0	15.0
Unconfined compression strength q_u	kN/m^2	90.9	70.5

from Fig. xx for C_c, H, u_{cy} and p'_c as 1.06, 5.0 m, 0.5 p'_c and 110 kN m^{-2}, then we obtain 33.9 cm for S_{post}. It is assumed here for calculation using Eq. (2) that the normalized excess pore pressure in the clay layer, u_{cy}/p'_c is 0.5. Therefore, the amount of settlement will be less than this value, 05, when $u_{cy}/p'c$ is smaller than 0.5. The amount of settlement of 33.9 cm obtained using Eq. (2) is less than the 47.6 cm obtained using Eq. (1).

3 Post-earthquake settlement of cohesive soil deposits in Ibaraki

3.1 Outlines of post-earthquake settlement of residential areas beside the River Dykes

Another case study in Ibaraki is that where clay deposits near the river dyke have undergone post-earthquake settlement causing damage to many private residences. The horizontal profile of the river dyke and the residential areas portrayed in Fig. 4 indicate a thick clay layer with xx m deposits on the sand layer with the higher ground water level (GWL). clay is characterized by the following. Highly sensitive with high liquidity index High water content with high undrained strength Post-earthquake settlement is expected to result from consolidation followed by dissipation of excess pore pressures generated by a strong earthquake. The average index properties of clay are presented in Table 1. This

Characteristic (ii) does not make sense from geotechnical perspective. It is extremely unusual although characteristic (i) does make sense. This clay has high structures of clay particles, which readily undergo deformation by disturbance, leading to radical deformation and settlement. For comparison with Kujikawa clay, Table 1 also includes index properties of Joban clay, deposited beneath the Kanda area of Joban express highway, which was constructed almost 35 years ago. Both clays are similar: both are highly plastic and become considerably soft with water contents beyond the liq-

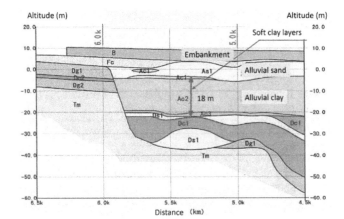

Figure 4. Geological profile of the objective site.

uid limit, although it cannot be said that the water content of either clay is very high.

A set of typical observed settlement vs. time relations continuously before and after the earthquake is presented in Fig. 5. Settlement before the earthquake is caused mainly by land subsidence because of ground water abstraction. However, settlement after the earthquake is very complex but perhaps caused by combining the following:

i. land sinking by crustal movement,

ii. immediate settlement followed by liquefaction of upper sand deposits, which is not time-dependent, and

iii. time-dependent settlement of clay followed by dissipation of excess pore pressure generated by earthquakes.

Those two components are present in addition to land subsidence.

The current paper, however, addresses issue (iii) because it emphasizes an examination of post-earthquake time-dependent settlement of clay deposits, which corresponds to the latter part of the settlement vs. time curves depicted in Fig. 6. The rates of settlement before and after the earthquake are quite noticeable. The settlement has been exacer-

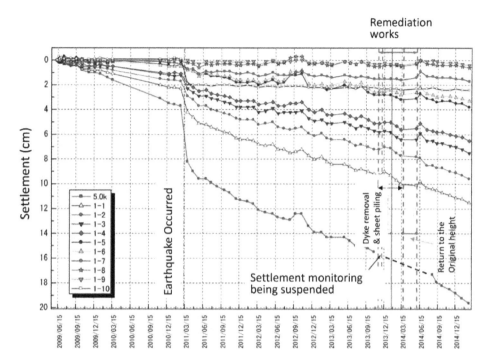

Figure 5. Settlement vs. time curves of river dykes.

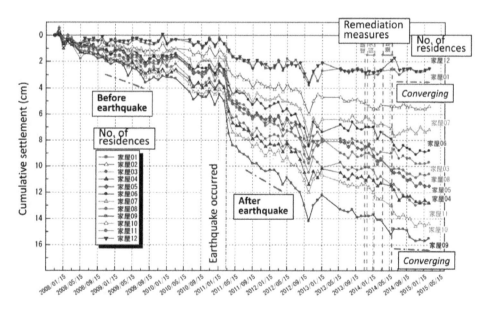

Figure 6. Settlement vs. time curves for residences.

bated by earthquakes because of the post-earthquake settlement of clay deposits caused by delayed dissipation of excess pore pressure in clay deposits after the earthquake. Generally speaking, such settlement takes a long time to cease. Therefore, it is necessary for engineers and researchers to ascertain when the rate of residual settlement after the earthquake becomes negligible for residents not to feel at risk in everyday life.

3.2 Results from settlement monitoring and numerical analysis

Since 2007, the Governmental Office in charge of river management in the objective area has been monitoring variations of settlement with time inside and outside the river dykes. Results of settlement monitoring indicate the following:

i. Settlement of residences accumulated up to 14 cm for the prior six years starting in 2007.

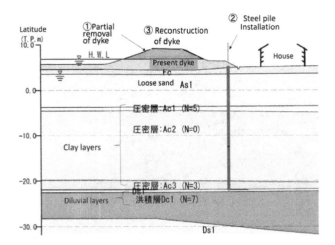

Figure 7. Remediation for reducing post-earthquake settlement of soft clay deposits.

ii. Settlement increased to around 2 cm after the earthquake in 2011.

According to elasto-visco-plastic two-dimensional numerical analysis for settlement of river dykes and residences conducted in addition to settlement monitoring:

  i. The predicted total settlement of river dykes will be greater than 90 cm.

  ii. Settlement must persist for more than 30 years.

  iii. Inclination of residences also continues for a long time.

4   Effects of damage mitigation countermeasures

To prevent the effects of river dykes on post-earthquake settlement of residences, the following countermeasures were undertaken (see Fig. 7).

  i. Remove 2.2 m of dykes to reduce dyke self-weight.

  ii. Thereafter install sheet piles aside from the river dykes into the hard stratus, called A_{c3} layers, to intercept the effects of river dyke self-weight on residences.

  iii. Return the dyke height to the original height with 4.1 m.

As might be readily apparent from the results depicted in Figs. 5 and 6, the following behavior was observed after carrying out the countermeasures stated above.

  i. Remaining excess pore pressure on grounds both outside and inside the dykes became less than that before partial removal of dykes.

  ii. Installation of sheet piles is associated with increased ground settlements beneath the dykes, but does not increase settlement of the ground outside the dykes, but rather might decrease settlement of the ground of residents.

iii. Filling up to the original height of dykes decreased settlement of the ground of residents because of the intercept of stress distribution of dykes to residential ground.

As a summary of the tendencies stated above, probably the countermeasures adopted herein have been effective to date for reducing post-earthquake settlement of residents near the river dykes. However, monitoring of settlements should be conducted because the influence of groundwater abstraction on continuous settlement of residential areas remains uncertain.

5   Conclusion

The paper presents an outline of the present situation of ground sinking, land subsidence, and settlement induced by tectonic movements on a global scale and ground movements on a local scale induced by the Tohoku Earthquake in 2011. Particularly, the paper describes a specific examination of post-earthquake settlement of clay layers observed in some locations in the Tohoku and Kanto regions. This paper describes two case histories related to this category of settlement: One in Miyagi for a methodology for predicting settlement and another in Ibaraki for effective countermeasures for settlement. Investigation of the two case histories suggests that careful attention should be devoted to the fact that such settlement is time-dependent and that monitoring of settlement should therefore continue for a long period. These phenomena show different tendencies from those of two kinds of settlement originating from tectonic movement, followed by liquefaction.

Acknowledgements. The authors are grateful for financial support from a Grant-in-Aid for Scientific Research from the Ministry of Education, Culture, Sports, Science and Technology (FY2014–FY2017, Project No. 26281055), Japan, whose representative is Makoto Tamura, Associate Professor of Ibaraki University, Japan.

References

Imakiire, T. and Koarai, M.: Wide-area land subsidence caused by "the 2011 OFF the Pacific Coast of Tohoku Earthquake", Soils Found., 52, 842–855, 2012.

Joint Editorial Committee in Tohoku Branch for the Report on the Great East Japan Earthquake Disaster 2013, Part 3 Geohazards, Ch. 8, Land sinking and inundation, 3-212–3-218, 2013 (in Japanese).

Kazama, M.: Geotechnical subjects seen to the damage of the 2011 off the Pacific Coast of Tohoku Earthquake, Proc. 11th Symp. on Ground Improvement, Journal of Material Science, Japan, 1–20, 2014 (in Japanese).

Matsuda, H., Nhan, T. T., Nakahara, D., Thien, D. Q., and Tuyen, T. H.: Post-cyclic recompression characteristics of a clay subjected to undrained uni0directional and multi-directional cyclic shears, Proc. 10th US National Conf. on Earthquake Engineering, Earthquake Eng. Research Institute, Anchorage, AK, 2014.

Ohara, S. and Matsuda, H.: Study on the settlement of saturated clay layer induced by cyclic shear, Soils Found., 28, 103–113, 1988.

Yasuhara, K. and Andersen, K. H.: Recompression of normally consolidated clay after cyclic loading, Soils Found., J. JGS, 31, 83–94, 1991.

Yasuhara, K. and Matsuda, H.: 5. Dynamic properties of cohesive soils, J. of Japan Geotechnical Society, 46, 59–64, 1998 (in Japanese).

Yasuhara, K. and Usui, T.: Potential application of geosynthetics for reconstruction following coastal land subsidence induced by the Great East Japan Earthquake, Geosynthetics Journal, Japan Chapter of IGS, 27, 69–77, 2012 (in Japanese).

Yasuhara, K., Murakami, S., Toyota, N. and Hyde, A. F. L.: Settlements in fine-grained soils under cyclic loading, Soils Found., J. of JGS, 41, 25–36, 2001.

21

Subsidence monitoring with geotechnical instruments in the Mexicali Valley, Baja California, Mexico

E. Glowacka[1], O. Sarychikhina[1], V. H. Márquez Ramírez[2], B. Robles[3], F. A. Nava[1], F. Farfán[1], and M. A. García Arthur[1]

[1]Centro de Investigacion Cientifica y Educacion Superior de Ensenada, Ensenada, Mexico
[2]UNAM Campus, Centro de Geociencias, Juriquilla, Querétaro, Mexico
[3]Instituto Mexicano de Tecnología de Agua, Jiutepec, Morelos, Mexico

Correspondence to: E. Glowacka (glowacka@cicese.mx)

Abstract. The Mexicali Valley (northwestern Mexico), situated in the southern part of the San Andreas fault system, is an area with high tectonic deformation, recent volcanism, and active seismicity. Since 1973, fluid extraction, from the 1500–3000 m depth range, at the Cerro Prieto Geothermal Field (CPGF), has influenced deformation in the Mexicali Valley area, accelerating the subsidence and causing slip along the traces of tectonic faults that limit the subsidence area. Detailed field mapping done since 1989 (González et al., 1998; Glowacka et al., 2005; Suárez-Vidal et al., 2008) in the vicinity of the CPGF shows that many subsidence induced fractures, fissures, collapse features, small grabens, and fresh scarps are related to the known tectonic faults. Subsidence and fault rupture are causing damage to infrastructure, such as roads, railroad tracks, irrigation channels, and agricultural fields.

Since 1996, geotechnical instruments installed by CICESE (Centro de Investigación Ciéntifica y de Educación Superior de Ensenada, B.C.) have operated in the Mexicali Valley, for continuous recording of deformation phenomena. Instruments are installed over or very close to the affected faults. To date, the network includes four crackmeters and eight tiltmeters; all instruments have sampling intervals in the 1 to 20 min range.

Instrumental records typically show continuous creep, episodic slip events related mainly to the subsidence process, and coseismic slip discontinuities (Glowacka et al., 1999, 2005, 2010; Sarychikhina et al., 2015).

The area has also been monitored by levelling surveys every few years and, since the 1990's by studies based on DInSAR data (Carnec and Fabriol, 1999; Hansen, 2001; Sarychikhina et al., 2011).

In this work we use data from levelling, DInSAR, and geotechnical instruments records to compare the subsidence caused by anthropogenic activity and/or seismicity with slip recorded by geotechnical instruments, in an attempt to obtain more information about the process of fault slip associated with subsidence.

1 Introduction

The Mexicali Valley is located, in the southern part of the San Andrés fault system, within the southern part of the Salton Trough, on the border between the North America and Pacific tectonic plates. The Valley is characterized by recent volcanic hydrothermal processes, active tectonics, and high seismicity. Moreover large earthquakes concentrate along the major, Imperial and Cerro Prieto faults, while scattered seismicity, (mainly swarms), and deformation are observed in the Pull-apart Cerro Prieto Basin (Lomnitz et al., 1970; Nava

and Glowacka, 1994; Suárez-Vidal et al., 2008). Fluid extraction began in the Cerro Prieto Geothermal Field (CPGF) in 1973, and brine injection therein began in 1989; these processes have been influencing deformation, stress, and seismicity of the area (Majer and McEvilly, 1981; Glowacka and Nava, 1996; Fabriol and Munguía, 1997; Glowacka et al., 1999, 2005; Trugman et al., 2014).

The subsidence area is limited to the area between the Imperial, Saltillo, Cerro Prieto and Morelia faults (Fig. 1). This zone, also known as Cerro Prieto basin, is larger than

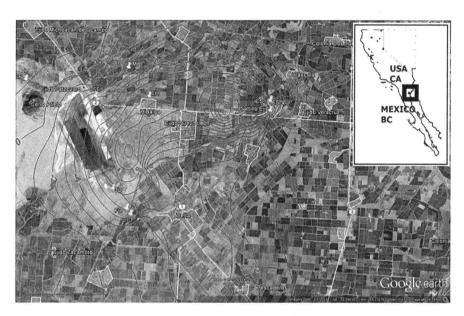

Figure 1. Instruments (yellow pins) and the subsidence rate blue isolines (cm yr^{-1}) for 1997–2006 (modified from Glowacka et al., 2011). Red lines indicate tectonic faults. The inset shows the geographical location of the study area.

the extraction zone, suggesting the existence of a recharging region. Since the sunken area is limited by tectonic faults, Glowacka et al. (1999, 2005, 2010a), suggest that these faults constitute a boundary of the subsiding region, due to differential compaction and/or due to poor permeability in the direction perpendicular to the faults that acts as a groundwater barrier.

The CPGF has 720 MW production capacity, or 1/3 of electricity production in Baja California; however, the anthropogenic subsidence caused by deep extraction of fluids in the CPGF causes damage to roads, railways and irrigation systems, increasing the natural hazards in this area that is quite vulnerable because of its tectonic situation.

The history of subsidence at the CPGF area has been well documented. Geodetic studies in the Mexicali Valley began in the 1960's. Leveling surveys have been done in the area of the CPGF and the Mexicali Valley since 1977 (Velasco, 1963; Lira and Arellano, 1997; Glowacka et al., 1999, 2012). Subsidence in the CPGF has also been measured via DInSAR (Differential Synthetic Aperture Radar Interferometry) by Carnec and Fabriol (1999) and Hanssen (2001) using ERS1/2 images acquired during 1993–1997 and 1995–1997, respectively, and interpreted as an anthropogenic effect of fluid extraction.

Sarychikhina et al. (2011) applied the DInSAR technique using C-band ENVISAR ASAR data acquired between 2003 and 2006, to determine the extent and amount of land subsidence in the Mexicali Valley near the CPGF. Recently Sarychikhina et al. (2015) applied DInSAR technique, together with leveling results and geotechnical instruments data and modeling, to estimate seismic and aseismic deformation in Mexicali Valley for the period 2006–2009. The current subsidence rate, evaluated from DInSAR data is of the order of 12 cm yr^{-1} in the production area and 18 cm yr^{-1} for the recharging zone (Sarychikhina et al., 2015; Sarychikhina and Glowacka, 2015a).

2 Geotechnical instruments monitoring

To study the spatial and temporal distribution of crustal deformation in the Mexicali Valley, CICESE installed a network of geotechnical instruments, starting in 1996. Since then, the REDECVAM (Red de Deformaciones de la Corteza en el Valle de Mexicali) network has included four creepmeters (wide range extensometer) and nine tiltmeters installed over different periods in different places; all instruments have sampling intervals in the 1 to 20 min range (Nava and Glowacka, 1999; Glowacka et al., 2002, 2010b, c; Sarychikhina et al., 2015). Creepmeters (Geokon model 4420) were installed mainly on vertical planes perpendicular to faults, in order to record vertical displacement. Biaxial tiltmeters (Applied Geomechanics, models 711, 712, and 722) were installed in shallow vaults or wells, close to faults.

Data from a creepmeter and a tiltmeter installed on the Saltillo fault (Fig. 1) show fault vertical displacement rate at ∼ 5.3 cm yr^{-1} until 2003 and ∼ 7.3 cm yr^{-1} since then. The distance-time relationship between changes in extraction at the CPGF and the displacement rate change found at the Saltillo fault suggests that the fault is probably affected by extraction through diffusive transmission of pore pressure changes, with a characteristic hydraulic diffusivity (Glowacka et al., 2010a).

Vertical displacement at the Saltillo fault consists of continuous creep and episodic slip events, sometimes concen-

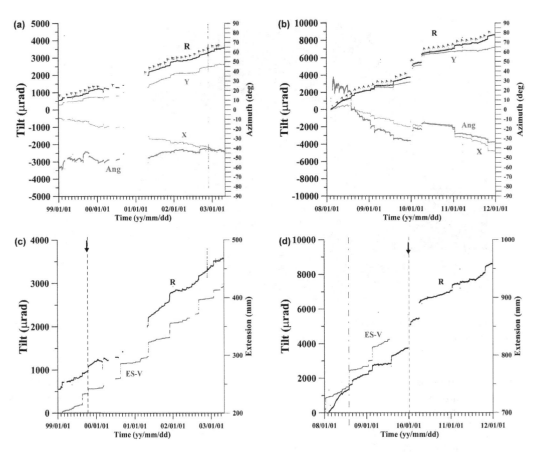

Figure 2. Tilt observations. X and Y are the East-West and North–South components of tilt, respectively, and R is the resultant tilt magnitude in the azimuth direction, Ang, for 1999–2003 (**a**) and 2003–2012 (**b**). X, Y and R are referred to the left axes, while azimuth Ang is referred to the right axes. Azimuth is also shown by the direction in which brown arrowheads point, with North straight up. Extension ES-V and magnitude R for 1999–2003 (**c**) and 2008–2012 (**d**). The vertical arrow indicates the $M = 7.2$ hector Mine earthquake in (**c**), and $M = 5.8$ 2009 earthquake in (**d**). The dotted-line indicates slip event in August 2008.

trated in suites (Nava and Glowacka, 1999). Episodic slips have 1 to 3 cm magnitudes and 1–3 days duration, separated by months of monotone creep, and release about 50 % of total slip. The episodic fault slip in the Saltillo fault appears mainly as slip-predictable, normal, aseismic slip (e.g. Glowacka et al., 2001, 2010a). Using reservoir model, geological structure, and subsidence data, Glowacka et al. (2010a), speculatively proposed the depth range of slip events to be from 1 to 2.5 km.

Some of the slip events are triggered by distant earthquakes. This includes, for example, Hector Mine Earthquake (of magnitude 7.2 and at a distance 260 km away 15 on 1999), as described by Glowacka et al. (2002), and, probably, by the Canal de las Ballenas earthquake (of magnitude 6.9 and a distance ~ 400 km away 2009), as described by Glowacka et al. (2015). These studies suggest that distant earthquakes can influence fault slip related to the subsidence. Both, sporadic and triggered slip events confirm slip predictable fault behavior.

Detailed analysis of tiltmeter and creepmeter data done by Sarychikhina et al. (2015), identified which part of subsidence observed by DInSAR in the Mexicali valley during 2006–2009 was caused by local seismicity. Glowacka et al. (2015) estimated that during the 2006–2009 period of relatively high seismicity, the anthropogenic subsidence was of the order of 80 % of the total subsidence, which is close to estimates done by Glowacka et al. (2005), Camacho Ibarra (2006), and Sarychikhina et al. (2015).

The goal of this paper is to find if there are differences, in the signal characteristics observed on the creepmeter and/or tiltmeter records, between continuous creep, sporadic slip, triggered slip and coseismic slip recorded on the Saltillo fault.

3 Results

In the following we will analyze records from ES-V creepmeter and ES-I biaxial tiltmeter installed very close to each other on the Saltillo fault. A biaxial tiltmeter (ES-I) was installed in 1998, very close to the creepmeter and recorded ground tilt until 2003 (Fig. 2a), and after a small reconstruction, again since 2008 (Fig. 2b). E–W (X) and N–S (Y) inclination recorded on ES-I tiltmeter are presented on Fig. 2a (1999–2003) and Fig. 2b (2008–2012).

For the tilt we calculate the resultant magnitude, R, and the azimuth, Ang, as

$$R = \left(X^2 + Y^2\right)^{1/2}, \qquad (1)$$
$$\text{Ang} = \arctan(\Delta X / \Delta Y), \qquad (2)$$

where X and Y are the measured tilts in the N–S and E–W directions, respectively; $\Delta X = X \cos\theta + Y \sin\theta$, and $\Delta Y = Y \cos\theta - X \sin\theta$, and θ is the correction for the azimuth orientation of the Y axis from North. $\theta = -10°$ for ES-I.

The tilt magnitude, R, and the ES-V vertical extension recorded by the creepmeter, are shown in Fig. 2c (1999–2003) and Fig. 2d (2008–2012).

Figure 3 shows R and Ang for the August 2008 slip episode (Fig. 3a), the slip event triggered by the $M = 7.2$ Hector Mine earthquake (Fig. 3b) and by the local $M = 5.8$, 30 December 2009, earthquake (Fig. 3c). While episodic slip events and triggered slip events show very similar behavior of R and Ang, the slip related to the 5.8 local earthquake has different behavior.

4 Discussion

Figure 2c and d show R and creepmeter extension ES-V observed between 1999–2003 and 2008–2012, respectively. The similarity between R and ES-V seen on both Fig. 2 suggests that tilt increases when vertical displacement increases on the fault, as observed during continuous creep and slip events.

There is not such an evident tendency for azimuthal behavior (Fig. 2a and b); the azimuth oscillated during 1999–2000, slightly increased during 2001–2003, and diminished with time during 2008–2012, except during December 2009–April 2010, which will be discussed later.

However, a significant azimuth change can be seen between Fig. 2a and b. Tilt tends to be more north-oriented for the 2008–2012 period. This phenomenon can be related to the subsidence amplitude increase in the recharge area observed for 1993–2009 by Sarychikhina and Glowacka (2015b) (this book).

Three kinds of deformation are shown in Fig. 3. The typical slip event shown on Fig. 3a is characterized by an R increase, related to the extension (subsidence) increase and azimuth decrease, with about one day duration. Triggered by the $M = 7.2$ HME, the slip event shown on Fig. 3b is characterized by abrupt R increase, extension (subsidence) increase, and azimuth decrease. This is followed by the slow deformation with the same orientation, similar to the regular slip event, but with smaller magnitude, expected since

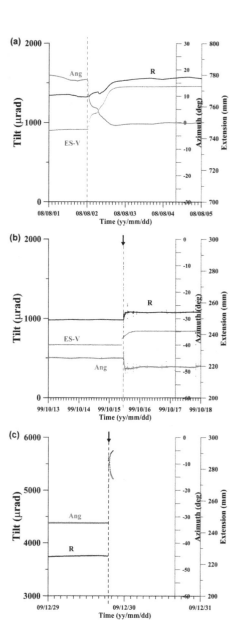

Figure 3. Tilt observations vs. time for episodic slip (**a**), triggered slip (**b**) and coseismic slip (**c**). R is the resultant tilt magnitude in the azimuth direction Ang. R is referred to the left axes, while azimuth Ang is referred to the right axes. Extension ES-V is shown for the period when the creepmeter was working. The vertical arrow indicates the $M = 7.2$ 1999 hector Mine earthquake (**b**) and the $M = 5.8$, 2009 earthquake (**c**).

the slip predictable character of deformation suggested by Glowacka et al. (2002). The deformation shown on Fig. 3c was caused by a local $M = 5.8$ earthquake and about 1 h of deformation was recorded after the earthquake before the instrument went out of range. Within the time precision of the tiltmeter (one sample every 4 min) the immediate R increase followed by postseismic R increase can be observed. However, the azimuth shows immediate increase and postseismic decrease.

Although creepmeters were not working during the $M = 5.8$, 2009 earthquake, Sarychikhina et al. (2015), using data from InSAR and leveling, showed there was subsidence (about 20 cm) and slip associated with this earthquake. They suggested that the observed deformation could be caused by triggered aseismic slip on the Saltillo fault or by earthquake-triggered soft sediments subsidence.

5 Conclusions

Our results show that the tilt magnitude is proportional to the extension recorded by a creemeter, which may eventually allow to substitute results when one instrument is missing.

Since magnitude and azimuth behaviors are very similar for episodic and triggered slip we suggest that they occur along the same fault segment and the same depth.

However, the different tilt azimuthal behavior for local earthquakes suggests that the slip on the fault has a different behavior, possibly including a horizontal component or/and that the slip was deeper than for episodic slip. More dense instrumentation with a higher frequency of measurements could be useful to confirm this suggestion.

Acknowledgements. This research was sponsored in part by CONACYT, project 105907 and CICESE internal funds.

References

Camacho Ibarra, E.: Análisis de la deformación vertical del terreno en la región de confluencia del sistema de fallas Cerro Prieto – Imperial en el periodo 1962–2001, MSc Thesis, Centro de Investigación Científica y Educación Superior de Ensenada, Ensenada, B.C., Mexico, 135 pp., 2006

Carnec, C. and Fabriol, H.: Monitoring and Modeling Land Subsidence at the Cerro Prieto Geothermal Field, Baja California,Mexico, Using SAR Interferometry, Geophys. Res. Lett., 26, 1211–1214, 1999.

Fabriol, H. and Munguia, L.: Seismic Activity At The Cerro Prieto Geothermal Area (Mexico) From August 1994 To December 1995, And Relationship With Tectonics And Fluid Exploitation, Geophys. Res. Lett., 24, 1807–1810, 1997.

Glowacka, E. and Nava, F. A.: Major earthquake in Mexicali valley, Mexico, and Fluid Extraction at Cerro Prieto Geothermal Field, Bull. Seismol. Soc. Am., 86, 93–105, 1996.

Glowacka, E., González, J., and Fabriol, H.: Recent Vertical deformation in Mexicali Valley and its Relationship with Tectonics, Seismicity and Fluid Operation in the Cerro Prieto Geothermal Field, Pure Appl. Geophys., 156, 591–614, 1999.

Glowacka, E., González, J. J., Nava, F. A., Farfán, F., and Díaz de Cossío, G.: Monitoring Surface Deformations in the Mexicali Valley, B.C., Mexico, Proceedings of Tenth International Symposium on Deformation Measurements, Orange, California, USA, 175–183, 2001.

Glowacka, E., Nava, F. A., Diaz de Cossio, G., Wong, V., and Farfan, F.: Fault slip, Seismicity and Deformation in the Mexicali

Valley (B.C., Mexico) after the 1999 Hector Mine Earthquake ($M = 7.1$), Bull. Seismol. Soc. Am., 92, 1290–1299, 2002.

Glowacka, E., Sarychikhina, O., and Nava, F. A.: Subsidence and stress change in the Cerro Prieto Geothermal Field, B.C., Mexico, Pure Appl. Geophys., 162, 2095–2110, 2005.

Glowacka, E., Sarychikhina, O., Suárez, F., Nava, F. A., and Mellors, R.: Anthropogenic subsidence in the Mexicali Valley, B.C., Mexico, and the slip on Saltillo Fault, Environ. Earth Sci., 59, 1515–1523, 2010a.

Glowacka, E., Sarychikhina O., Suárez, F., Nava, F. A., Farfan, F., Batani, G., and Garcia-Arthur, M. A.: Anthropogenic subsidence in the Mexicali Valley, B.C., Mexico, caused by the fluid extraction in the Cerro Prieto geothermal Field and the role of faults, Proceedings, World Geothermal Congress, Bali, Indonesia, 25–29, 2010b.

Glowacka, E., Sarychikhina, O., Nava, F. A., Suarez, F., Ramirez, J., Guzman, M., Robles, B., and Farfan, F.: Guillermo Diaz De Cossio Batani, Continuous monitoring techniques of fault displacement caused by geothermal fluid extraction in the Cerro Prieto Geothermal Field (Baja California, Mexico), in: Land Subsidence, Associated Hazards and the Role of Natural Resources Development, edited by: Correón-Freyre, D., Cerca, M., and Galowey, D., Land Subsidence, Associated Hazards and the Role of Natural Resources Development, Proceedings of EISOLS 2010, 17–22 October 2010, Querétaro, Mexico, IAHS Publ., 339, 326–332, 2010c.

Glowacka, E., Sarychikhina, O., Vazquez-Gonzalez, R., Nava, F. A., García-Hernández, A., Pérez, A., Aguado, C., López-Hernández, M., Farfan, F., Diaz De Cossio, G., and Orozco, L.: Catálogo de eventos de slip registrados con los instrumentos geotérmicos y piezómetros en la cuenca de Cerro Prieto, Reunión Anual Unión Geofísica Mexicana, 31, p. 121, 2011.

Glowacka, E., Sarychikhina, O., Robles, B., Suarez Vidal, F., Ramírez Hernández, J., and Nava, F. A.: Estudio geológico para definir la línea de hundimiento cero y monitorear la subsidencia de los módulos 10, 11 y 12 en el Valle de Mexicali, en el distrito de riego 014, Reporte Final, Rio Colorado, B.C., 2012.

Glowacka, E., Sarychikhina, O., Márquez Ramírez, V. H., Nava, F. A., Farfán, F., and García Arthur, M. A.: Deformation Around the Cerro Prieto Geothermal Field Recorded by the Geotechnical Instruments Network REDECVAM During 1996–2009, WGC 2015, Proceedings CD, Melbourne, Australia, 2015.

González, J., Glowacka, E., Suárez, F., Quiñones, J. G., Guzmán, M., Castro, J. M., Rivera, F., and Félix, M. G.: Movimiento reciente de la Falla Imperial, Mexicali, B. C. Ciencia para todos Divulgare, Universidad Autónoma de Baja California, 6, 4–15, 1998.

Hanssen, R. F.: Radar Interferometry: Data Interpretation and Error Analysis, Kluwer Academic Publishers, Dordrecht, the Netherlands, 328 pp., 2001.

Lira, H. and Arellano, J. F.: Resultados de la nivelación de precisión realizada en 1997, en el campo geotérmico Cerro Prieto, Informe Técnico RE 07/97, Comisión Federal de Electricidad, Residencia de Estudios, México, 28 pp., 1997.

Lomnitz, C., Mooser, F., Allen, C. R., Brune, J. N., and Thatcher, W.: Seismicity and tectonics of the northern Golf of California region, Mexico, preliminary results, Geof. Int., 10, 37–48, 1970.

Majer, E. L. and Mcevilly, T. V.: Seismological Studies at the Cerro Prieto Geothermal Field, 1978–1982, Proc. Fourth Symp. on the Cerro Prieto Geothermal Field, Baja California, Mexico, Comisión Federal de Electricidad, Guadalajara, Mexico, 145–151, 1981.

Nava, F. A. and Glowacka, E.: Automatic identification of seismic swarms and other spatio-temporal clustering from catalogs, Comput. Geosci., 20, 797–820, 1994.

Nava, F. A. and Glowacka, E.: Fault slip triggering, healing, and viscoelastic afterworking in sediments in the Mexicali_Imperial Valley, Pure Appl. Geophys., 156, 615–629, 1999.

Sarychikhina, O. and Glowacka, E.: Application of DInSAR Stacking Method for Monitoring of Surface Deformation Due to Geothermal Fluids Extraction in the Cerro Prieto Geothermal Field, Baja California, Mexico, Proceedings, World Geothermal Congress, Melbourne, Australia, 2015a.

Sarychikhina, O. and Glowacka, E.: Spacio-temporal evaluation of aseismic grond deformation in the Mexicali Valley (Baja California, Mexico) from 1993 to 2010, using differential SAR interferometry, NISOLS, Nagoya, Japan, 2015b.

Sarychikhina, O., Glowacka, E., Mellors, R., and Suárez-Vidal, F.: Land subsidence in the Cerro Prieto Geothermal Field, Baja California, Mexico, from 1994 to 2005. An integrated analysis of DInSAR, leveling and geological data, J. Volcanol. Geoth. Res., 204, 76–90, 2011.

Sarychikhina, O., Glowacka, E., Robles, B., Nava, F. A., and Guzmán, M.: Estimation of Seismic and Aseismic Deformation in Mexicali Valley, Baja California, Mexico, in the 2006–2009 Period, Using Precise Leveling, DInSAR, Geotechnical Instruments Data, and Modeling, Pure Appl. Geophys., 172, 3139–3162, 2015.

Suárez-Vidal, F., Mendoza-Borunda, R., Naffarrete-Zamarripa, L. M., Ramírez, J., and Glowacka, E.: Shape and dimensions of the Cerro Prieto pull-apart basin, Mexicali, Baja California, Mexico, based on the regional seismic record and surface structures, Int. Geol. Rev., 50, 636–649, 2008.

Trugman, D. T., Borsa, A. A., and Sandwell, D. T.: Did stresses from the Cerro Prieto Geothermal Field influence the El Mayor-Cucapah rupture sequence?, Geophys. Res. Lett., 41, 8767–8774, 2014.

Velasco, J.: Levantamiento Gravimétrico, Zona Geotérmica de Mexicali, Baja California. Technical Report, Consejo de Recursos Naturales no Renovables, 55 pp., 1963.

22

Has land subsidence changed the flood hazard potential? A case example from the Kujukuri Plain, Chiba Prefecture, Japan

H. L. Chen[1], Y. Ito[2], M. Sawamukai[3], T. Su[2], and T. Tokunaga[2]

[1]School of Environmental Science and Engineering, Zhejiang Gongshang University, Hangzhou, 310018, China
[2]Department of Environment Systems, School of Frontier Sciences, The University of Tokyo, Tokyo, 277-8563, Japan
[3]VisionTech Inc., Tsukuba-city, Ibaraki, 305-0045, Japan

Correspondence to: H. L. Chen (hualichen57@126.com)

Abstract. Coastal areas are subject to flood hazards because of their topographic features, social development and related human activities. The Kujukuri Plain, Chiba Prefecture, Japan, is located nearby the Tokyo metropolitan area and it faces to the Pacific Ocean. In the Kujukuri Plain, widespread occurrence of land subsidence has been caused by exploitation of groundwater, extraction of natural gas dissolved in brine, and natural consolidation of the Holocene and landfill deposits. The locations of land subsidence include areas near the coast, and it may increase the flood hazard potential. Hence, it is very important to evaluate flood hazard potential by taking into account the temporal change of land elevation caused by land subsidence, and to prepare hazard maps for protecting the surface environment and for developing an appropriate land-use plan. In this study, flood hazard assessments at three different times, i.e., 1970, 2004, and 2013 are implemented by using a flood hazard model based on Multicriteria Decision Analysis with Geographical Information System techniques. The model incorporates six factors: elevation, depression area, river system, ratio of impermeable area, detention ponds, and precipitation. Main data sources used are 10 m resolution topography data, airborne laser scanning data, leveling data, Landsat-TM data, two 1 : 30 000 scale river watershed maps, and precipitation data from observation stations around the study area and Radar data. The hazard assessment maps for each time are obtained by using an algorithm that combines factors with weighted linear combinations. The assignment of the weight/rank values and their analysis are realized by the application of the Analytic Hierarchy Process method. This study is a preliminary work to investigate flood hazards on the Kujukuri Plain. A flood model will be developed to simulate more detailed change of the flood hazard influenced by land subsidence.

1 Introduction

Flooding in coastal areas is one of the most devastating natural hazards, which causes considerable property damage and personal injury (Doornkamp, 1998; Hunt and Watkiss, 2011). Different from other areas, coastal lowland areas can be influenced by both natural process and anthropogenic activities which make the area at higher risk of floods than others. At the same time, there is a growing awareness of anthropogenic impact and positive/negative feedbacks on flooding (IPCC, 2013). Rapid expansion in human settlements increases imperviousness which in turn increases peak discharge and surface runoff. Widespread land subsidence caused by excessive groundwater withdrawal or engineering activities further amplifies the harmful effects of flooding by modifying the ground elevation, stream slopes and flow pathways. On the other hand, some activities such as construction of detention ponds can buffer the flow to reduce the risk of flooding. Factors related to flood hazards are changing in spatial and temporal scales which increase the difficulties of accurate flood hazard assessment.

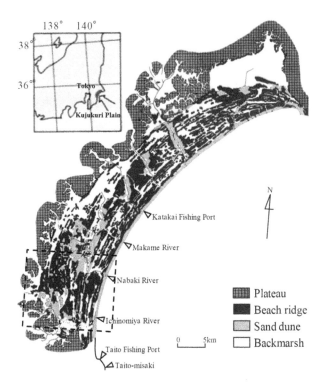

Figure 1. Geomorphological map of the Kujukuri Plain (Obanawa et al., 2010). Square marked by dashed line indicates the study area.

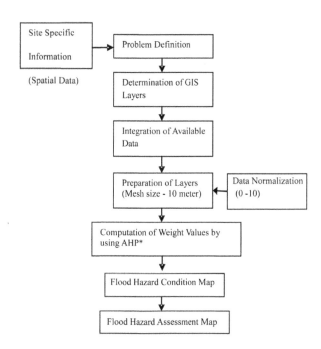

Figure 2. Flow chart for making the flood hazard map. * AHP – Analytical Hierarchy Process.

2 Methodology

The Kujukuri Plain (Fig. 1), Chiba Prefecture, Japan, suffers from flood hazards. Many natural processes and anthropogenic activities have changed the surface environment of the Plain, such as agricultural development, urban and industrial development, change of drainage patterns, and deposition and/or erosion of river beds. Due to the exploitation of groundwater, extraction of natural gas dissolved in brine, and natural consolidation of the Holocene and landfill deposits, widespread land subsidence has occurred in the Plain (Environmental and Community Affairs Department, Chiba Prefecture 1970–2007). All of these factors could more or less increase the risk of flooding in the Kujukuri Plain. Therefore, constructing flood hazard potential maps for different periods is meaningful to study the change of the spatial and temporal patterns of the hazard potential for discussing future land-use planning of the Kujukuri Plain.

2 Methodology

A semi-quantitative method was selected to investigate the flood hazard assessment due to the scarcity of good quality data in the study area (Chen et al., 2015). Figure 2 shows the steps used in this study including data gathering, Geographic information system (GIS) manipulation, and Multicriteria Decision Analysis. The main data used are airborne laser scanning data, leveling data, Landsat-TM data, two river watershed maps with 1 : 30 000 scale, and radar precipitation data. Based on these data, five layers expressing condition factors were constructed, i.e., river system, elevation, depression area, ratio of impermeable area, and detention ponds. Precipitation was the main factor to trigger floods. A layer for the triggering factor, i.e., precipitation distribution, was created from radar precipitation data. The next step was to assign weights to each layer. The Analytic Hierarchy Process method (Saaty, 1980) was applied to assign the weight values (Table 1). The pixel value of the condition and assessment map was calculated by overlapping the factor layers.

3 Analysis and results

As shown in Fig. 3, each layer was created for three periods, i.e., year 1970, 2004 and 2013. The river system data for 2004 and 2013 were similar because no significant change was found between two periods. Data quality of year 1970 were not good compared with those of year 2004 and 2013, especially the elevation and the precipitation data. For detention ponds layer, there was no such structure for year 1970. So, the condition and assessment maps were calculated only for year 2004 and 2013 (Fig. 4). Here, the higher values indicate the higher likelihood of floods. In the flood hazard assessment map, the studied area was divided into five classes by using the quantiles method which created nearly equal ratios of the coverage in each class, i.e., very high, high, moderate, low, and very low hazards potential. The assessment map (Fig. 4) shows that the southeastern, the northeastern and the central parts of the study area are expected to have highest flood hazard potential as a consequence of the combination of high precipitation with lower elevation difference

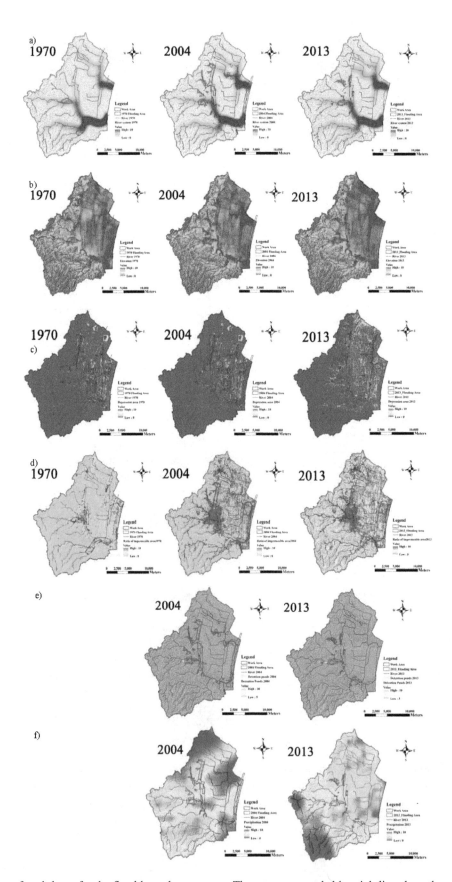

Figure 3. The images of each layer for the flood hazard assessment. The areas surrounded by pink line show the ones actually flooded. Figures in the left column are those for year 1970, the center for year 2004, and the right for year 2013. (**a**) River system layer, (**b**) elevation layer, (**c**) depression area layer, (**d**) ratio of impermeable area (Ratio of IA) layer, (**e**) detention ponds layer, and (**f**) precipitation layer.

Table 1. Assigned weights for the layers.

Layer	River system	Elevation	Depression area	Ratio of impermeable area	Detention ponds	Precipitation	Consistency ratio
Condition map	0.300	0.233	0.233	0.167	0.067	–	0.012
Assessment map	0.225	0.175	0.175	0.125	0.050	0.250	0.014

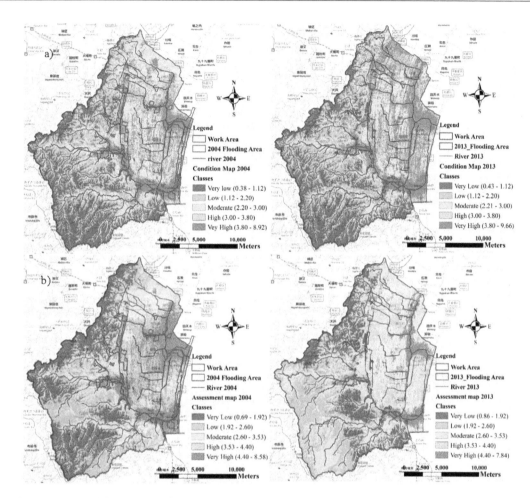

Figure 4. (a) Flood hazard condition map and **(b)** flood assessment map obtained by this study, for left (2004) and right (2013). The areas surrounded by pink line show the ones actually flooded (Chiba Prefecture, 2005, 2013).

and higher river density. The comparison between the actual flood areas reported by the local authorities (Chiba Prefecture, 2005, 2013) and the map constructed in this study showed that most of the flooded areas were located in the very high hazard potential area.

4 Discussion and summary

By comparing the results for three different periods, 1970, 2004 and 2013, the urban area enlarged significantly, the area effected by detention ponds became lager, while the change of the river system, the depression area and that of the elevation layer were small. For precipitation, it was difficult

Table 2. The areal percentage of each class for the condition and assessment maps.

Class name	Condition map		Assessment map	
	2004	2013	2004	2013
Very low	18.89 %	18.36 %	19.49 %	11.75 %
Low	20.53 %	20.44 %	20.48 %	22.21 %
Moderate	20.04 %	16.07 %	20.00 %	35.28 %
High	20.52 %	17.87 %	20.06 %	19.72 %
Very High	20.02 %	27.25 %	19.96 %	11.04 %

to compare because of differences in the resolution of the data and the rather large spatial difference of precipitation patterns. From Table 2, it was found that the percentage of very high hazard potential was increased from year 2004 to 2013 for the condition map with slight decreases of the high and moderate potential classes. However, other classes did not show big changes. Due to the large difference of spatial patterns of precipitation between year 2004 and 2013, the change in the assessment map was difficult to compare.

The areas with highest flood potential are in the southeastern, the northeastern and the central parts of the study area. The land cover change may increase the flood hazard potential. Construction of flood hazard maps with different periods described in this paper can assist decision makers and planners to evaluate the study area with respect to the temporal and spatial changes of different factors. This study can be applied as an early stage assessment of the flooding problem. A flood model will be developed to simulate more detailed change of the flood hazard influenced by land subsidence.

Acknowledgements. We would like to express our sincere gratitude to the Keiyo Natural Gas Association (Japan), the National Natural Science Foundation of China (41401539), and Science Research Foundation for the Returned Overseas Chinese Scholars (State Education Ministry of China, (2015) No. 1098) for supporting this study. Residential maps provided by ZENRIN CO., LTD are used as the CSIS Joint Research (No. 524) using spatial data provided by Center for Spatial Information Science, The University of Tokyo.

References

Chen, H. L., Ito, Y., Sawamukai, M., and Tokunaga, T.: Flood hazard assessment in the Kujukuri Plain of Chiba Prefecture, Japan, based on GIS and multicriteria decision analysis, Nat. Hazards, 78, 105–120, 2015.

Chiba Prefecture: Flooding report in Chiba Prefecture, Chiba, 12–12, 2005 (in Japanese).

Chiba Prefecture: Flooding report in Chiba Prefecture, Chiba, 18–18, 2013 (in Japanese).

Doornkamp, J. C.: Coastal flooding, global warming and environmental management, J. Environ. Manage., 52, 327–333, 1998.

Environmental and Community Affairs Department: Chiba Prefecture, Leveling results in Chiba Prefecture, reference date: February 1970–1 January 2007, 2007 (in Japanese).

Hunt, A. and Watkiss, P.: Climate change impacts and adaptations in cities: A review of the literature, Climatic Change, 104, 13–49, 2011.

IPCC, Climate change 2013: The physical science basis, Cambridge, edited by: Stocker, T. F., Qin, D., Plattner, G. K., Tignor, M. M. B., Allen, S. K., Boschung, J., Nauels, A., Xia, Y., Bex, V., Midgley, P. M., Cambridge, Cambridge University Press, 659–741, 2013.

Obanawa, H., Tokunaga, T., Rokugawa, S., Deguchi, T., and Nakamura, T.: Land subsidence at the Kujukuri Plain in Chiba Prefecture, Japan: Evaluation and monitoring environmental impacts, IAHS Publ., 339, 17–22, 2010.

Saaty, T. L.: The Analytic Hierarchy Process, Mc Graw Hill Company, New York, 1980.

23

Application of high resolution geophysical prospecting to assess the risk related to subsurface deformationin Mexico City

F. A. Centeno-Salas[1,3]**, D. Carreón-Freyre**[2]**, W. A. Flores-García**[3]**, R. I. Gutiérrez-Calderón**[3]**, and E. Luna-Sánchez**[1]

[1]Posgrado en Ciencias de la Tierra, Universidad Nacional Autónoma de México (UNAM), Mexico City, Mexico
[2]Laboratorio de Mecánica Geosistemas (LAMG), Centro de Geociencias, UNAM, Querétaro, Mexico
[3]Centro de Evaluación de Riesgo Geológico (CERG), Delegación Iztapalapa del Distrito Federal, Mexico

Correspondence to: D. Carreón-Freyre (freyre@geociencias.unam.mx)

Abstract. In the eastern sector of Mexico City the sub soil consists of high contrasting sequences (lacustrine and volcanic inter bedded deposits) that favor the development of erratic fracturing in the surface causing damage to the urban infrastructure. The high-resolution geophysical prospecting are useful tools for the assessment of ground deformation and fracturing associated with land subsidence phenomena.

The GPR method allowed to evaluate the fracture propagation and deformation of vulcano-sedimentary sequences at different depths, the main electrical parameters are directly related with the gravimetric and volumetric water content and therefore with the plasticity of the near surface prospected sequences. The active seismology prospection consisted in a combination of Seismic Refraction (SR) and Multichannel Analysis of Surface Waves (MASW) for the estimation of the velocity of the mechanical compressive (P) and the shear (S) waves. The integration of both methods allowed to estimate the geomechanical parameters characterizing the studied sequence, the Poisson Ratio and the volumetric compressibility.

The obtained mechanical parameters were correlated with laboratory measured parameters such as plasticity index, density, shear strength and compressibility and, GPR and seismic profiles were correlated with the mapped fracture systems in the study area. Once calibrated, the profiles allowed to identify the lithological contact between lacustrine and volcanic sequences, their variations of thicknesses in depth and to assess the deformation area in the surface. An accurate determination of the geometry of fracturing was of the most importance for the assessment of the geological risk in the study area.

1 Introduction

The ground-fracturing phenomenon related with land subsidence has increased recently in Mexico City due to natural and anthropogenic processes. One of the areas with the highest differential subsidence is the lacustrine plain of Iztapalapa (Carreón-Freyre, 2011). The high-resolution geophysical prospecting, such as Ground Penetrating Radar (GPR) and active seismology, are useful tools for the assessment of ground deformation. In the eastern sector of Mexico City the sub soil consists of high contrasting sequences (lacustrine and volcanic inter bedded deposits) that favor the development of erratic fracturing in the surface causing damage to the urban infrastructure.

The GPR method is based on the generation of short electromagnetic pulses generated by a transmitter that penetrate in the subsoil, and are reflected according with the properties of optics laws, toward the receiving antenna. This method allowed to evaluate the fracture propagation and deformation of vulcano-sedimentary sequences at different depths (Daniels et al., 1998). By the use of the Common Mid Point (CMP) arrangement was possible to measure the propagation velocity of electromagnetic waves through the contrasting layers and allowed to build high-resolution geological

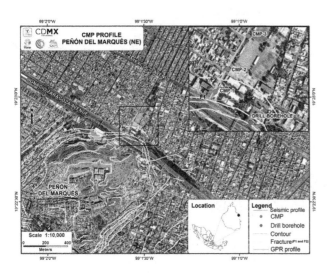

Figure 1. Location of the study area, east of Mexico City with spatial distribution of ground-fracturing and geophysical study.

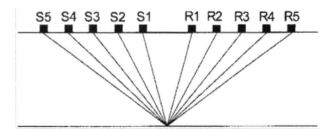

Figure 2. Common Mid Point (CMP) acquisition technique, where "S" denotes the transmitter location and "R" denotes the receiver location (Huisman, 2003).

sections (Grandjean et al., 2000). The energy is reflected to the surface by the contrasting electrical properties of each lithological sequence; the main electrical parameters are dielectric constant, ε, the magnetic permeability, μ, and the electrical conductivity, σ, which are directly related with the gravimetric and volumetric water content and therefore with the plasticity.

The MASW seismic method allows the record of the surface waves generated by a mechanical source in the surface (hammer) by the analysis of the dispersion curve of the fundamental mode, and eventually the higher modes. This curve can be inverted in a vertical section of shear strength because the surface waves are directly related by a factor of 0.97 with shear wave moreover, these take usually 70 % of the total seismic energy. The active seismology prospection consisted in a combination of Seismic Refraction (SR) and Multichannel Analysis of Surface Waves (MASW) for the estimation of the velocity of the mechanical compressive (P-) and the shear (S-) waves. The variation in depth of the shear velocity of seismic waves is directly related with some mechanical parameters of the prospected sequence such as: bulk modulus, Young's modulus, shear modulus and Poisson's ratio, used for the estimation of their deformation and potential fracturing.

2 High-resolution geophysical studies: GPR and MASW

The study area is located in the Basin of Mexico, east of the city, within a transition zone between volcanic and lacustrine materials (Fig. 1). The volcanic deposits come from a volcanic cone called "Peñón del Marques" composed by pyroclastic rocks and lava. Lake sediments are mainly clays, with large amounts of water content retained in their struc-

ture and with high plasticity values. The geophysical profiles were acquired at the base of the volcanic structure, over a high deformed and fractured area. GPR and seismic profiles are parallel, separated by 50 m. In the same area a geotechnical borehole was drilled for the characterization of the lithological sequence. In Fig. 1 the location of the study area is shown.

2.1 GPR prospecting

GPR fundamentals are based in the electromagnetic induction, where time is a decisive variable in conjunction with the spatial coordinates x, y, z. The induction field is of the same frequency as the transmitter current and varies in time within the prospected media, following electromagnetic laws, it creates secondary currents emerging from the prospected media to be measured by a receiver (Nabighian, 1988).

This is a high resolution method because it is capable to get 64 samples per meter, allowing to identify areas of deformation and fracture by reflection profiles. The Common Mid Point (CMP) technique was used to estimate the propagation velocity of the electromagnetic waves in different media and allowed the estimation of other physical properties. The CMP technique consists on separating the transmitter and receiver at different constant distances, as shown in Fig. 2. Three CMP profiles were acquired to characterize the spatial variation of the geological materials by the analysis of their velocity. Figure 3 shows the field data, processing, and interpretation of velocities from CMP. A GPR profile of 200 m length (radargram) was acquired complement the punctual measurements. The radargram shown in Fig. 4 was acquired with a 100 MHz antenna with 32 scans m^{-1}, each scan contains 1024 samples. The horizontal resolution is about 3 cm and the vertical resolution is less than 1 cm. This resolution allowed the characterization of the lithological variations related with the deformation and fracturing of the studied sequence through the change in the waveform.

The radargram shows three main layers: (a) a layer of volcanic sand 2 m width; (b) a lacustrine layer corresponding to clayey materials of variable thickness; and (c) a pyroclastic deposit is the subspace corresponding to volcanic sand until 10 m depth. The radargram shows a distortion of the se-

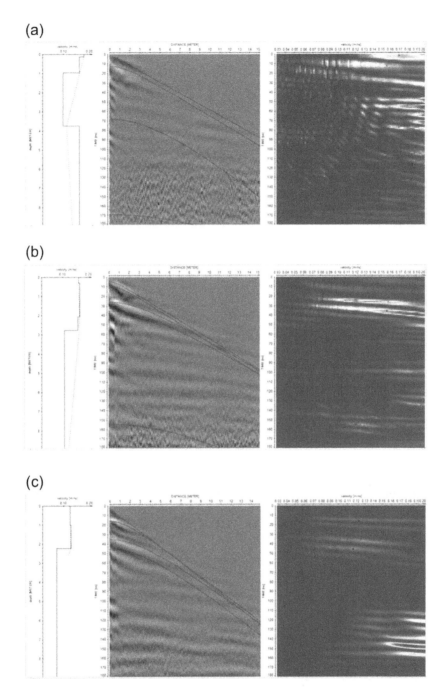

Figure 3. Data and results of the three CMP, on the left side the velocity, the field data on central side, and the processing on right side. **(a)** First CMP; **(b)** second CMP; **(c)** third CMP.

quence because of the high plasticity of the clayey materials, and fracturing recorded in the lower sandy layer.

2.2 MASW and Refraction Seismic (SR) profiles

The MASW method deals with surface waves in the low frequencies (1–30 Hz) and uses a shallow range of investigation (approximately 20 m depth). These superficial waves are related with S-wave by a factor of 0.97. Shear velocity is directly linked to the material stiffness and is one of the most critical mechanical parameters. The shear-wave velocity (V_s) is the best indicator for the detection of fracture zones because of the dramatic decrease of velocities.

The SR method involves the analysis of the travel times of first wave arrivals through the prospected layer to surface. The interpretation of the seismic data involves resolving the number of layers of the sequence, according to the velocity

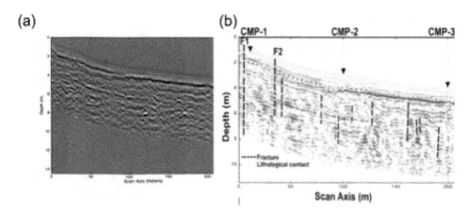

Figure 4. Reflection Profile of GPR. (**a**) Profile without geological interpretation; (**b**) profile with geological interpretation and CMP data correlation.

Table 1. Mechanical properties inferred from wave velocities of body.

Lithologhy	P-wave velocity ($m\,s^{-1}$)	S-wave velocity ($m\,s^{-1}$)	Bulk module ($kg\,cm^{-2}$)	Shear module ($kg\,cm^{-2}$)	Young module ($kg\,cm^{-2}$)	Poisson module	Water content (%)
Volcanic sequence (L_1)	500	270	65 000–90 000	20 000–34 000	180 000–260 000	0.04–0.16	11–47
Volcanic sequence (L_2)	280–400	90–180	15 000–30 000	15 000–30 000	100–600 000	0.48	226–258
Volcanic sequence (L_3)	480–552	300	65 00–90 000	65 000–90 000	180 000–260 000	0.04–0.16	14–50

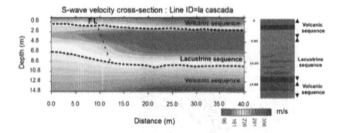

Figure 5. S-wave model, acquired from MASW method and a stratigraphic correlation with the borehole.

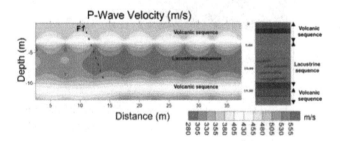

Figure 6. P-wave model, acquired from the determination of the first arrivals method and a stratigraphic correlation with the borehole. SR Method.

variation of each layer, and the travel-time from a given refractor to the ground surface.

MASW and SR profiles were acquired using 16 geophones of 12.5 Hz acquisition frequency, spaced every 2.5 m with a total length of 40 m. The S-wave velocity is determined by analyzing the Rayleigh wave and, P-wave velocity by calculating the first arrivals. Figures 5 and 6 show the S-wave model and P-wave model respectively. In Table 1 the relationship of P-wave and S-wave velocity is presented.

The rate of the two body waves (P and S) determine the spatial distribution of the different mechanical parameters. Equation (1) describe the compression waves and Eq. (2) the shear wave.

$$V_p = \sqrt{\frac{\lambda + 2\mu}{\rho}} \tag{1}$$

$$V_s = \sqrt{\frac{\mu}{\rho}}, \tag{2}$$

with $\lambda = \kappa - \frac{2}{3}\mu$.

The parameter λ is known as the first Lamé constant, which has no physical sense. The V_p and V_s parameters are the P-wave and S-wave velocities, μ is the modulus of rigidity (or shear), κ is the bulk modulus (or incompressible) and ρ is the density. These mechanical parameters result from the theoretical basis of the dynamic elasticity that studies the behavior of a solid upon the application of a force, both compression and shear, in one or in all three spatial dimen-

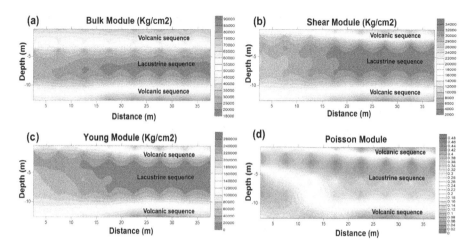

Figure 7. Distribution of mechanical parameters obtained from modeling of the P- and S-wave (SR Method): (a) spatial model of Bulk module; (b) spatial model of shear modulus; (c) spatial model of Young's modulus; (d) model spatial of Poisson constant.

Table 2. Velocities obtained from the CMP data.

Lithologhy	CMP-1 Depth velocity		CMP-2 Depth velocity		CMP-2 Depth velocity	
	(m)	(m ns^{-1})	(m)	(m ns^{-1})	(m)	(m ns^{-1})
Volcanic sequence (L$_1$)	0–1	0.16	0–2.8	0.15	0–2.3	0.13
Volcanic sequence (L$_2$)	1–3.8	0.085	2.8	0.09	2.3	0.05
Volcanic sequence (L$_3$)	3.8	0.16	–	–	–	–

sions (Velis, 2007). Seismic velocities are directly related with these mechanical properties for a prospected media.

The bulk modulus, rigidity, and Young modulus, can be determined either from static or dynamic experiments, involving the travelling of seismic waves through the soil. The bulk modulus of a soil measures the resistance to uniform compression. It is defined as the ratio of the infinitesimal pressure increase to the resulting decrease of the volume. The shear modulus is the coefficient of elasticity for a shearing force. It is defined as the ratio of shear stress to the unit displacement per sample length. Mathematically is the tangential force exerted and the related strain change. The elastic constant, called the Poisson's ratio (σ) measure the compressibility of a material perpendicular to the applied stress, or the ratio of vertical and horizontal strains. This can be expressed in terms of properties measured in the field, as velocities of P-waves and S-waves and in terms of μ and λ, as shown below (Aki, 2002).

$$\sigma = \frac{\lambda}{2(\lambda + \mu)} \tag{3}$$

with

$$\mu = V_s^2 \rho; \quad \lambda = V_p^2 \rho - 2\mu = V_p^2 \rho - 2V_s^2 \rho$$

$$\sigma = \frac{V_p^2 \rho - 2V_s^2 \rho}{2(V_p^2 \rho - 2V_s^2 \rho + V_s^2 \rho)} \tag{4}$$

$$= \frac{V_p^2 - 2V_s^2}{2(V_p^2 - 2V_s^2)}$$

$$= \frac{(V_p/V_s)^2 - 2}{2(V_p/V_s)^2 - 2}.$$

If V_s is small then Poisson's ratio equals 0.5, indicating either a fluid, because shear waves do not pass through fluids, or a material that maintains constant volume regardless of stress, also known as an ideal incompressible material.

The Young modulus, E, describes the elastic properties of a solid undergoing tension or compression in only one direction. Young's modulus is a measure of the ability of a material to withstand changes in length when under lengthwise tension or compression (see Eq. 5).

$$E = 3\kappa(1 - 2\sigma) \tag{5}$$

If seismic data are available, the spatial distribution of mechanical parameters of the prospected sequence can be modeled to create pseudo-geologic sections as shown in Fig. 7. The distribution of the obtained mechanical parameters (κ, μ, σ and E) was correlated with the log of the studied sequence and with physical measurements from borehole samples.

The Poisson constant distribution, varying from 0.38 to 048, and S-wave velocities, varying from 90 to 180 m s^{-1}, are correlated with the lacustrine layer having the larger water content, approximately 200 %, in the sequence. The low values of E modulus, κ modulus and μ modulus show that the lacustrine material is highly deformable (i.e. low bearing capacity). The volcanic sand layer shows higher values of κ, μ and E, less deformable, having lower water content values, 30 % in average.

The distribution of the mechanical parameters (Fig. 7 and Table 2) also allowed identifying the transition zone by the increase the thickness of the lacustrine sequence which contact correspond to the location of fracture F_1.

3   Conclusions

The results of applying the GPR, MASW and SR methods simultaneously show the persistence of the clayey lacustrine material with variable thickness encased by volcanic materials. This material was characterized in the laboratory a present gravimetric large water contents varying from 200 to 300 %.

The contrasting mechanical properties between the two types of material may cause slippage over lithological contacts, causing the propagation of deformation and fracturing of the sequence, as shown in the GPR profile of Fig. 4. This can also be correlated with the distribution of the S-wave velocities presented in Fig. 5, where the lower values (about $100 \, \text{m s}^{-1}$) corresponds to the location of fracture F_1.

The application of seismology and electromagnetic high resolution methods allow to characterize deformation and fracturing in a subsidence area. GPR studies allowed the monitoring of the deformation zone and fracturing and to correlate lithological variations through CMP profiles. Seismology shows to be very useful to quantify the mechanical behavior of the studied sequences.

The physical and mechanical parameters of the high deformability of the lacustrine sequence showing large water content and low shear strength. Fracturing showed to be strongly related with the lithological contact between materials having contrasting mechanical parameters in heterogeneous subsidence areas such as the east zone of Mexico City.

Acknowledgements. The authors thank his colleagues at the Centre for Evaluation of Geological Risk and the Center of Geosciences for the field support and technical suggestions for improving this article. The authors acknowledge the support of the DGAPA-UNAM research project No. IN114714.

References

Aki, K. and Richards, P.: Quantitative seismology: theory and methods: W. H. Freeman and Co. University Science Book, 2nd Edn., 67–288, 2002.

Carreón-Freyre, D.: Identificación y Caracterización de los Diferentes Tipos de Fracturas que Afectan el Subsuelo de la Delegación Iztapalapa en el Distrito Federal, Informe Técnico, Ingeniería Geologica, AI México, 3–9, 2011.

Daniels, D., Gunton, D., and Scott H.: Introduction to subsurface radar, IEEE Proceedings, 135, 278–320, 1998.

Grandjean, G., Gourry, J., and Bitri, A.: Evaluation of GPR techniques for civil-engineering applications: study on a test site, J. App. Geophys., 45, 141–156, doi:10.1016/S0926-9851(00)00021-5, 2000.

Huisman, J., Hubbard, S., Redman, J., and Anna, P.: Measuring Soil Water Content with Ground Penetrating Radar: A Review, Vadose Zone J., 2, 476–491, 2003.

Nabighian, N.: Electromagnetic Methods in Applied Geophysics: Theory, Society of Exploration Geophysics, Tulsa, OK, USA, 1, 131–141, 1998.

Velis, D.: Improved AVO fluid detection and lithology discrimination using Lame petrophysical parameters "$\lambda\rho$", "$\mu\rho$", and "λ/μ fluid stack" from P and S inversions, 67tn Annual International Meeting, SEG, Expanded Abstracts, 183–186, 1997.

24

Subsidence by liquefaction-fluidization in man-made strata around Tokyo bay, Japan: from geological survey on damaged part at the 2011 off the Pacific Coast of Tohoku Earthquake

O. Kazaoka[1], S. Kameyama[2], K. Shigeno[3], Y. Suzuki[3], M. Morisaki[1], A. Kagawa[1], T. Yoshida[1], M. Kimura[1], Y. Sakai[1], T. Ogura[1], T. Kusuda[a], and K. Furuno[a]

[1]Research Institute of Environmental Geology, Chiba (RIEGC), Japan
[2]Environmental Protection division of Chiba Prefectural Government, Japan
[3]Meiji Consultante Co., Ltd, Japan
[a]Former member of Research Institute of Environmental Geology, Chiba

Correspondence to: O. Kazaoka (osamu.kazaoka@gmail.com)

Abstract. Geological disaster by liquefaction-fluidization happened on southern part of the Quaternary Paleo-Kanto submarine basin at the 2011 Earthquake off the Pacific Coast of Tohoku. Liquefaction-fluidization phenomena occurred mainly in man-made strata over shaking 5+ intensity of Japan Meteorological Agency scale. Many subsided spots, 10–50 m width, 20–100 m length and less than 1 m depth, by liquefaction-fluidization distributed on reclaimed land around northern Tokyo bay. Large amount of sand and groundwater spouted out in the terrible subsided parts. But there are little subsidence and no jetted sand outside the terrible subsided part. Liquefaction-fluidization damaged part at the 1987 earthquake east off Chiba prefecture re-liquefied and fluidized in these parts at the 2011 great earthquake. The damaged area were more wide on the 2011 earthquake than the 1987 quake. Detailed classification maps of subsidence by liquefaction-fluidization on the 2011 grate earthquake were made by fieldwork in Chiba city around Tokyo bay. A mechanism of subsidence by liquefaction-fluidization in man-made strata was solved by geological survey with continuous large box cores on the ACE Liner and large relief peals of the cores at a typical subsided part.

1 Introduction

Strong shaking hit the Kanto district and the Tohoku district at the 2011 off the Pacific Coast of Tohoku Earthquake. Strong shaking over 5+ intensity of Japan Meteorological Agency scale distributed widely in southeast part of the Neogene Kanto Basin (JMA, 2011; Fig. 1). The shaking 5+ intensity of JMA scale correlate with about 8 intensity of MSK scale. Liquefaction-fluidization phenomena caused on reclaimed land by hydraulic fill in the strong shaking area at the Neogene Kanto Basin (RIEGC, 2011c; Fig. 1).

2 Distribution of liquefaction-fluidization site on Boso Peninsula

Liquefaction-fluidization sites were distributed on north part of the Boso peninsula, southeastern Neogene Kanto Basin at the earthquake (RIEGC, 2011c; Fig. 2). Many of these liquefaction- fluidization sites had been caused liquefaction-fluidization phenomena at the 1987 east off Chiba prefecture earthquake (RIEGC, 2011a; RIEGC,2011c; Fig. 3). Each liquefaction-fluidization sites at the 2011 earthquake is more wide than the liquefaction-fluidization sites at the 1987 earthquake.

Liquefaction-fluidization sites concentrate on belts with 0.5 km wide, NE-SW extending on reclaimed land along

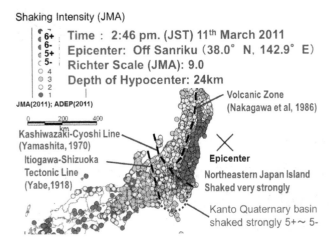

Figure 1. Distribution of JMA intensity on the 2011 Earthquake off the Pacific Coast of Tohoku and geological structure.

Figure 2. Distribution of liquefaction-fluidization phenomena and JMA intensity by the 2011 Earthquake.

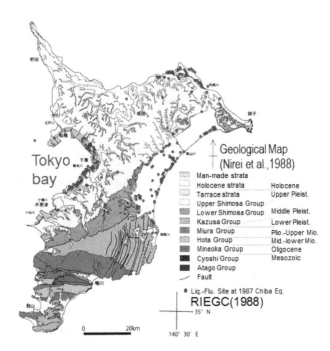

Figure 3. Distribution of liquefaction-fluidization phenomena by the 1987 east off Chiba prefecture earthquake.

there are little subsidence and no jetted sand outside the terrible subsided part. Detailed classification maps of subsidence by liquefaction-fluidization on the grate earthquake were made by fieldwork in Chiba city around Tokyo bay (RIEGC, 2011c; Fig. 6).

3 Mechanism of subcidence by liquefaction-fluidization in manmade-strata

A mechanism of subsidence by liquefaction-fluidization in man-made strata was solved by geological survey with continuous large box cores on the ACE Liner (Fig. 7) and continuous boring cores in Chiba city (Fig. 6). Continuous box core samples from surface to 5–7 m depth could be taken at the each 3–5 m length from little subsided part to terrible subsided part (Fig. 8). Detailed litho-stratigraphy, sedimentary structure, dencity and hardness of strata were studied on the continuous box core samples, continuous boring core samples and large relief peels of the cores (RIEGC, 2014). These data indicate that:

1. The thickness of man-made strata is about 12 m.

2. Man-made strata is composed of Dumped Association, Upper Filling Association and Lower Filling Association (RIEGC, 2014; Figs. 9, 10). Two Filling Associations were made by sand pump method from bottom sediments in the Tokyo bay. Upper Filling Association consists of lowermost, lower, upper and uppermost bundle.

3. Litho-facies of each man-made strata is as follows.

Dumped Association: This association is composed of 1.5–2.2 m thick sandy silt to silty fine sand layers with silt-

Tokyo bay (RIEGC, 2011a, c; Fig. 4). These belts may be on the thick part of the Alluvium formation after the Wurm glacial stage (Kazaoka, 2011; RIEGC, 2011b; Fig. 5).

Liquefaction-fluidization phenomena caused partially with subsidence on the reclaimed land of northern Tokyo bay at the 2011 earthquake. Many subsided spots, 10–50 m width, 20–100 m length and less than 1 m depth, by liquefaction-fluidization distributed on reclaimed land around northern Tokyo bay. Large amount of sand and groundwater spouted out in the terrible subsided parts. But

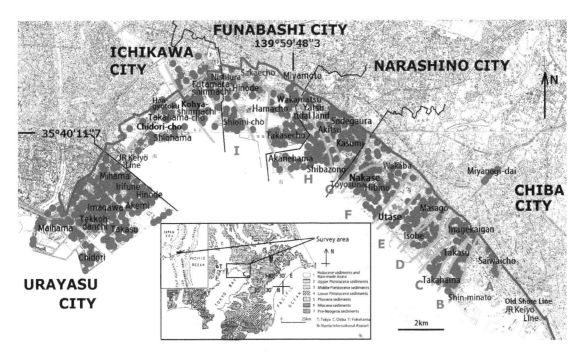

Figure 4. Distribution of liquefaction-fluidization phenomena. Blue spot means liquefaction-fluidization site (Kazaoka et al., 2011).

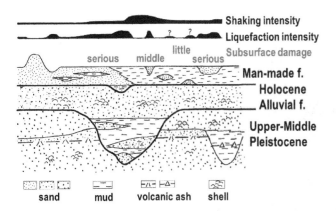

Figure 5. Idealized geological cross section on reclaimed land of Tokyo bay. Intensity of shaking and liquefaction show line wideth (Kazaoka, 2011).

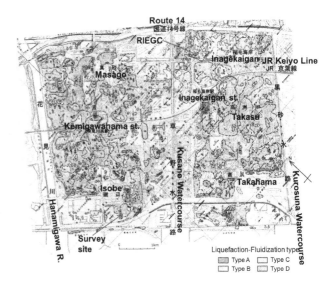

Figure 6. Classification of subsidence by liquefaction-fluidization on the 2011 earthquake in the reclaimed land in Chiba city. A Type: subsidence > 0.3 m, B type: 0.3 m < subsidence < 0.1 m, C Type: 0.1 m < subsidence < few centimeter, D Type: no subsidence.

stone brocks and rock gravels. Sand dike of yellowish brown sand and gray sand distribute rarely (RIEGC, 2014; Fig. 11).

Uppermost Bundle of Upper Filling Association: this bundle is composed of 0.2–0.8 m thick yellowish brown laminated fine to medium sand layers. Upper part of this bundle lost partially primary sedimentary structures and loose. The feature described above suggests that the losting structure part is liquefaction-fluidization part. The base of this bundle consists of laminated coarse-very coarse sandy shell fragment layers.

Upper Bundle of Upper Filling Association: this bundle is composed of 0.4–1.8 m thick gray laminated medium sand layers. Shell fragment layers often interbeded in this sand

layers. The sand layers lost widely primary sedimentary structures and very loose or relatively dence.

Lower Bundle of Upper Filling Association: this bundle is composed of 0–1.8 m thick gray silt layers.

Lowermost Bundle of Upper Filling Association: this bundle is composed of 0.7–1.8 m thick gray shelly medium sand layers. Shell fragment layers often interbedded in the shelly sand layers. In upper part of this bundle, sand layer lost

Figure 7. Continuous sampling by ACE liner and continuous man-made strata samples on this survey site.

Figure 8. Distribution of subsidence and sampling points by ACE liner. Ground surface subside a little at the sampling points No. 4, 5, 9 and 10. The surface subsided 0.4 m height at the sampling points No. 1, 2, 6 and 7. Roofs were deformed by subsidence.

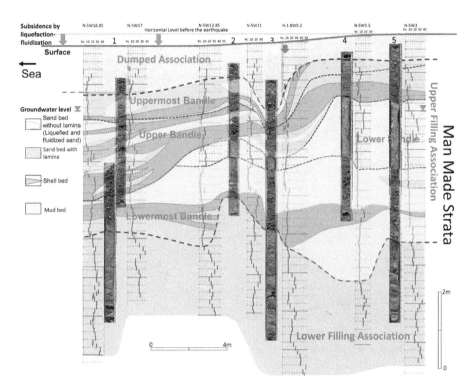

Figure 9. Geological cross section from No. 1 to No. 5 sampling points of Fig. 8.

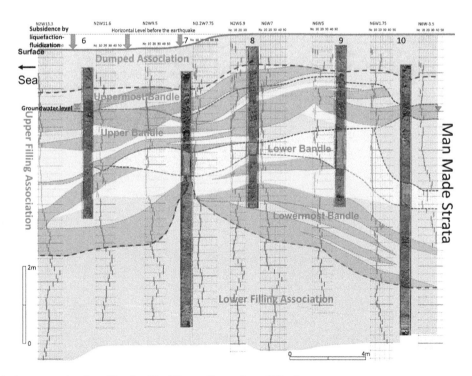

Figure 10. Geological cross section from No. 6 to No. 10 sampling points of Fig. 8.

Figure 11. Jetted yellowish gray sand along fisher on No. 4 point of Fig. 2.

widely primary sedimentary structures. The medium sand injected in the upper silt layers.

Lower Filling Association: this association is composed of 0.5–3.5 m thick yellowish gray laminated relatively dence matrix free good sorted fine-medium sand layers. The sand layers lost partially primary sedimentary structure. The liquefaction-fluidization part is cut by the Upper Filling Association.

4. Liquefaction-fluidization parts are in lowermost bundle, upper bundle and uppermost bundle of the Upper Filling Association.

5. Subsidence part is entirely identical with distribution of liquefaction-fluidization part in upper bundle of the Upper Filling Association.

Acknowledgements. We thank Yuji Takanashi and Yutaka Yazawa, former Chief of Chiba Environmental Research Center and Hiroaki Hiura, Chief of Chiba Environmental Research Center for helpful to study and survey liquefaction-fluidization.

References

Japan Meteorological Agency: The seismological bulletin of the Japan Meteorological Agency for March 2011, Japan Meteorological Agency, 321 pp., 2011.

Kazaoka, O.: Liquefaction-fluidization damage and geological structure of Man-made strata, Symposium for Disaster in man-made strata and 2011 Tohoku Earthquake, Japanese Society of Geo-pollution Science, Medical Geology and Urban Geology, 1–21, 2011. Kazoka, O., Furuno, K., Kagawa, A., Kusuda, T., Sakai, Y., Yoshida, T., Kato, A., Yamamoto, M., and Takanashi, Y.: Distribution of Geological Disaster by Liquefaction-Fluidization Phenomena at The 2011 off the Pacific coast of Tohoku Earthquake on Tokyo Bay Reclaimed Land, Jour. Geo-Pol. Sci., Med. Geol. and Urban Geol., 7, 10–21, 2011.

Kazaok, O., Kameyama, S., Morisaki, M., Shigeno, K., Suzuki Y., Kagawa A., Yoshida T., Kimura, M., Sakai, Y., and Ogura, T.: Characteristic of Man-made strata in liquefaction-fluidization site at the 2011 off the Pacific coast of Tohoku Earthquake: from geologaical survey on reclaimed land in Isobe district around Tokyo bay, Chiba city, Proceedings of the 24th Symposium on Geo-Environments and Geo-Technics, Japanese Society of Geo-pollution Science, Medical Geology and Urban Geology, 9–14, 2014.

Nirei, H., Kusuda, T., Suzuki, K., Kamura, K., Furuno, K., Hara, Y., Satoh, K., and Kazaoka, O.: The 1987 East off Chiba Prefecture Earthquake and its Hazard, Memoir of Geological Society of Japan, Geological Society of Japan, no. 35, 31–46, 1990.

RIEGC: Liquefaction-Fluidization Phenomena on Boso Peninsula at the 2011 Earthquake off the Pacific Coast of Tohoku, Central Japan: Part 1. RESEARCH REPORT G-8, Chiba Environmental Research Center, 1-1-1-8, 2011a.

RIEGC: Liquefaction-Fluidization Phenomena on Boso Peninsula at the 2011 Earthquake off the Pacific Coast of Tohoku, Central Japan: Part III, RESEARCH REPORT G-8, Chiba Environmental Research Center, 3-1-3-26, 2011b.

RIEGC: Liquefaction-Fluidization Phenomena on Boso Peninsula at the 2011 Earthquake off the Pacific Coast of Tohoku, Central Japan: Part IV. RESEARCH REPORT G-8, Chiba Environmental Research Center, 4-1-4-69, 2011c.

RIEGC: Liquefaction-Fluidization Phenomena on Boso Peninsula at the 2011 Earthquake off the Pacific Coast of Tohoku, Central Japan: Part V. RESEARCH REPORT G-8, Chiba Environmental Research Center, 5-1-5-8, 2012.

RIEGC: Liquefaction-Fluidization Phenomena on Boso Peninsula at the 2011 Earthquake off the Pacific Coast of Tohoku, Central Japan: Part VI., Chiba Environmental Research Center, 4P (http://www.pref.chiba.lg.jp/wit/chishitsu/ekijoukahoukoku/documents/6-2.pdf), 2014.

25

Analysis of the variation of the compressibility index (Cc) of volcanic clays and its application to estimate subsidence in lacustrine areas

D. Carreón-Freyre[1], **M. González-Hernández**[2], **D. Martinez-Alfaro**[3], **S. Solís-Valdéz**[1],
M. Vega-González[1], **M. Cerca**[1], **B. Millán-Malo**[4], **R. Gutiérrez-Calderón**[2], **and F. Centeno-Salas**[3]

[1]Centro de Geociencias de la UNAM Juriquilla. Queretaro, Mexico
[2]Centro de Evaluación de Riesgos Geológicos CERG, Iztapalapa, Mexico City, Mexico
[3]Posgrado en Ciencias de la Tierra, UNAM Juriquilla, Queretaro, Mexico
[4]Centro de Física Aplicada y Tecnología Avanzada, UNAM Juriquilla, Queretaro, Mexico

Correspondence to: D. Carreón-Freyre (freyre@geociencias.unam.mx)

Abstract. An analysis of the deformation conditions of lacustrine materials deposited at three sites in the volcanic valley of the Mexico City is presented. Currently geotechnical studies assume that compressibility of granular materials decreases in depth due to the lithostatic load. That means that the deeper the sample the more rigid is supposed to be, this assumption should be demonstrated by a decreased Compression Index (Cc) in depth. Studies indicate that Mexico City clays exhibit brittle behaviour, and have high water content, low shear strength and variable Cc values. Furthermore, groundwater withdrawal below the city causes a differential decrease in pore pressure, which is related to the physical properties of granular materials (hydraulic conductivity, grain size distribution) and conditions of formation. Our results show that Cc for fine grain materials (lacustrine) can be vertically variable, particularly when soils and sediments are the product of different volcanic materials. Lateral and vertical variations in the distribution of the fluvio-lacustrine materials, especially in basins with recent volcanic activity, may be assessed by Cc index variations. These variations can also be related to differential deformation, nucleation and propagation of fractures and need to be considered when modelling land subsidence.

1 Introduction

Mexico City is the best known example of land subsidence in Mexico (Fig. 1). The city is located within the geological province of the Transmexican Volcanic Belt (Ferrari et al., 2012), over a desiccated lacustrine plain bounded by faults and volcanic edifices. The near-surface lithology in the Mexico basin is highly heterogeneous and includes fluvio-lacustrine sediments with particle sizes varying from gravel, sand, and silts to clays, with interbedded layers of pyroclastic rocks and lava flows (Carreón-Freyre, 2010, Carreón-Freyre et al., 2011). Analyses of groundwater withdrawal and of the associated land subsidence have been documented systematically in Mexico City since the 1940s (Carrillo, 1947). As a consequence of descending water level, the Basin of Mexico has been divided in several clay bearing fluvio-lacustrine sub-basins, which include Mexico City, Tlahuac-Chalco, Texcoco, Xochimilco, and Zumpango basins.

Although subsidence in Mexico City is thought to be related to high compressibility rates of the clayey sequences when the pore pressures decrease (Mesri et al., 1976), there is a lack of systematic studies to explain the compositional and mechanical behavior variations in clayey sequences and their role on the estimation of vertical deformation. The Compression Index (Cc; Lambe and Whitman, 1969) is obtained from the void ratio vs log of normal effective stress plot in consolidation tests (ASTM, 1994), and was developed for the estimation of settlements for fine-grained deposits. The evaluation of vertical and lateral variations in the Compression Index Cc may facilitate the assessment of the potential of land subsidence in clay-bearing sequences. Several authors (see Haigh et al., 2013; and references therein) found that Cc

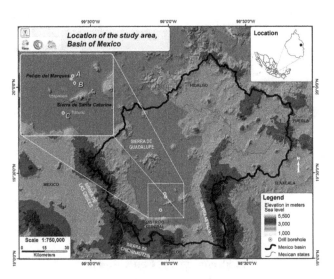

Figure 1. Location of the study areas in the Basin of Mexico (black polygon), Iztapalapa and Tlahuac. The geotechnical borehole locations are shown by blue circles: (A) La Cascada, (B) Lucio Blanco, (C) Tlahuac. Lacustrine planes are shown in green and volcanic structures in yellow and red. Grey lines indicate administrative boundaries.

can be correlated with water content, Atterberg Limits, and void ratio variations (e). The presented analysis of the Cc was made on silt and clayey deposits because of the strong interdependence of water content and compressibility with mineralogy, mechanical behaviour, and structure.

Three silty and clayey sequences, located in Iztapalapa and Tlahuac localities (Fig. 1), are compared to identify the conditions of formation and evolution that led to their physical and mechanical properties. The systematic characterization of these sequences included compressibility, grain size distribution, mineralogy of clays (analyzed by X-ray diffraction), water content, and other physical parameters.

2 Lacustrine deposits in the Iztapalapa and Tlahuac-Chalco sub-basins

The near surface stratigraphy of Iztapalapa and Tlahuac areas comprise fluvial and lacustrine sequences (silt and clay bearing) interbedded with tephra and lava (Zeevaert, 1953). The main volcanic structures are the Sierra de Santa Catarina volcanic complex, that separates the Iztapalapa and Thalhuac sub-basins and, the Peñón del Marques located at the north of the Iztapalapa plain. Weathering and hydrothermal alteration of basaltic rocks and volcanic ashes generate different types of clays (Hillier, 1995; Velde, 1995; Wada, 1987). The influence of mineralogy composition of clays on the mechanical behavior of lacustrine materials in Mexico City has been widely studied (Diaz-Rodriguez et al., 1998; Peralta y Fabi, 1989). The physical characteristics and composition of the lacustrine deposits have been documented from geotechnical and geophysical surveys (Carreón-Freyre and Cerca 2006;

Hernández-Marin et al., 2005; Carreón-Freyre et al., 2003). Clays are metastable materials, their forming processes are sequential and the existence of each type is restricted to specific temperature and time spans. The environmental conditions of a lake are a consequence of its salinity, pH, water depth, temperature, and water residence time (with restricted drainage). These conditions varied dramatically in very short periods because of the intense local volcanic activity (mafic or silicic in composition, effusive and/or pyroclastic rocks) in the sub-basins of the Basin of Mexico and their interaction with the lake deposits. The water level of the former lakes also played an important role to differentiate lacustrine conditions in adjacent sub-basins, such as Iztapalapa and Tlahuac.

A common clay mineral assumed to predominate in the sediments of the Basin of Mexico is montmorillonite, from the smectite family (Diaz Rodriguez, et al., 1998; Peralta y Fabi, 1989), that is characterized by: (1) 2 : 1 crystalline layered structure (2 tetrahedral sheets to 1 octahedral sheet) (Velde, 1995); (2) high values of plasticity and compressibility index (Hernandez-Marin and Carreón-Freyre, 2002); (3) large amounts of water retained in structure (generally more than 100 %); and, (4) high solids density ($2.5\,\mathrm{g\,cm^{-3}}$). Smectite formation is favored in lake areas with restricted drainage and low saline-alkaline conditions (Hillier, 1995). In such environments, iron (Fe) oxidizes and accumulates. When the lake level increases, iron reduces and becomes soluble forming green clay sequences (Carreón-Freyre et al., 2006). Mafic minerals (e.g., olivine and pyroxene) from volcanic rocks may form chlorite and talc. Moreover, alteration of pumicite-rich volcanic ashes produces allophane and imogolite, clay minerals of low crystalline order having a gel-like structure, low vertical compressibility, and susceptibility to shear deformation (Wesley, 2001). These materials have a characteristic amorphous spectrum (with no specific clay reflexions) of X-Ray Difraction (XRD) analysis (Carreón-Freyre et al., 2006). If the environmental conditions favor dehydration, these materials become gibbsite, halloysite, and talc (Velde, 1995). Heterogeneity of mineral content (mica, pyroxene, and plagioclase) suggests a close similarity with the volcanic source and less chemical equilibrium. The close relation between mineralogical conditions of the clayey materials, their ability to retain water, and their compressibility has been doumented by Haigh et al. (2013), Robinson (1999) and Warren and Rudolph (1997).

3 Characterization of the studied sequences

The three study sites were selected after detailed mapping of the fluvio-lacustrine sediments and volcanic rocks. Geological mapping was complemented with a topographic survey of the sites with evidence of ground surface deformation, where vertical displacements varied from 50 cm to 3 m. The different sites have approximately the same elevation. Samples

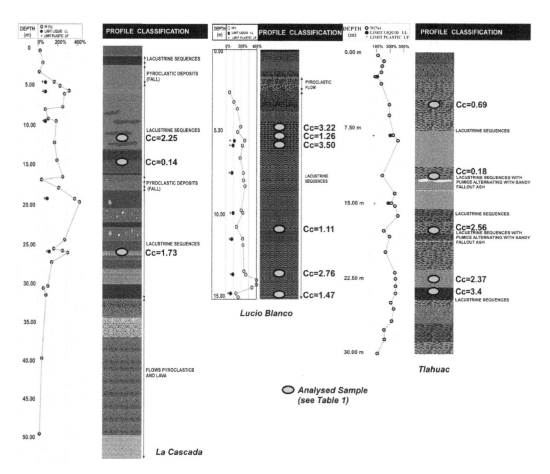

Figure 2. Lithological correlation of three fluvio-lacustrine and volcanic sequences in the Iztapalpa (La Cascada and Lucio Blanco) and Tlahuac sub-basins, obtained from geotechnical boreholes. Green ellipses indicate the location in depth of non-altered samples used for physical and mechanical characterization (see Fig. 1 for location).

were collected through three geotechnical boreholes named La Cascada (50 m depth), Lucio Blanco (15 m depth), and Tlahuac (30 m depth) (Figs. 1 and 2).

3.1 Physical properties and clay mineralogy

Physical properties of the collected samples were determined in the Laboratorio de Edafología and the Laboratorio de Mecánica de Geosistemas (LAMG), of the Centro de Geociencias, UNAM. These determinations included: gravimetric water content (W_{grav} %), grain-size distribution by sieving and fine-grain distribution using the Bouyoucos densitometer, organic matter content (%), and solid density (S_s) by the pycnometer method. Particles with diameters smaller than 2 microns were segregated from samples and dried at room temperature. They were crushed in an agate mortar and sieved with a 200 mesh. Carbonates were removed by washing with buffer solution of sodium acetate-acetic acid (Buhrke et al., 1998). Subsequently the samples were washed with deionized water and mild sonic vibration was applied. Particle dispersion was propitiated by the addition of sodium hexametaphosphate. Samples for XRD analysis were pre-

pared according to the standard methodology proposed by Buhrke et al. (1998); and were analyzed at the Laboratorio de Difracción de Rayos X, Centro de Fisica Aplicada y Tecnologia, UNAM. The microstructure of some samples were also analyzed using a Scanning Electron Microscope (SEM) and an Transmission Electron Microscope (TEM) in the Laboratorio de Fluidos Corticales, Centro de Geociencias.

3.2 Mechanical properties: compressibility

More than 30 borehole samples were tested by the standard incremental consolidation method in order to determine their compressibility (Cc) and overconsolidation ratio (OCR, overconsolidated having values > 1), but only 14 representative samples are discussed in this work (Table 1 and Figs. 2 and 3). Samples were classified according to the Unified Soil Classification System (USCS). Liquid limit (LL), plastic limit (PL), and plasticity Index (PI) were determined. Each type of soil has a certain water retention capacity and, therefore, its plasticity index may reflect their mineralogical condition. The pneumatic consolidometer used for the compressibility tests (ASTM, 1998) included measurements of

Table 1. Physical and mechanical characterization of the studied lacustrine sequences. S_s, solid density; W_{grav}, gravimetric water content; GSD, Grain Size Distribution: C, Clay, M, Silt, S, Sand; e_0 and e_f, initial and final void ratio, OCR, Overconsolidation Ratio, Cc, Compressibility Index, USCS, Unified Soil Classification System: MH, High Plasticity Silt, SM, Silty Sand.

Depth m	S_s g cm^{-3}	W_{grav} %	LL-PI/IP %	GSD C/M/S	$e_0 - e_f$	Δe	OCR	Cc	USCS
Tláhuac									
6.0	2.20	106	74–59/15	14/60/26	2.5–2.1	0.4	13	0.69	MH
12.15	2.21	187	76–57/19	19/43/38	3.5–3.3	0.2	2.8	0.18	MH
17.05	2.27	228	123–90/33	11/32/57	5.3–2.4	2.9	1	2.56	MH
22.6	2.23	260	95–67/28	15/62/23	5.2–2.5	2.7	1	3.13	MH
22.9	2.05	260	136–90/46	15/62/23	7.6–3.2	4.4	1	3.40	MH
La Cascada									
12.65	2.43	203	87–61/26	16/34/50	4.7–2.2	2.5	1.8	2.25	MH
14.65	2.53	45	–	7/15/78	0.8–0.7	0.1	3.7	0.14	SM
26.84	2.26	130	103–74/29	16/38/46	3.0–1.4	1.6	10	1.73	MH
L. Blanco									
4.7	2.39	190	–	12/29/59	5.1–2.8	2.3	12.5	3.22	SM
5.6	2.26	234	103–74/29	24/29/47	0.8–0.0	0.8	5.5	1.26	MH
5.85	2.21	222	96–72/24	46/31/23	4.9–1.8	3.1	6.5	3.5	MH
11.5	2.23	213	75–59/16	19/36/45	3.2–1.8	1.4	20.5	1.1	MH
14.1	2.23	302	101–67/34	18/37/45	5.9–2.2	3.7	1.4	2.76	MH
14.8	2.36	120	59–47/12	6/41/53	2.9–1.7	1.2	3.6	1.47	SM

variations of axial pressure, back pressure, pore pressure and displacement with electronic sensors (Hernandez-Marin and Carreón-Freyre, 2002). The description of its operation and data acquisition software is detailed in Flores et al. (2010). Compression Index (Cc) and Overconsolidation Ratio (OCR) were obtained from graphs of void ratio vs log effective stress.

Results obtained for representative samples (14 out of more than 30) of the three studied sequences are presented in Table 1. Mostly, sediments are high plasticity silts (MH) with some intercalations of uniform sand and gravel deposits of black colour and ash layers (sandy silts) that correspond to pyroclastic deposits. Water content varies from 0 to 100 % for pyroclastic deposits and from 200 to 300 % for lacustrine deposits (Fig. 2). Variations of colour from red to olive green indicate the oxidation or reduction of iron (Fe) present in clays. Likely, local variations of pH, temperature, and salinity produced different clayey materials within the same depositional environment. In the La Cascada samples green colours of the clayey sediments denote a reduction condition indicating a water cover. The Lucio Blanco site is located in a channel that connected with the Texcoco sub-basin from the north within the lacustrine area of Iztapalapa (Fig. 1). Variations of the water level in the channel might have varied conditions from dry to humid, and the presence of Fe and magnesium (Mg) from basaltic rocks propitiated the crystallization of smectite clays, such as montmorillonite. In the Tlahuac site, the sequence is composed mainly by pumice and ash (silicic py-

roclastic deposits) that rapidly generated fine grain materials, rich in silica (SiO_2), and with a low level of crystallization. The initial void ratio (e_0) for each layer indicates the potential compressibility and is directly related to the grain size and mineralogy.

Three samples from different lacustrine layers at 12, 14, and 27 m depth in La Cascada borehole are overconsolidated. The highest Cc value and the higher water content correspond to the 12 m sample and the plasticity index (IP) is similar to that of 27 m sample which has the lower Cc value (Table 1, Fig. 3b). The sequence is interpreted to be formed by clayey well crystallized lacustrine sediments.

In the Lucio Blanco borehole, six samples were tested from 4.7, 5.6, 5.85, 11.5, 14.1, and 14.8 m depth. The highest Cc value corresponds to the 5.85 depth but does not present the highest water content or IP. The lowest Cc value corresponds to a sandy material that is a pyroclastic flow deposit (Fig. 3c and d). Pyroclastic deposits in general behave as non-cohesive potentially collapsible material. The sequence was interpreted as a wetland area within a lake channel, the red colour (oxidation) of the lacustrine deposits indicated large variations of water level.

In the Tláhuac borehole, five samples were tested from 6, 12, 17, 22.6, and 22.9 m depth. The highest Cc value corresponds to the deepest layer and the two lowest to the shallower layers. Also in this case the most compressible layer does not present the highest water content, neither the highest IP (Fig. 3a). The sequences is interpreted to be mainly

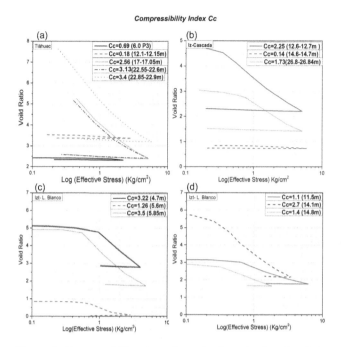

Figure 3. Compressibility plots of samples from the three borehole samples. (**a**) Five normally consolidated samples from 6 to 22.9 m depth in the Tlahuac site, only the material in the surface is over consolidated and have the smaller Cc value and the deeper samples have the higher Cc values; (**b**) three over consolidated samples from 12.6 to 26.8 m depth in La Cascada site, the lowest Cc value is placed at 14 m depth and the highest at 12 m depth; (**c and d**) six over consolidated samples from 4.7 to 14.8 m depth in the Lucio Blanco site, the lower Cc value corresponds to a 11.5 m depth sample and the higher at 5.85 m depth. Note also high variations in Cc for the same depth (e.g. at 12 m depth: 0.18, 2.25, and 1.1) for the different sequences.

formed by amorphous materials, formed by the alteration of ashes, that may explain the apparently contradictory values.

3.3 Mineralogy

For better understanding of the Cc variations we carried out XRD analysis in order to identify mineralogy; SEM and TEM imaging to characterize structure. Figure 4 shows some of the results for the Lucio Blanco and Tlahuac sites. The XRD records of the sample from 14 m depth in the Lucio Blanco site, show heterogeneous mineralogical composition of the entire sample: mica, illite, anorthite, silica, talc, and possibly nontronite. The clay fraction sample (after washing salt) indicates the presence of montmorillonite and cristobalite (silica) (Fig. 4a). Anorthite, a calcium plagioclase, and mica are minerals formed during volcanic eruptions that have not been completely altered. Cristobalite may have a volcanic origin or be part of the diatomaceous frustule (e.g., Marsal and Mazari, 1959). The montmorillonite suggests saturated conditions during its formation. The presence of talc indi-

cates alteration of the basic volcanic minerals at high salinity and temperatures below 100 °C (Velde, 1995).

The XRD record for the Tlahuac samples indicates the presence of vermiculite which also is related to the alteration of volcanic minerals (an example from 22 m depth is presented in Fig. 4a). Amorphous material and low-crystallized materials cannot be identified by XRD but SEM and TEM can identify such materials. The SEM image presented in Fig. 4b corresponds to a silty clay from 6 m depth (Fig. 4b1), and to a low crystallized material from 22 m depth (Fig. 4b1 and b2). The Fig. 4b3 was obtained using an TEM and shows the lack of structure. This material is similar to allophane and is a product of the alteration of acid volcanic ashes.

4 Discussion

The results of the physical characterization presented in this paper indicate that the conditions of formation and the evolution of lacustrine sequences control mechanical behaviour, and ultimately differential subsidence in the Basin of Mexico, and particularly in Mexico City. The silty-clay fraction has a highly heterogeneous composition including partially altered volcanic ash, illite, smectite, and allophane.

These materials are usually assumed to be highly compressible and with low resistance to shear. However, complex consolidation of these materials is also influenced by the displacement of free water, adsorbed water and/or intermolecular water, which holding force is dependent on the mineralogical characteristics of the solid phase. In such lacustrine materials having a metastable structure and mineralogy the mechanical behavior cannot be represented only by one plasticity or compressibility index. Lacustrine clay heterogeneity due to changes in depositional environment should be considered for a proper assessment of the compressibility index variations. This parameter is the most important because it is widely used for the estimation of land subsidence and, hence of the fracturing susceptibility of sediments.

A common assumption is that Cc decreases with depth due to lithostatic load; however, the analysis of the data presented in this work suggest a different mechanical behaviour for the lacustrine sediments. The mixture of silt and sand composed of silica, talc, and plagioclase with smectite clays (montmorillonite) presents variable consolidation conditions. Sandy lenses are less compressible, although the silt and clay content is higher. For the case of Tlahuac, it should be noted that the sand corresponds to pyroclastic materials composed of pumice that may degrade and change structure in hundreds or tens of years, and this explains the presence of low crystallized materials in the whole sequence. These materials show a lower plasticity than the smectites, higher contraction limits and a brittle behavior. The lateral and vertical compressibility variations are one of the causes of the substantial differential deformation observed in the study area. The high water

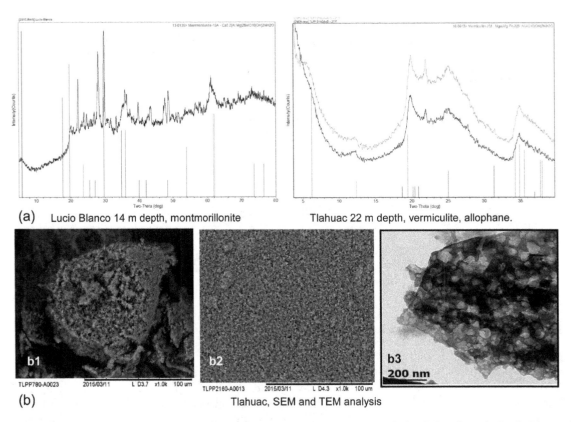

Figure 4. Characterization of the mineralogy and structure of clayey materials. **(a)** XRD analysis of clays from the Lucio Blanco site (14 m depth) showing the montmorillonite signature and from Tlahuac (22 m depth) showing the signature of vermiculite and low crystallized material; **(b)** SEM images from Tláhuac deposits: **(b1)**, is a silty material at 6 m depth; **(b2)**; is a low crystallized material at 22 m depth and, the **(b3)** image, taken using an TEM, shows the lack of structure in material similar to allophane.

content values suggest that these materials have low shear strength and fracture easily. Once fractured, the space created is filled with material coming from the surface creating a tensional state of stress that propitiates the propagation of fracturing to the surface.

Furthermore, the studied silt-clay sequences are commonly interbedded with fluvial and pyroclastic granular materials and/or volcanic rocks, which increase the mechanical heterogeneity of sequences and favours differential deformation. Fracturing of the subsurface in the Basin of Mexico, (e.g. Iztapalapa and Tlahuac) might be provoked by several factors, such as groundwater withdrawal and regional faulting; nevertheless, the gradual deformation related to differential consolidation between the different types of fluviolacustrine materials in volcanic areas also play an important role and should be better studied.

Acknowledgements. The authors acknowledge the work of characterization of samples in the laboratory of Mr. Ricardo Carrisoza, laboratory technician in the Laboratorio de Mecánica de Geosistemas (LAMG) at the Centro de Geociencias. The field support of Mr. Cesar Tovar CERG-Iztapalapa is also appreciated. The authors also thank support from the UNAM DGAPA Project PAPIIT No. 114174.

References

American Society for Testing Materials (ASTM): D2435-96 Standard method for one-dimensional consolidation properties of soils, ASTM Standards, USA, 5 pp., 1998.

Buhrke, V. E., Jenkis, R., and Smith, D. K.: A practical guide for the Preparation of specimens for X-ray Fluorescene and X-Ray diffraction Analysis, Wiley-VCH, New York, 333 pp., ISBN: 0-471-19458-1, 1998.

Carreón-Freyre, D.: Land subsidence processes and associated ground fracturing in Central Mexico, in: Land Subsidence, Associated Hazards and the Role of Natural Resources Development (Proceedings of EISOLS 2010, Querétaro, México), edited by: Carreón-Freyre, D., Cerca, M., and Galloway, D. L., Red Book Series Publication 339, IAHS Press, CEH Wallingford, UK, ISBN: 978-1-907161-12-4, ISSN: 0144-7815, 149–157, 2010.

Carreón-Freyre, D. and Cerca, M.: Delineating the near-surface geometry of the fracture system affecting the valley of Queretaro, Mexico: Correlation of GPR signatures and physical properties of sediments, Near Surface Geophysics, EAGE (European Assoc. of Geoscientists and Engineers), 4, 49–55, 2006.

Carreón-Freyre, D. C., Cerca, M., and Hernández-Marín, M.: Correlation of near-surface stratigraphy and physical properties of clayey sediments from Chalco Basin, Mexico, using Ground Penetrating Radar, J. Appl. Geophys., 53, 121–136, 2003.

Carreón-Freyre, D. C., Hidalgo-Moreno, C., and Hernández-Marín, M.: Mecanismos de fracturamiento de depósitos arcillosos en zonas urbanas. Caso de deformación diferencial en Chalco, Estado de México, Boletín de la Sociedad Geológica Mexicana, Número Especial de Geología Urbana, Tomo LVII, 2, 237–250, 2006.

Carreón-Freyre, D., González-Hernández, M., Cerca M., and Gutiérrez-Calderón, R. I.: Caracterización geomecánica de los suelos de Iztapalapa, México, para evaluar el fracturamiento causado por deformación diferencial, Proceedings of the 14th Pan-American Conference on Soil Mechanics and Geotechnical Engineering, 2–6 October 2011, Toronto, Canada, 2011.

Carrillo, N.: Influence of artesial wells in the sinking of Mexico City, en Volumen Nabor Carrillo "El hundimiento de la Ciudad de México y el Proyecto Texcoco", Comisión Impulsora y Coordinadora de la Investigación Científica Anuario, 47, 7–14, 1947.

Díaz-Rodríguez, A., Lozano-Santacruz, R., Dávila-Alcocer, V. M., Vallejo, E., and Girón, P.: Physical, chemical, and mineralogical properties of México City sediments: a geotechnical perspective, Can. Geotech. J., 35, 600–610, 1998.

Ferrari, L., Orozco-Esquivel, T., Manea, V., and Marina, M.: The dynamic history of the Trans-Mexican Volcanic Belt and the Mexico subduction zone, Tectonophysics, 522–523, 122–149, doi:10.1016/j.tecto.2011.09.018, 2012.

Flores, C. O., Gómez, R. E., Hernández, M. S., and Carreón Freyre, D.: Automatización de un consolidómetro neumático, Memorias de la XXV Reunión Nacional de Mecánica de Suelos e Ingeniería Geotécnica, Mexico, ISBN: 978-968-5350-23-5, 8–16, 2010.

Haigh, S. K., Vardanega, P. J., and Bolton, D.: The plastic limit of clays, Géotechnique, 63, 435–440, doi:10.1680/geot.11.P.123, 2013.

Hernández-Marín, M., Carreón-Freyre., D., and Cerca Martínez, M.: Mechanical and physical properties of the montmorillonitic and allophanic clays in the near-surface sediments of Chalco Valley, Mexico: Analysis of contributing factors to land subsidence, Proceedings of the 7th International Symposium on Land Subsidence SISOLS 2005, 23–28 October 2005, Shanghai, China, ISBN: 7-5323-8209-5, Vol. I, 276–285, 2005.

Hillier, S.: Erosion, sedimentation and sedimentary origin of clays, in: Composition and mineralogy of Clay Minerals, edited by: Velde, B., Springer-Verlag, Berlin, Heidelberg, New York, 162–214, ISBN: 3-540-58012-3, 1995.

Lambe, W. T. and Whitman, R. V.: Soil Mechanics, John Wiley and Sons Inc., NY, 553 pp., 1969.

Marsal, R. J. and Mazari, M.: El subsuelo de la Ciudad de México, Instituto de Ingeniería, U.N.A.M. I, II, Mexico, 505 pp., 1959.

Mesri, G., Rokhse, A., and Bonor, B. F.: Composition and compressibility of typical samples of Mexico City clay, Geotechnique, 25, 527–554, 1976.

Peralta y Fabi, R.: Sobre el origen de algunas propiedades mecánicas de la formación arcillosa superior del Valle de México, Simposio sobre Tópicos Geológicos de la Cuenca del Valle de México, Mexico, 282 pp., 1989.

Robinson, R. G.: Consolidation analysis with pore pressure measurements, Géotechnique, 49, 127–132, 1999.

Velde, B.: Origin and Mineralogy of Clays, Clays and the environment, Springer-Verlag, Berlin, Heidelberg, New York, 334 pp., ISBN: 3-540-58012-3, 1995.

Wada, K.: Minerals formed and mineral formation from volcanic ash by weathering, Chem. Geol., 60, 17–28, 1987.

Warren, C. J. and Rudolph, D. L.: Clay minerals in basin of Mexico lacustrine sediments and their influence on ion mobility in groundwater, J. Contam. Hydrol., 27, 177–198, 1997.

Wesley, L. D.: Consolidation behaviour of allophane clays, Géotechnique, 51, 901–904, 2001.

Zeevaert, L.: Estratigrafía y problemas de ingeniería en los depósitos de arcilla lacustre de la Ciudad de México, Memoria del Congreso Científico Mexicano, 5, 58–70, 1953.

26

Sinking coastal cities

G. Erkens[1,2]**, T. Bucx**[1]**, R. Dam**[3]**, G. de Lange**[1]**, and J. Lambert**[1]

[1]Deltares Research Institute, Utrecht, the Netherlands
[2]Utrecht University, Utrecht, the Netherlands
[3]WaterLand Experts, Amsterdam, the Netherlands

Correspondence to: G. Erkens (gilles.erkens@deltares.nl)

Abstract. In many coastal and delta cities land subsidence now exceeds absolute sea level rise up to a factor of ten. A major cause for severe land subsidence is excessive groundwater extraction related to rapid urbanization and population growth. Without action, parts of Jakarta, Ho Chi Minh City, Bangkok and numerous other coastal cities will sink below sea level. Land subsidence increases flood vulnerability (frequency, inundation depth and duration of floods), with floods causing major economic damage and loss of lives. In addition, differential land movement causes significant economic losses in the form of structural damage and high maintenance costs for (infra)structure. The total damage worldwide is estimated at billions of dollars annually.

As subsidence is often spatially variable and can be caused by multiple processes, an assessment of subsidence in delta cities needs to answer questions such as: what are the main causes? What is the current subsidence rate and what are future scenarios (and interaction with other major environmental issues)? Where are the vulnerable areas? What are the impacts and risks? How can adverse impacts be mitigated or compensated for? Who is involved and responsible to act?

In this study a quick-assessment of subsidence is performed on the following mega-cities: Jakarta, Ho Chi Minh City, Dhaka, New Orleans and Bangkok. Results of these case studies will be presented and compared, and a (generic) approach how to deal with subsidence in current and future subsidence-prone areas is provided.

1 Introduction

Currently, global mean absolute sea- level rise is around $3\,\mathrm{mm\,yr^{-1}}$, and projections until 2100 based on Intergovernmental Panel on Climate Change (IPCC) scenarios expect a global mean absolute sea-level rise in the range of 3–$10\,\mathrm{mm\,yr^{-1}}$ (Church and White, 2011; Slangen, 2012). However, currently observed subsidence rates in coastal megacities are in the range of 6–$100\,\mathrm{mm\,yr^{-1}}$, and projections until 2025 expect similar subsidence rates (Fig. 1).

In coastal cities around the world, land subsidence increases flood vulnerability (flood frequency, inundation depth, and duration of floods), and hence contributes to major economic damage and loss of lives. Land subsidence is additionally responsible for significant economic losses in the form of structural damage and high maintenance costs; it affects roads and transportation networks, hydraulic infrastructure, river embankments, sluice gates, flood barriers, pump-ing stations, sewage systems, buildings, and foundations. The total damage associated with subsidence worldwide is estimated at billions of dollars annually.

There are no indications that neither subsidence nor the resulting damage will reduce in the near future. In fact, both are likely to increase. Ongoing urbanization and population growth in delta areas, in particular in coastal mega-cities, continues to fuel economic development in subsidence-prone areas. Consequently, economic development drives both the growing demand for groundwater, thereby increasing subsidence rates, and the growth of the total value of assets at risk. These impacts are aggravated on the long term in coastal areas, by expected future climate change impacts, such as sea-level rise, increased storm surges, and changes in precipitation.

In this paper, we focus on land subsidence in the urban environment, rather than land subsidence in rural agricultural areas, where the drivers may be similar, but the impact very

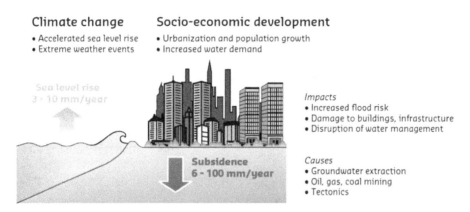

Figure 1. Drivers, processes and impacts of land subsidence in coastal cities. Land subsidence can exceed global absolute sea-level rise (SLR) with a factor 10.

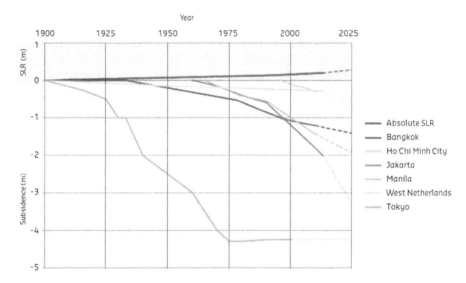

Figure 2. Subsidence history (cumulative) in a series of coastal cities around the world. Absolute sea level rise is depicted as reference. Subsidence can differ considerably within a city area, depending on groundwater levels and subsurface characteristics. Values provided here can be seen as average for the local subsidence hotspots. Some cities are currently seeing an acceleration of subsidence as a result of economic growth. Tokyo stands out as an example where subsidence has stopped after successful mitigation measures were implemented. The caption of Table 1 provides references.

Table 1. Subsidence in coastal cities. Estimated additional mean cumulative subsidence until 2025 (mm) are linear interpolations of the current rates, notwithstanding any policy changes. Sources: Bangkok: MoNRE-DGR (2012), Aobpaet et al. (2013); Ho Chi Min City: van Trung and Minh Dinh (2009); Jakarta: Bakr (2011); Manila: Eco et al. (2011); West Netherlands: van de Ven (1993); Tokyo: Kaneko and Toyota (2011).

City	Mean cumulative subsidence in period 1900–2013 (mm)	Mean current subsidencerate (mm yr^{-1})	Maximum subsidence rate (mm yr^{-1})	Estimated additional mean cumulative subsidence until 2025 (mm)
Jakarta	2000	75–100	179	1800
Ho Chi Minh City	300	up to 80	80	200
Bangkok	1250	20–30	120	190
New Orleans	1130	6	26	> 200
Tokyo	4250	≈ 0	239	0

different. Figure 2 and Table 1 show that land subsidence rates widely vary from city to city. In many cases, the underlying processes and the relative contribution of the different drivers is not well understood. Similar to the level of technical understanding, policy formulation and governmental engagement in cities is equally diverse. Whereas some cities are in an early state of research and policy development on land subsidence, others have already implemented measures mitigating subsidence and the resulting damage. The observed different stages in development mean that cities can learn from each other, thereby avoiding re-inventing the wheel. Cities that actively pursue a policy on subsidence have valuable experiences to share with cities that have just started to address their subsidence.

This is exactly the thought behind the assessment that was carried out for this research. We compared five cities regarding their state of subsidence research and policy development: Jakarta, Ho Chi Minh City, Dhaka, New Orleans and Bangkok. The assessment aimed at getting insight into the processes causing subsidence in the urban environment, obtaining a (generic) research agenda for this topic, and listing best practice cases. Results of these case studies will be presented and a (generic) approach how to cope with subsidence in current and future subsidence-prone areas is provided.

2 Results of the review

For the quick assessment we used published reports on both the technical and the policy aspects of subsidence in the focus cities. In addition, we interviewed local scientist and policymakers to obtain their perspective. It became quickly clear that all cities tried to answer similar questions. We compiled the interview results into seven interrelated questions (Fig. 3). They include questions such as: what are the main causes for subsidence? How much is the current subsidence rate and what are future scenarios? Where are the vulnerable areas? What are the impacts and risks? How can adverse impacts be mitigated or compensated for? Who is involved and responsible to act? How to monitor the effect of the implemented measures? The interrelation between the questions is indicated with the arrows in Fig. 3. The indicated interrelation does not necessarily mean that each question needs to be answered in a specific order, but it merely indicates that each answer may be valuable input for a next question.

In this paper we follow this framework (in seven steps) and illustrate how these questions are addressed in example cities, thereby discussing both technical and policy aspects of subsidence. In this way, this framework could serve as a blue print for cities to shape their policy and research agenda regarding subsidence.

1. How much subsidence is there?

2. What is causing subsidence?

3. How much subsidence is predicted?

4. What impact has subsidence?

5. Who is responsible?

6. What are solutions?

7. Monitoring and evaluation of measures

Figure 3. Seven questions that need to be addressed to pursue a successful policy to coop with subsidence. This is loosely based on the policy cycle, a popular framework to analyse policy development.

3 Measuring and monitoring

The first step towards a successful strategy for subsidence is to establish if a certain area is actually subsiding. This may not me evident from the field, particular if subsidence is non-differential and no structural damage (cracks, tilting) is observed in buildings or infrastructure. Typically, the loss of elevation, which may have been observed, is interpreted as the result of climate-driven sea level rise instead of the result of subsidence.

To determine land subsidence rates, accurate measuring techniques are required. Continuous subsidence monitoring provides the necessary insight into changes – ranging from minor to very significant changes – in the topography of the urban area. These observations are also essential to validate subsidence prediction models in a later stage.

The following geodetric observation methods are being used:

– optical levelling;

– Global Positioning System (GPS) surveys;

– Laser Imaging Detection and Ranging (LIDAR);

– Interferometric synthetic aperture radar (InSAR) satellite imagery.

Following early work with systematic optical levelling nowadays GPS surveys and remote sensing techniques (LIDAR and InSAR) are deployed with impressive results. In contrast to surveys, LIDAR and InSAR images give a spatially resolved subsidence signal. InSAR images date back to the 1990s and can now be used to establish subsidence since that time. Application of this technique in soft soil areas is for the moment limited to the build-up environment, as a result of the need for stable reflectors. Ideally, multiple observation techniques are combined, for instance absolute measurements from GPS and Optical Levelling Spatially can be combined with remotely sensed, relative displacement measurements from InSAR. In this way, spatially resolved subsidence maps with respect to a global reference frame can be produced. InSAR measurement can therefore not replace periodic and systematic ground surveys, as they remain essential for ground truthing subsidence rates derived by remote sensing and as an independent source for validating subsidence prediction models.

Systematic observation of elevation forms the base for subsidence monitoring systems. Monitoring results can be used to develop a so-called dynamic digital elevation model (dDEM). This is not a static, one-time only (preferably high resolution) recording of the local topography, but an elevation model that can be corrected and updated from time to time, and that can be used in hydraulic models for flood prediction and urban water management.

All techniques mentioned above measure land surface elevation change, but give no information on the source of the subsidence. Subsidence benchmarks or extensometers can provide in-situ information of ground movement, as they record the volume reduction across a certain stretch in the subsurface, or even of individual geological layers. Ideally the benchmarks or extensometers need to be connected to surface movement observations, for instance by using a combined extensometer and continuous GPS station (e.g. Wang et al., 2014). Monitoring total subsidence at these "super-sites", where a terrestrial network of site specific measurement stations is combined with remote sensing, forms the backbone of a spatially resolved subsidence measurement system (Allison et al., 2014). To support subsidence modelling, hydraulic heads of different aquifer systems and the phreatic groundwater level need to be monitored at these super sites as well. Measurements of geotechnical parameters at the same site provide additional necessary input for model studies.

4   Unravelling the subsidence signal

Subsidence can have natural as well as anthropogenic causes. The natural causes include tectonics, loading by ice sheets, by sediments, of by the ocean/sea (isostatic adjustment), and natural sediment compaction (autocompaction). Anthropogenic causes include compression of shallow soft layers by loading (with buildings for instance), or as a result of drainage and subsequent oxidation and consolidation of organic soils and peat. Alluvial or coastal sediments consisting of alternating layers of sand, clay, and peat are specifically compressible and vulnerable for oxidation. This is related to the physical characteristics of these sediments and makes low-lying coastal and delta areas specifically prone to subsi-

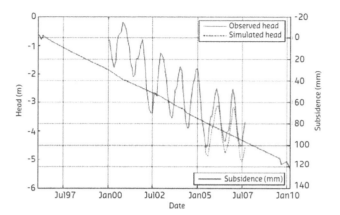

Figure 4. A distinct relation between falling hydraulic heads and subsidence in Ho Chi Minh City (Vietnam). This is indicative of an important contribution of groundwater over-exploration to subsidence, although it is not necessarily the only component contributing to the total subsidence signal.

dence. In deeper layers subsidence is caused by extraction of resources such as oil, gas, coal, salt, and groundwater.

In most of the large delta cities where land subsidence is severe (Jakarta, Ho Chi Minh City, Bangkok, Dhaka, Shanghai, and Tokyo), the main cause is extraction of groundwater. Rapidly expanding urban areas require enormous amounts of water for domestic and industrial water supply. This need often leads to over-exploitation of groundwater resources, especially when surface waters are seriously polluted (Jakarta, Dhaka). Dhaka (Bangladesh) is an example of a city that started to discover that it subsided after the flood frequency increased. In this rapidly expanding city data on subsidence and its impacts are currently largely lacking. Large-scale extractions cause groundwater levels to fall by 2–3 m yr^{-1}. At present, 87 % of the supplied water is from groundwater extraction, and it has been acknowledged that a shift to using surface water instead is necessary. However, treating the polluted surface water is much more technically complex and expensive than extracting groundwater.

Although groundwater extraction is often not the sole source of subsidence, studies in many cities have revealed a distinct relation between falling groundwater levels and subsidence, indicative of an important contribution of aquifer compaction (Fig. 4). The resulting spatial pattern of subsidence and its progress over time are strongly related to the local composition of the subsurface and the number and positions of groundwater abstraction wells.

New Orleans (USA) is a prominent example of a city where an array of processes contributes to the total subsidence of the city. The Mississippi Delta subsides as a result of natural processes, such as autocompaction, faulting, sediment loading and isostacy (e.g. Törnqvist et al., 2008; Yu et al., 2012). Within the urban area of New Orleans, there additionally is anthropogenic induced subsidence as a result of drainage of shallow soft soils (Stuurman and Erkens, 2015)

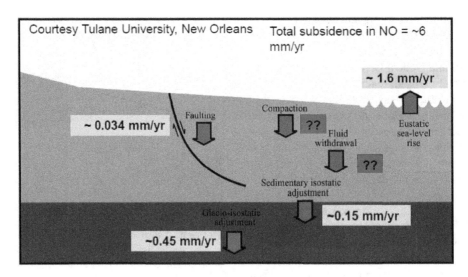

Figure 5. Subsidence components in the urban area of New Orleans. Values are derived from studies of Tulane University, New Orleans, and are indicative. The total subsidence rate is derived from InSAR measurements (Dixon et al., 2006). It shows that in the urban area natural subsidence forms the smaller portion of the total subsidence and that human induced subsidence dominates.

and extraction of deeper groundwater in confined aquifers, for industrial use mainly (Dokka, 2011). After drainage of the organic rich soils, they start to oxidize and lose volume, and this process will continue to cause subsidence as long as organic material is available in the drained subsoil.

The average measured subsidence rate in the city of New Orleans (including the urban area of Jefferson and St Bernard Parishes) is $6 \, \text{mm} \, \text{yr}^{-1}$ (Dixon et al., 2006). Many studies try to quantify one or more of the different components contributing to the total measured subsidence. Figure 5 shows how this may look for New Orleans, when components are quantified step by step (source: Tulane University, New Orleans). In-situ observation data may provide an independent valuable source of information to unravel the total subsidence signal, as argued in Sect. 3. Another approach to unravel the subsidence signal is inverse modelling, whereby with the use of a careful inversion scheme, the available knowledge on the geology and hydrological dynamics of a system can be quantitatively constrained with subsidence observations (e.g. Fokker et al., 2007).

From Fig. 5 also follows that in the urban area of New Orleans, human induced subsidence has a much larger contribution to the total subsidence signal than natural subsidence. This is often the case, as natural subsidence rates are mainly limited to tens of millimeters per year, to millimeters per year in exceptional cases. Human induced subsidence rates can easily reach centimeters per year, to even tens of centimeters per year (e.g. Jakarta). For policy development this is an important notion: it is worthwhile to implement measures to reduce human-induced subsidence.

5 Modelling subsidence to make predictions

In step three, once the causes for land subsidence have been established (see Sect. 4), predictions can be made to get insight in future land subsidence. Land subsidence modelling and fore-casting tools are being progressively developed that enable quantitative assessment of medium- to long- term land subsidence rates, and determination of multiple causes. Modelling tools are ideally complemented with monitoring techniques (i.e., GPS leveling, the use of InSAR -monitoring techniques), see Sect. 3.

Because land subsidence is in many places closely linked to excessive groundwater extraction, we focus in this paper on modelling of aquifer compaction. One of the most widely used computer program to simulate vertical compaction in models of regional ground-water flow is MODFLOW SUB-WT (Leake and Galloway, 2007). MODFLOW SUB-WT is developed by the US Geological Survey and uses changes in groundwater storage in subsurface layers (aquifers and aquitards) and accounts for temporal and spatial variability of geostatic and effective stresses to determine layer compaction.

In soft soils, such as unconsolidated Holocene layers of peat and clay, the classical consolidation theory by Terzaghi is unable to explain observed consolidation behaviour. These lithology form the aquitards and interbed units in confined aquifer complex systems, albeit often more consolidated, which start to compact after groundwater is extracted from the confined aquifers. Creep deformation is one of the typical processes that occur when the effective stress is increased in clay or peat soils. The creep deformation (also known as secondary strain) of soils is a secondary consolidation process that leads to a reduction in void ratio at con-

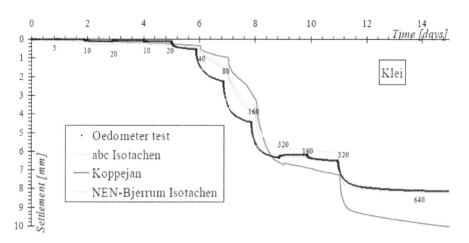

Figure 6. The performance of a series of models used to calculate settlement compared to Oedometer test results. Prediction made by classic models such as Koppejan fit the measurements less well compared to models based on the isotachs method, such as NEN-Bjerrum or abc-Isotachs.

stant effective stress, and consequently, to the development of an apparent pre-consolidation pressure (Den Haan, 1994). It is seen as visco-plastic behaviour and is considered a slow process, compared to primary or elastic consolidation. The inclusion of creep behaviour in numerical descriptions of the consolidation process has a long history, which is excellently described in Bakr (2015). An important aspect of the creep based models is that, due to secondary compression, there is a family of stress-strain curves rather than a single curve describing the relationship between stress and strain. Each of these curves, called "time lines" (i.e. isochrones), corresponds to a different duration of the applied load in a standard oedometer test. For soft soils, model predictions that make use of the isochrones method tend to match the oedometer test results best, specifically on longer time periods (Fig. 6).

Deltares Research Institute modified the US Geological Survey SUB-WT module by including isotachs (line of equal speed) based consolidation predictions. This model, MOD-FLOW SUB-CR (SUBsidence Creep), is used to determine medium- to long-term land subsidence trends under different scenarios of groundwater usage. In this way, the consequences of groundwater extraction for urban flood management become clear.

Because the SUB-CR model works with isotachs to calculate consolidation, it differs from the SUB-WT model in two ways:

- It predicts on the longer term more consolidation, thus subsidence, in clay and peat layers, as creep is a slow and largely irreversible component of subsidence

- Creep may continue for some time even after the hydraulic heads increased, introducing a time lag in consolidation.

As a result of these differences, aquifers with many fine-grained interbeds, creep forms a considerable part of the to-

tal amount of settlement over time and should not be neglected. An example is the subsidence predictions conducted for Jakarta, Indonesia, using isotachs-based consolidation calculations. Bakr (2015) calculates the subsidence occurring in four future groundwater management scenarios for Jakarta. The four scenarios are:

1. drawdown for all aquifers are kept zero till 2100 by maintaining piezometric levels at their values of 2010 (no change);

2. drawdown for all aquifers increase 5 m every 5 years from 2010 till 2030 (business as usual);

3. piezometric heads are recovered for all aquifers by 2015 to their values of 1995 (recovery),

4. piezometric heads are recovered for all aquifers by 2015 to the maximum level of all aquifers in 1995 (full recovery).

In Table 2, we report the predicted cumulative subsidence for Jakarta as calculated by Bakr (2015), calculated with the inclusion of creep. The results indicate that (i) if the hydraulic head declines continue with the current rate (scenario 2) parts of North Jakarta will sink an additional 3.9 m, and (ii) even if hydraulic heads remain the same (scenario 1) or are restored (scenarios 3 and 4) subsidence continues, up to 2.3 or 2.4 m in the recovery scenarios in 2100. This residual subsidence is the result of both delayed pore water pressure dissipation and visco-plastic creep compaction. This means that even if effective stresses do not change, land subsidence will continue till all layers reach hydrostatic equilibrium and creep compaction of all layers vanishes by time due to aging (Bakr, 2015).

This has important implications for policy development in the city of Jakarta. The significant predicted subsidence in the

Table 2. Cumulative subsidence (m), modelled including creep behaviour, for 4 groundwater management scenarios for Jakarta, Indonesia, by Bakr (2015). Because of the slow creep rates, subsidence continues even after hydraulic heads are restored (scenarios 3 and 4).

Year	Scenario 1 (no change)	Scenario 2 (business as usual)	Scenario 3 (recovery)	Scenario 4 (rapid recovery)
2020	1.97	2.48	1.74	1.73
2025	2.08	2.75	1.80	1.77
2030	2.18	2.92	1.85	1.81
2100	3.01	3.91	2.43	2.30

business as usual scenario justifies a subsidence mitigation policy. But the forecasted subsidence values in the recovery scenarios indicate that for the remaining residual subsidence an adaptation strategy must be developed too. Because of these far reaching implications for policy development, it is important that subsidence predictions are as accurate as possible. Although the inclusion of creep behaviour in prediction models aimed to increase accuracy of predictions, they are also sensitive for the geo(hydro)logical schematisation used in the model. This is because the fine-grained interbeds and aquitards are most sensitive to creep, and their exact distribution both in vertical and lateral direction determines the model outcome. The 3-D distribution of fine-grained deposits in the subsurface, and their geo-mechanical properties, are therefore key to reliable subsidence predictions for cities.

6 Impact and damage

With subsidence predictions for different management scenarios (step 3, see Sect. 5), damage estimates (step 4 in the framework, Fig. 3) provide additional information for policy decisions. The estimation of costs associated to subsidence is very complex. Subsidence is a "hidden threat" because in practice, costs appear on financial sheets as part of ad hoc investments or planned maintenance schemes, but are not labelled as subsidence-induced damage. Dedicated damage estimates of subsidence can help to raise awareness among policymakers and initiate policy development.

Generally, two (very different) types of damage as a result of subsidence can be recognised: (i) increased flood risk (due to increased flood frequency, floodwater depth, and duration of inundation) and more frequent rainfall-induced floods due to ineffective drainage systems, and (ii) damage to buildings, foundations, infrastructure (roads, bridges, dikes), and subsurface structures (drainage, sewerage, gas pipes, etc.). The former is mainly the result of non-differential subsidence, which is characteristic for large subsidence bowls that exist when groundwater or hydrocarbons at greater depth are extracted. Examples of cities that have increased flood risk as a result of subsidence include Jakarta, Ho-Chi-Minh and Bangkok. The second type of damage, to structures, is the result of differential subsidence. This commonly happens when fault systems are (re)activated, or when the subsidence is

the result of shallow processes (loading or drainage of soft soils). Examples of cities in which structures are damaged include New Orleans, Venice (Italy) and Amsterdam (the Netherlands). Note that the construction site preparation and construction costs in soft-soil areas should be considered as subsidence-related costs, as these are mainly incurred to prevent consolidation. On the longer term, however, cumulative subsidence of soft soils may also increase flood risk as for instance happened in the Netherlands (subsidence over the last ~ 1000 years) or in New Orleans (subsidence over the last ~ 150 years). The extent of the damage is different in the two cases: increased flood risk usually applies to a larger area than structural damage that applies to single structures or parts of the network. The owner of the problem is also different: it is the local government who is investing in reducing flood risk, whereas local communities, (utility) companies or even home owners pay for the damage to (infra)structures.

Making an estimation of costs associated with subsidence is notoriously complex. Some bulk estimates are available. For instance, in China, the average total economic loss due to subsidence is estimated at around USD 1.5 billion per year, of which 80–90 % is from indirect losses. In Shanghai, over the period 2001–2010, the total loss cumulates to approximately USD 2 billion. In the Netherlands, new estimates based on subsidence modelling, try to unravel the bulk costs. For instance, it is calculated that damage to foundations (as a result of subsidence) has been more than EUR 5 billion thus far, and might reach EUR 40 billion in 2050 (although this is a theoretical maximum, Hoogvliet et al., 2012). The communities in soft soil areas in the Netherlands spend EUR 0.25 billion per year more on maintenance than the communities on supportive soils. This values consists of EUR 0.17 billion per year maintenance for roads and water networks and EUR 0.08 billion per year for sewage systems (Lambert et al., 2014). The total damage associated with subsidence worldwide is unknown, but estimated based on the aforementioned values suggest billions of dollars annually. Because of ongoing economic and urban development, the potential damage costs of subsidence will increase considerably in the future, especially in subsidence-prone areas such as flood plains.

Damage estimates form the core of cost-benefit analyses. For subsidence, cost-benefit analyses will help to systematically calculate and compare benefits and costs of a decision or government policy on the short and long term. Being a gradual process, usually mitigation measures for subsidence are costly on the short term, but cost-effective on the longer term. Cost-benefit analyses could provide this insight in a quantitative way.

7 Measures and monitoring

Once the damage caused by subsidence is quantified (Sect. 6), the responsible actors (step 5 of the framework, Fig. 3) can work out a policy on subsidence (step 6), that should be evaluated after implementation (step 7). In this section we focus on action necessary for steps 6 and 7.

There are generally two policy strategies for subsiding cities: mitigation and adaptation – analogue to the climate change policy discussions. A successful strategy, however, probably includes both. Mitigation only works for human-induced subsidence (Sect. 4). Typical mitigation measures include restrictions of groundwater extraction, artificial recharging aquifers, or raising (phreatic) water levels in areas with organic rich soils, thereby reducing oxidation of organic matter. Building with lighter materials decreases the load on soft soils, thereby decreasing consolidation and subsidence (Lambert et al., 2014).

For the human induced subsidence that cannot be mitigated, either because of technical difficulties (for instance the use of lighter building materials in high rise buildings), or because of financial reasons (i.e. the mitigation costs are too high), an adaptation strategy should be considered. This is also true for residual subsidence after a successful mitigation of subsidence (see Sect. 5) or for natural subsidence, where mitigation is not possible.

Adaptation must focus on reducing the impact of subsidence, for instance by decreasing the vulnerability of a certain asset to the negative impacts of subsidence. For increased flood risk as a result of subsidence, adaptation measures include the strengthening or heightening of embankments, building on mounds or piles, or conduct spatial planning in such a way that new constructions are only built on elevated areas. For damage to structures, adaptation strategies may include the use of flexible pipes and cables (specifically for connection points), the use of better foundations for structures, or again careful spatial planning, whereby building is limited to areas with supportive soils (for instance channel belt deposits within a delta).

Adaptation strategies are commonly applied in subsiding coastal cities, for instance most of them have network of embankments that reduces the flood risk. Cities that pursue an active policy on subsidence mitigation are less common, but successful examples do exist. In Tokyo, after taking regulations measures restricting the groundwater use were imposed in the early 1960s, the groundwater levels began to rise as a result (Fig. 7). Subsidence came to hold 10 years later as a result of the delayed response in the compacting layers (see

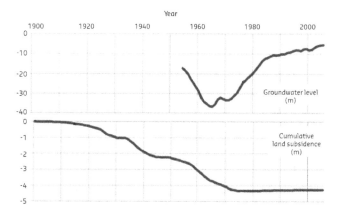

Figure 7. Land subsidence and groundwater levels in the Tokyo area (Japan), modified after Kaneko and Toyota (2011). The effect of the reduction of groundwater extraction on groundwater levels is clearly visible. Note that land subsidence completely stops 10 years after the groundwater level recovery started.

also Sect. 5). The restrictions on groundwater use meant that a replacement water source had to be found. Dams were constructed in several river basins that were designated for water resources development. During the 1970s and 80s numerous dams were built to provide storage to avoid future water scarcity and to supply the growing cities with sufficient water. Beginning in the 1960s an additional investment in waste water treatment was initiated.

Shanghai in China is another example of a city with a successful subsidence mitigation strategy. Following the increased understanding of the close relationship between groundwater extraction and land subsidence in Shanghai (e.g. Shi et al., 2008), groundwater levels were restored with active recharge techniques. Although this approach reduced the further lowering of groundwater tables and limited subsidence, it did not completely eliminate the effects of subsidence on infrastructure, roads, and buildings. The Shanghai case shows that, with active and substantial recharge, sustainable groundwater use is achievable, without severe subsidence, provided that average yearly pumping rates are in balance with the average yearly recharge.

In Bangkok, Thailand, regulation of and restrictions on groundwater extraction have successfully reduced severe land subsidence. A specific law (the Groundwater Act) was enacted in 1977. The most affected areas were designated as Critical Zones, and the government was given more control over private and public groundwater use in these areas. Groundwater use charges were first implemented in 1985 and have gradually increased. Currently about 10 % of the total water use is supplied by groundwater extraction, and this mainly used by the industry in Bangkok. In most urban areas, subsidence is now reduced to $1\,\mathrm{cm\,yr^{-1}}$, with local increased subsidence rates of $2\,\mathrm{cm\,yr^{-1}}$ in the aforementioned industrial sites.

Jakarta (Indonesia) and Ho-Chi-Minh City (Vietnam) are considering similar subsidence mitigation strategies. In the Greater Jakarta area, metropolitan authorities and technical agencies are advocating the reduction of groundwater extraction in vulnerable areas. The goal is to completely phase out the use of groundwater by taxing groundwater consumption. This would require developing an alternative water supply for large industrial users or relocation of large groundwater users, outside the so-called "critical zones". The number of "unregistered" users is still a problem. To some extent, spatial planning measures have been applied to avoid subsidence-prone areas, but the fast growth of informal settlements has made many of these plans obsolete. Recently, the Jakarta province government started to clear out the water management structures to reduce flood risk. In 2015, the Governor of Jakarta announced the reduction of the usage of deep groundwater in all government and public buildings, as a first step in the transition to piped water supply. The expected delayed response of subsidence to groundwater head recovery (Bakr, 2015; Sect. 5) asks for accurate subsidence prognosis. They form a vital component for any integrated flood management and coastal defence strategy (Dam, 2012).

Although land subsidence in Ho Chi Minh City has been observed since 1997, there is still – similar to Jakarta – considerable disagreement about the underlying processes and impacts. This is partly due to poor land level and groundwater extraction monitoring data (Ho Chi Minh City Flood and Inundation Management, 2013). Restrictions of groundwater extraction have been initiated, but it is too early to observe any effects.

In the Netherlands, with arguably the longest history of human induced subsidence in the world (since 1000 AD), the focus has been on adaptation strategies for more than nine centuries. In the coastal peatlands, after $\sim$ 1200 AD, adaptation measures included improving drainage (digging canals), the closing of (tidal) creeks and rivers, raising dikes and creating polders, and the improvement of foundations of buildings and infrastructure. Only in the last 50 years, with ever increasing damage to structures, mitigation measures were implemented. Nowadays, groundwater is sustained as shallow as possible in the peatlands. This means careful land use planning (less productive grassland and nature development in the most sensitive areas and considering alternative crops elsewhere) and the inlet of fresh water in polders in dry periods. Complete mitigation of subsidence is probably not possible, because that would end agriculture in a major part of western and northern Netherlands. The associated high economic losses are socially and culturally not acceptable. In the northern part of the Netherlands, gas extraction results in significant subsidence. Here policy similarly developed towards mitigation measures, albeit on a shorter time scale. Gas extraction started in the 1960's, but until about 2010 the governmental response to subsidence was limited to adaptation of the surface water management system. After 2010, subsidence was accompanied by more frequent and power-ful induced seismicity (earthquakes). The resulting damage of houses and other constructions forced the government in 2014 to start additional mitigation measures in the form of a significant reduction of gas exploitation in the most critical fields. Again, full mitigation was very difficult as stopping of the gas exploitation would endanger the national energy supply and would reduce the gas revenues by several billion of euro's per year. In addition, even if the gas exploitation was completely phased out, the subsidence and earthquakes are likely to continue. Concluding, for the Netherlands full mitigation of subsidence is far more expensive than implementing adaptation measures and an adaptation strategy combined with limited mitigation is a much more feasible option.

For all measures taken to reduce subsidence or its impacts, it is important that the effectiveness of these efforts is monitored. This implies that a subsidence monitoring network (see Sect. 3) need to be installed before the measures are implemented. The monitoring data form an important contribution to any subsidence monitoring network that has been established in step 1 (Sect. 3). Preferably, the monitoring data and analytical results (of the various modelling tools) are stored in a central database.

8 Concluding remarks

- In urban areas, human induced land subsidence dominates the total subsidence signal.

- Land surface elevation measurements need to be combined with in-situ measurements in order to be able to unravel the total subsidence signal.

- There are two types of damage as a result of subsidence: increased flood risk (with non-differential subsidence) and damage to structures (with differential subsidence).

- Analogue to climate change policies, a successful policy on subsidence consists of adaptation measures (reducing the damage and vulnerability) and mitigation measures (actively reducing subsidence).

- Delayed response of aquitards and interbed compaction may introduce unwanted additional subsidence after implementing mitigation measures, which is presently unaccounted for.

Acknowledgements. This study is based on research conducted by scholars working in the cities mentioned in this paper. They shared data and insights from which this study greatly benefitted. They are all gratefully acknowledged.

References

Allison, M., Yuill, B., Törnqvist, T., Amelung, F., Dixon, T., Erkens, G., Stuurman, R. J., Milne, G., Steckler, M., Syvitski, J., and Teatini, P.: Coastal subsidence: global risks and research priorities, EOS Transactions, 2014.

Aobpaet, A., Caro Cuenca, M., Hooper, A., and Trisirisatayawong, I.: InSAR time series analysis of land subsidence in Bangkok, Thailand, Int. J. Remote Sens., 34, 8, 2013.

Bakr, M.: Land subsidence in North Jakarta – preliminary analysis results, part of Jakarta Coastal Defencse Strategy (JCDS) study, JCDS Atlas, 2011.

Bakr, M.: Influence of Groundwater Management on Land Subsidence in Deltas: A Case Study of Jakarta (Indonesia), Water Resour. Manage., 29, 1541–1555, doi:10.1007/s11269-014-0893-7, 2015.

Church, J. A. and White, N. J.: Sea-Level Rise from the Late 19th to the Early 21st Century, Surv. Geophys., 32, 585–602, 2011.

Dam, R.: Jakarta Coastal Defencse Strategy (JCDS) study, Activity Report: Land subsidence and adaptation/mitigation strategies, JCDS Bridging Phase, 2012.

Den Haan, E. J.: Vertical compression of soils, PhD thesis, Technical University of Delft, Delft, the Nehterlands, 1994.

Dixon, T. H., Amelung, F., Ferretti, A., Novali, F., Rocca, F., Dokka, R., Sella, G., and Kim, S.-W.: Subsidence and flooding in New Orleans – A subsidence map of the city offers insight into the failure of the levees during Hurricane Katrina, Nature, 441, 587–588, 2006.

Dokka, R. K.: New Orleans The role of deep processes in late 20th century subsidence of New Orleans and coastal areas of southern Louisiana and Mississippi, J. Geophys. Res., 116, 2011.

Eco, R. C., Lagmay, A. A., and Bato, M. P.: Investigating ground deformation and subsidence in northern Metro Manila, Philippines using Persistent Scatterer Interferometric Synthetic Aperture Radar (PSInSAR), American Geophysical Union, Fall Meeting, San Francisco, CA, USA, 5–9 December 2011, G23A-0822, 2011.

Fokker, P. A., Muntendam-Bos, A.-G., and Kroon, I. C.: Inverse modelling of surface subsidence to better understand the Earth's subsurface, First Break, 25, 8, 2007.

Ho Chi Minh City Flood and Inundation Management: Final Report, Volume 2: IFRM Strategy, Annex 3: Land Subsidence, Royal Haskoning-DHV and Deltares, 2013.

Hoogvliet, M., van de Ven, F., Buma, J., van Oostrom, N., Brolsma, R., Filatova, T., Verheijen, J., and Bosch, P.: Schades door watertekorten en – overschotten in stedelijk gebied – Quick scan van beschikbaarheid schadegetallen en mogelijkheden om schades te bepalen, Deltares report 1205463-000, 128 pp., 2012 (in Dutch).

Kaneko, S. and Toyota, T.: Long-Term Urbanization and Land Subsidence in Asian Megacities: An Indicators System Approach, in: Groundwater and Subsurface Environments: Human Impacts in Asian Coastal Cities, 2011.

Lambert, J. W. M., van Meerten, J. J., Woning, M. P., and Eijbersen, M. J.: Verbeterde onderhoud strategie infrastructuur in slappe bodemgebieden, Deltares report 1209950-000, 31 pp., 2014 (in Dutch).

Leake, S. A. and Galloway, D. L.: MODFLOW ground-water model, User guide to the Subsidence and Aquifer-System Compaction Package (SUB-WT) for water-table aquifers: US Geological Survey, Techniques and Methods 6–A23, 42 p., 2007.

MoNRE-DGR (Ministry of Natural Resources and Environment, Department of Groundwater Resources): The study of systematic land subsidence monitoring on critical groundwater used area project, Project report, 2012.

Shi, X., Wua, J., Yea, S., Zhangb, Y., Xuea, Y., Weic, Z., Lic, Q., and Yud., J.: Regional land subsidence simulation in Su-Xi-Chang area and Shanghai City, China, Engineering Geology, 100, 27–42, 2008.

Slangen, A. B. A.: Towards regional projections of twenty-first century sea-level change based on IPCC SRES scenarios, Clim. Dynam., 38, 5–6, 2012.

Stuurman, R. J. and Erkens, G.: New Orleans' soft soils need water and solid management, J. Hydrogeol., in prepraration, 2015.

Törnqvist, T. E., Wallace, D. J., Storms, J. E. A., Wallinga, J., van Dam, R. L., Blaauw, M., Derksen, M. S., Klerks, C. J. W., Meijneken, C., and Snijders, E. M. A.: Mississippi Delta subsidence primarily caused by compaction of Holocene strata, Nat. Geosci., 1, 173–176, 2008.

van der Ven, G. P.: Man-made lowlands, history of water management and land reclamation in the Netherlands, Uitgeverij Matrijs, Utrecht, the Netherlands, 1993.

van Trung, L. and Minh Dinh, H. T.: Monitoring Land Deformation Using Permanent Scatterer INSAR Techniques (case study: Ho Chi Minh City), 7th FIG Regional Conference, Vietnam, 2009.

Wang, G., Yu, J., Kearns, T. J., and Ortega, J.: Assessing the Accuracy of Long-Term Subsidence Derived from Borehole Extensometer Data Using GPS Observations: Case Study in Houston, Texas, J. Surv. Eng., 140, 2014.

Yu, S.-Y., Törnqvist, T. E., and Hu, P.: Quantifying Holocene lithospheric subsidence rates underneath the Mississippi Delta, Earth Planet. Sci. Lett., 331–332, 21–30, 2012.

27

Subsidence at the "Trébol" of Quito, Ecuador: an indicator for future disasters?

T. Toulkeridis[1], **D. Simón Baile**[1], **F. Rodríguez**[1], **R. Salazar Martínez**[1], **N. Arias Jiménez**[2], **and D. Carreon Freyre**[3]

[1]Universidad de las Fuerzas Armadas ESPE, Sangolquí, Ecuador
[2]Empresa Metropolitana de Alcantarrillado y Agua Potable de Quito, Quito, Ecuador
[3]Universidad Nacional Autónoma de México, Queretaro, Mexico

Correspondence to: T. Toulkeridis (ttoulkeridis@espe.edu.ec)

Abstract. A sinkhole of great proportions was produced in one of the most trafficked zones of Quito. Constructed in the late sixties, this area is of high importance in solving the traffic jams of the capital city. The sinkhole called "El Trebol" started to be generated in the form of a crater, reached finally dimensions of approximately 120 m in diameter and some 40 m of depth, where at its base the river Machangara appeared. The generation of this sinkhole paralyzed the traffic of the south-central part of the city for the following weeks and therefore the state of emergency was declared. Soon the cause of the sinkhole was encountered being the result of the lack of monitoring of the older subterranean sewer system where for a length of some 20 m the concrete tunnel that canalized the flow of the river collapsed generating the disaster. The collapse of this tunnel resulted from the presence of a high amount of trash floating through the tunnel and scratching its top part until the concrete was worn away leaving behind the sinkhole and the fear of recurrence in populated areas. The financial aspects of direct and indirect damage are emphasized.

1 Introduction to the event and its causes

On Monday, 31 March 2008, around 14:00 LT, a giant subsidence in form of a sinkhole has been generated in a site called "El Trebol", where 25 000 m³ of earthy material disappeared. This area is the most important southern inter-connector of the city to the Valley "De los Chillos" in the eastern end of Quito, but serves as well to connect the southern to the northern side of the city, where an extremely high amount of vehicles transit daily. This site was constructed four decades before its collapse. The sinkhole started in form of a crater with a diameter of approximately 30 m, amplifying its size constantly due to the instability of the slopes, which being wet and saturated with water of the high precipitations in that area of those days prior to the event. Further presence of more subterranean water has been noticed in the southern slope of the sinkhole, when the diameter reached some 120 m in diameter and a depth of 40 m, determined from the top part down to the previously covered river. That river was detected after the visible collapse of the concrete channel-

ing which leads the river Machangara for a distance of some 20 m. Fortunately, no victims where reported. The result of this subsidence has been a traffic collapse and the declaration of the state of emergency of the Municipality of Quito. This event forced the authorities of Quito first to find out the causes of this subsidence and afterwards to search for further areas with similar problems and vulnerabilities in order to avoid future disasters where potentially people could be involved.

Three hypothesis were taken into consideration as the most potential cause for this disaster: (a) extreme discharge, (b) erosional process and chemical activity through time and finally (c) activity o subterranean waters, of which it resulted to be a combination of all three causes in different proportions.

Table 1. INAMHI's Meteorological Automatic Stations closest to El Trébol area.

Code	Station name	Latitude	Longitude	Altitude (m)
M003	Izobamba	−0.366089	−78.555061	3085.00
M024	Iñaquito	−0.175000	−78.485278	2789.12
M002	La Tola	−0.231667	−78.370333	2503.00

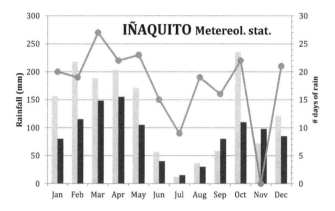

Figure 1. Total monthly rainfall, average multiannual rainfall (of the available data) number of days with rain per month (higher than 0.1 mm) of the closest meteorological station to El Trebol being Iñaquito. From National Institute of Meteorology and Hydrology (2010).

2 Hydrological characteristics

The hydrographic network of the South Valley of Quito is classified as dendritic, having the Machángara as its main river course. The Machángara rises in the steep foothills of volcano Atacazo and crosses the valley from south-west to north-east running parallel to the basin until it reaches the Trébol area (Panecillo) where it takes a turn to the East, gets deeper and flows to the valleys of Cumbayá and Tumbaco. Another important drainage is the so-called "Quebrada Grande" that has its source in the northwest foothills of volcano Atacazo and for a strecht it runs parallel to the Machángara river until it becomes its tributary. When it exits to the valley, the same river caudal varies between 3 m^3 s^{-1} during dry season and 170 m^3 s^{-1} during the wet season. Furthermore, the Machángara river is the main sewage receiver of Quito.

Regarding rainfall, the distribution along the year shows the prevalence of two periods with abundant rainfall being these, February–May and October–November respectively, (EMAAP-Q, 2006; Table 1, Fig. 1). On the other hand, as for the multiannual distribution of rainfall it is important to highlight that the wet season between January and March 2008 was considered the strongest of the last 20 years with accumulated rainfall during the first 3 months of 2008 being much higher that the recorded during any other heavy rainy seasons

of 1989, 1993 and 2000 (Salazar et al., 2009). In particular, a peak of 21.2 L m^{-2} was registered on 31 March 2008 during a heavy rainfall that lasted 4 h from 1 to 5 p.m. (INAMHI, 2008; El Comercio, 2008; Salazar et al., 2009). Furthermore, within the whole year 2008 and coinciding in the three meteorological stations closest to "El Trébol", March recorded continuous raining since it was the month with the highest number of days of rain, with 27 and 29 out of 31 days in Iñaquito and Izobamba respectively (Fig. 1.)

3 Historical background of the area

The site where El Trébol was constructed is about 300–600 m. downstream of the confluence of Jerusalem and El Tejar streams, with Machángara River. Today, the channeled stream of Jerusalem is the Boulevard 24 de Mayo Avenue and its extension to the old bus station, current Qumandá Park. The channeled stream of El Tejar, after passing below the historic center, merges and flows with the stream of Manosalvas down in the low neighborhood of San Juan, in the area that is the center of mass transit transfer, now known as La Marin, before reaching Machángara River.

As a result of uncontrolled growth of the city, this natural drainage system was conducted on sewer channels below the construction of roads for vehicular traffic, and historically have been recorded 12 sinkholes (Fig. 2a) of different magnitudes, defined as "declines or collapses of roadway in the filler material streams, caused by faulty sewers". The responsibility gets under the fill type initially performed with debris and garbage, as well as the age, type of construction and degree of deterioration of the physical work of the channels, sewers and collectors (Pewter, 1986; Fig. 2b). The location of these sinkholes is defined about the old bed of the streams filled so we can assume that on its way continue to occur, especially in periods of daily rainfall above the historical average. In the area of EL Trébol, the first serious collapse was registered on 31 March 2008 and took ten years after the river and fill channels were made for the construction of the infrastructure to solve the problems of the traffic jams. There was a second sink lower rates on 10 January 2014 (Fig. 2c).

4 Direct and indirect financial damage

The day after sinkhole, the Ecuadorean Government created a line of credit of USD 60 million to help the city and start the reconstruction (La Hora, 2008). At the same day, the Quito's Major and the City's Council created an emergency fund of USD 200 000. This emergency fund was increased to about USD 1 million.

The day after drainage vault collapsed, rebuilding process started with a team of 5 hydraulic excavators, 5 power shovels, 10 roll-off trucks, 4 system equipment of mobile industrial lighting. In the rebuilding process 210 workers teams were at the site 24/7 (Explored, 2009). The reconstruction of

a)   b)

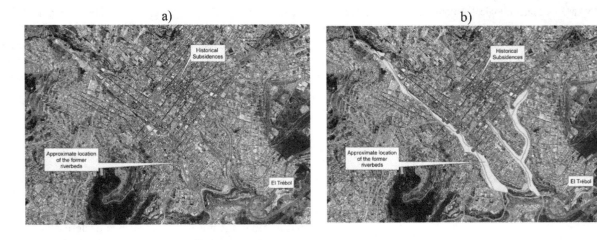

c)

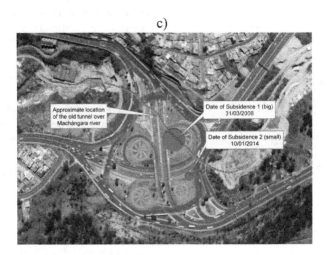

Figure 2. (a–c) Historical subsidences at Machángara river, El Trébol sector.

the cloverleaf interchange took 8 months to full traffic recovery, yet the entire reconstruction of the site took 22 months (Explored, 2010). El Trebol total reconstruction cost reached over USD 13 million, yet it did not required to use the full line of credit; details of expenditures are presented in Table 2.

However, this cost did not reflect the real cost because it does not take in account the externalities implicit with the sinkhole. Externalities associated to how users were affected by the sinkhole and cloverleaf interchange reconstruction. For instance, local authorities closed schools during the first week after the event, cost of students losing classes are not included neither teachers income lost, who were in per hour contracts.

We analyzed those costs of users who were affected by and were not compensated for. Because of lack of official information, we concentrated our efforts to estimate the cost of losing time during the reconstruction of the cloverleaf interchange and drainage vault, as well as the additional cost in gasoline of users of this crucial cloverleaf interchange.

Table 2. List of the costs of the reconstruction in USD (Source: MDMQ, 2015).

New tunnel	7 575 872.21
Rental of equipment	352 622.93
Land movement	151 526.15
Labor	801 546.35
Construction materials	435 495.03
Services	219 061.16
Other costs	31 236.78
EMOP	2 000 000.00
Vida para Quito	2 000 000.00
Total	13 567 360.61

In case of additional cost in gasoline, there were 80 000 of vehicles circulate and use El Trebol every day, in addition of 400 inter-parish buses of public transportation (La Hora, 2008). In order to estimate the value lost by users, we concentrated in private transportation under the assumption that a car-owner who uses his car to go to his job and back home

Table 3. Vehicle type and different gasoline prices in USD.

Vehicle	Gasoline price	
type	USD gallon^{-1}	Total cost*
17 456	2.00	11 210 941.44
62 544	1.48	40 168 258.56
Total gasoline		51 379 200.00

* Adding USD 0.07

Table 4. Cost category expressed in delay of 25 and 35 min.

Cost category	25 min	35 min
CPVTOTAL	34 185 200.00	47 859 280.00
CPSTOTAL	15 668 216.67	21 935 503.33
CPAvVITOTAL	7 121 916.67	9 970 683.33
CPúTOTAL	11 395 066,67	15 953 093.33
Total cost	68 370 400.00	95 718 560.00

fills his car gas tank once a week. This assumption seems reasonable for most of car owners. Yet, we did not include public transportation because we do not know how many times a transportation unit fills bus gas tank in a week. Based on public estimations, we used a value 0.07 additional dollar that an owner has to pay extra to fill his gas tank. It is reasonable also to assume that a car owner spends USD 20 week^{-1} filling his car tank. These USD 0.07 seems a low bound, but still reasonable. Then, we multiply the value of one gallon of gasoline adding these USD 0.07.

Based on AIHE statistics (AIHE, 2015), we know that 21 % of the car owners use a "super premium" gasoline which costs USD 2.00 gallon^{-1} and 78 % uses "extra" gasoline with a price of USD 1.48 gallon^{-1}. Based on the number of vehicles which circulated at that time "El Trebol" every day, 80 000 cars day^{-1}, we can say that approximately 17 000 cars used "super premium" gasoline and over 62 000 used "extra" gasoline type. We estimated that additional cost in gasoline for private car owners was USD 85 million for those 8 months of traffic problems at "El Trebol" (Table 3).

We did not include capital depreciation, even now, it is public knowledge that keep a car running while waiting depreciate its value faster than normal conditions. We did not include both public transportation and car owners who own a diesel engine car.

Concerning the cost of time lost (an opportunity cost), we estimate its value from per hour salary multiply for the additional time that users had to spend during the reconstruction process. We considered that a time between 25 up to 35 min is reasonable to believe users lost during their travel to workplaces or going back their homes. A user lost of USD 2.2–3.12 h^{-1} seems to be reasonable. This value is multiplied by the total time of site reconstruction. Since user came from

Table 5. Cost category.

Cost category	Amount (USD)	Ratio
Reconstruction cost	13 567 360.61	
Additional gasoline cost	51 379 200.00	0.264063290
Opportunity cost 25 min	68 370 400.00	0.198439100
Opportunity cost 35 min	95 718 560.00	0.141742214

different directions, we estimate the lost value separately. According to media reports, 48 000 came private own vehicles from "Los Chillos" Valley (CPVTOTAL), 22 000 came from southern part of Quito (CPSTOTAL), 10 000 were coming toward the valley or southern part of the city (CPAvVITOTAL) and 400 units of public transportation (CPúTOTAL). Regarding public transportation, we assume that each unit was carrying 40 passengers each trip, which it is a low bound because rush hour, these units can be a full capacity (around 72 passengers).

Finally we estimate users' opportunity cost multiplying per hour lost times the time of "El Trebol" reconstruction, which was total of 8 months. The users' opportunity cost for each category is presented in Table 4.

Users lost a considerable amount of time when reconstruction took place, adding all users (the aggregate value) it turned out that the real cost of the sinkhole increases significantly. As the table shows, under the assumption that users lost only 25 min, the opportunity cost reaches over USD 68 million during the 8 months of reconstruction, and under the assumption that user lost up to 35 min, the opportunity cost reaches over USD 95 million. As a result the real cost (under 25 min assumption) reached more than USD 133 million, and under 35 min assumption reached more than USD 160 million as real cost (Table 5).

5 Conclusions and recommendations

As result of this event, we consider as a priority to take corresponding actions to prevent future collapses. Taking in consideration the alignment of the actual and past sinkholes, these alignments need to be reinforced in order to avoid future disasters in that area. As demonstrated in our study, the real costs of damages are much higher in the indirect damage of such sinkhole events rather in the reconstruction of the disaster site itself. Unfortunately, the enforcement of the potential subsidence areas did not take place yet, as demonstrated by a new sinkhole in 2014 in a zone where the vulnerability has been previously emphasized.

References

AIHE: Estadísticas, Asociación de la Industria Hidrocarburífera del Ecuador, Quito, Asociación de la Industria hidrocarburífera del Ecuador (AIHE), Quito, Ecuador, available at: http://www.aihe.org.ec/index.php?option=com_content&view= article&id=122:estadisticas&catid=67&Itemid=142, 2015.

Cowen, T. (Ed.): Public Goods and Market Failures, Transaction Publishers, New Brunswick, NJ, USA, 1992.

Explored: Obra vial desaparece en la capital después de 40 años, Explored Noticias Ecuador publicado en 1 enero 2009, http://www.explored.com.ec/noticias-ecuador/nuevo-colector-ya-funciona-en-el-trebol-437565.html, 2009.

National Institute of Meteorology and Hydrology (INAHMI): Meteorological yearbook, Nr. 48, Quito, Ecuador, 2010.

La Hora: El Trebol se cayó en Quito, Diario La Hora Nacional, sección País, Domingo 6 de abril de 2008, available at: http://www.lahora.com.ec/index.php/noticias/show/703827/1/El_Tr/%C3/%A9bol_se_cay/%C3/%B3_en_Quito.html#.VT4MtJ_NHw, 2008.

Municipio del Distrito Metropolitano de Quito (MDMQ): Balance de los Estudios Urbanos (1985–2005) de la Coperación IRD-Municpio de Quito, in: Movilidad, elementos esenciales y riesgos en el Distrito Metropolitano de Quito, edited by: Demoraes, F., Co-edición MDMQ-IFEA-IRD, Quito, 2005.

O'Rourke, T. D. and Crespo, E.: Geotechnical properties of cemented volcanic soil, J. Geotech. Eng., 114, 1126–1147, 1988.

Peltre, P.: Quebradas y riesgos naturales en Quito, in: Riesgos Naturales en Quito, edited by: Pierre, P., Corporación Editora Nacional, Quito, Ecuador, 45–89, 1989.

28

Physical experiments of land subsidence within a maar crater: insights for porosity variations and fracture localization

M. Cerca[1], L. Rocha[2], D. Carreón-Freyre[1], and J. Aranda[1]

[1]Centro de Geociencias, Universidad Nacional Autónoma de México, Campus Juriquilla, Querétaro, Qro., 76230, México
[2]Posgrado en Ciencias de la Tierra, Centro de Geociencias, Universidad Nacional Autónoma de México, Campus Juriquilla, Querétaro, Qro., 76230, México

Correspondence to: M. Cerca (mcerca@geociencias.unam.mx)

Abstract. We present the results of a series of physical models aiming to reproduce rapid subsidence (at least 25 m in 30 years) observed in the sediments of a maar crater caused by extraction of groundwater in the interconnected adjacent aquifer. The model considered plausible variations in the geometry of the crater basement and the measured rate of groundwater extraction (1 m per year in the time interval from 2005 to 2011) in 15 wells located around the structure. The experiments were built within a rigid plastic bowl in which the sediments and rocks of the maar sequence were modeled using different materials: (a) plasticine for the rigid country rock, (b) gravel for the fractured country rock forming the diatreme fill and, (c) water saturated hollow glass microbeads for the lacustrine sedimentary fill of the crater. Water table was maintained initially at the surface of the sediments and then was allowed to flow through a hole made at the base of the rigid bowl. Water extraction provoked a sequence of gentle deformation, fracturing, and faulting of the surface in all the experiments. Vertical as well as lateral displacements were observed in the surface of the experiments. We discuss the results of 2 representative models. The model results reproduced the main geometry of the ring faults affecting the crater sediments and helps to explain the diversity of structures observed in relation with the diatreme geometry. The surface of the models was monitored continuously with an optical interferometric technique called structured light projection. Images collected at nearly constant time intervals were analyzed using the ZEBRA software and the obtained interferometric pairs permitted to analyze the full field subsidence in the model (submilimetric vertical displacements). The experiments were conducted at a continuous flow rate extraction and show a also a linear subsidence rate. Comparison among the results of the physical models and the fault system associated to subsidence in the maar show that fault geometry in the sedimentary sequence imitates closely the geometry of the volcanic basement.

1 Introduction

Rincón de Parangueo is a Quaternary volcanic maar, a crater formed by a phreatomagmatic eruption, located in the northern portion of the Michoacán-Guanajuato volcanic field in the central sector of the Trans Mexican Volcanic Belt (Fig. 1, location of the maar: 20°25.839′ N, −101°14.917′ W). Volcanic maars are craters located above a feeder magmatic dike that poses into contact magma with groundwater causing a gaseous explosion and resulting in a rock filled fracture with a conic shape, called diatrema (Lorenz, 1986). Although the literature on maars is extensive the studies of post-eruption subsidence in the lacustrine sequences filling the craters are rare (e.g., Suhr et al., 2006; Pirrung et al., 2008). Rincón de Parangueo was one of four maar craters nearby the locality of Valle de Santiago village that used to have a perennial lake and nowadays is partially occupied by a brine lake. During the 1980's progressive desiccation began as a consequence of water overdraft from the regional aquifer that used to feed the lake (Escolero-Fuentes and Alcocer-Duran, 2004;

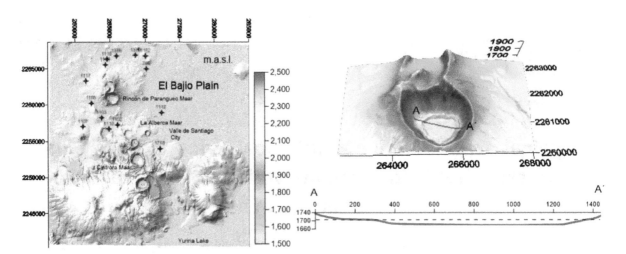

Figure 1. Digital elevation model obtained by LIDAR data from INEGI (Institute of Informatics and Geography; Mexico) showing the volcanic cluster in the area of Rincón de Parangueo maar. Perspective view of the crater and the annular fault system affecting the bottom. Topographic profile crossing the crater showing the initial lake surface (dashed line) and the present day leveling (red line).

Aranda-Gómez et al., 2013). Detailed geologic mapping in the sediments of the lake bed shows clear evidence of rapid active subsidence contemporaneous to desiccation (Aranda-Gómez et al., 2010, 2013; Aranda-Gómez and Carrasco-Nuñez, 2014; Cerca et al., 2014). Deformation is characterized by a main ring-shaped normal fault accompanied by gravitational sliding and gliding of mud blocks, and folding at the foot of the slides. Sinking of the maar crater is associated with the formation of vertical fractures and lateral mass movements to the depocenter, resulting in the formation of tensile fractures. Likewise, normal faulting exerts a major influence in the location of today depocenter.

Sinking at the bottom of the crater produced a ring-fault system which is exposed near the former coast of the lake, on a pair of parallel 15 m high topographic scarps. Attitudes of laminations in the mud exposed in the scarp are consistent with down sagging of the sediment caused by sinking of the lake's depocenter. Tilt angles of the laminations vary between 10 and 35° towards the basin. Normal faults exposed at the topographic scarp produced in some areas deformation domains that are consistent with the presence of listric detachment structures, which produced rollover anticlines with axes located near the former coast of the lake. Lake sediments are composed by a sequence presenting rhythmic layering with changes of colors between beige and dark brown; beige layers are composed carbonate micritic mud with crystals of 2–5 μm (Kienel et al., 2009). The dark brown layers are composed by clay, silt, and organic material. Also interbedded can be observed more prominent layers with black coloration constituted mainly silt and fine sand. In cores, the lower part of the observed sequence is composed of deformed layers affected by fractures with a small displacements. Mechanically, the whole sequence is characterized by

low compressible silty sediments, and the gravimetric water content increases gradually with depth up to more than 40 %.

Water withdrawal in the aquifer system adjacent to the volcanic field has been pointed out as the cause for deformation and desiccation of the lake since the 1980's decade (Escolero-Fuentes and Alcocer-Duran, 2004; Aranda-Gómez et al., 2013; and references therein). Fragmentation of the country rock due to diatreme formation is likely to be contained within the maar–diatreme and less likely to propagate down to deeper levels into the feeder dike. Nevertheless, alignment and overlap of the numerous craters at the study area indicates the presence in the subsurface of multiple parallel and orthogonal feeding dykes. Coalescence of the shallow structures suggested by vent distribution might be the cause of a high hydraulic connectivity below the volcanic edifices with the regional aquifer. Faults and fractures within the country rock enhance the hydraulic conductivity and might serve as drainage lines into the regional aquifer system. Thus, the maar may be divided into two significant hydrogeological units, a deforming lacustrine sedimentary unit, acting as an aquitard and, the high fractured diatreme hydraulically connected with the regional aquifer system. Nowadays groundwater extraction and evaporation exceed the inflow of groundwater into the crater. Therefore, the bottom of the crater will either be dry except for runoff from precipitation or develop into a brine lake where accumulation of salts will take place in the depocenter. The drawdown of the groundwater table will continue indefinitely because the region depends on groundwater for human consumption.

The present day elevation at the bottom of the crater (1868 m a.s.l.) is lower than the adjacent Bajio plain (1710–1730 m a.s.l). The variations in the piezometric levels were recorded by the National Water Commission (CNA) in selected wells around the Parangueo maar for the time period

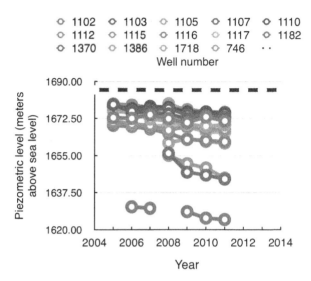

Figure 2. Piezometric level evolution for the time interval between 2005 and 2011 for selected extraction wells measured around the Rincón de Parangueo maar. Location of the wells is indicated in Fig. 1.

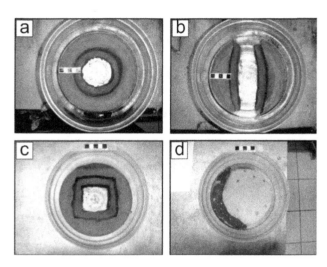

Figure 3. Photographs of the experimental setup showing the plasticine molded to produce variations in the geometry of the rigid country rock. The bowl was then filled to the level of the plasticine with gravel and covered with hollow glass microbeads to obtain a plain surface. (a) Ring faults circular, (b) parallel faults simulating a magmatic dike shape, (c) ring faults square, (d) asymmetric faults set up.

between 2005 and 2011. The observations were made twice in a year at both dry and raining seasons, we present here the measurements made during the dry season (Fig. 2). All piezometers record a similar depleting evolution of 1 m per year, but the cluster of wells 1370, 1182 and 746, located northeast of the Rincón de Parangueo maar record a deeper piezometric level. In summary, the piezometric data suggest an significant hydraulic gradient towards the northeast and the data can be interpolated to a lineal decrease in the water table for the last 30 years. In this work we aim at reproducing the subsidence observed in the Rincón de Parangueo maar by constructing a series of scaled physical models.

2 Experimental subsidence bowl

Experiments were conducted using a plastic bowl simulating the overall half-sphere shape of the maar crater. Plasticine was used to mold and vary systematically the different geometries of volcanic basement (rigid country rock) assumed to be below the sedimentary sequence. Variations in the geometry included ring circular faults, ring square faults, parallel faults, and asymmetric basement (Fig. 3). The internal part of the bowl was filled with gravel to simulate the highly fractured country rock in the volcanic conduct. Posteriorly, the bowl was filled up to an horizontal level with a fine grain size and well sorted, slightly cohesive granular media (hollow glass microbeads) completely saturated in water, simulating the initial conditions of the lake maar in the 1980's when the phreatic level reached the surface. In nature, porosity changes during subsidence are related to effective stresses. The water head decrease in the maar was of about 30 m in the last 35 years. For scaling we have measured

porosity of samples and assumed the initial porosity. Experiments are a simplification of nature and assume that: (1) the lacustrine material is relatively homogeneous, (2) the voids of the lacustrine sediments are filled with water below the piezometric surface and, (3) changes in the effective stress produced by the drop of piezometric level are the main cause for the subsidence in the crater maar. The water table is also assumed to be parallel to the surface of the lacustrine sediments at the beginning of the process in nature and in the model

$$\Delta\sigma_{\mathrm{ef}}^{*} = \frac{\Delta\sigma_{\mathrm{ef}}\mathrm{model}}{\Delta\sigma_{\mathrm{ef}}\mathrm{nature}}, \tag{1}$$

where σ_{ef} is the change in the effective stress and σ_{ef}^{*} is the relation among the variations in the effective stresses in model and nature, in this case the dimensionless scaling ratio is of the order of 5.33×10^{-4}. If the total stress is constant through time the differences in the effective stresses are related to the pore pressure drop, which in turn is proportional to the change in the groundwater level.

The experimental time for all models was about 20 min with a flow rate maintained constant at $1.57 \times 10^{-7}\,\mathrm{m^3\,s^{-1}}$ that was induced by opening of a 2 mm in diameter tube at the base of the bowl. Subsidence in the physical models was monitored at constant time intervals of 15 s using an optical system that allows obtaining subtle changes (submilimetric) in the experimental relief using the ZEBRA software (based on the work by Barrientos et al., 2008). A structured light (black and white fringes) pattern was projected on the surface of the model and the digital images recorded by a CCD sensor. Comparison between pairs of images (deformed state

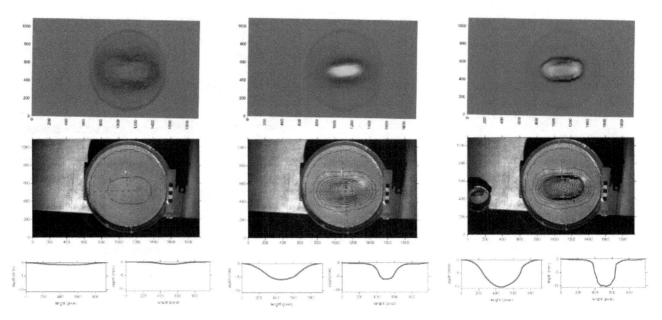

Figure 4. Example of the model results at the start, middle and end of the experiment. The images show the digital elevation model obtained with the ZEBRA software, a photograph of the model with overlapped contours and the topographic profile obtained in two orthogonal directions.

and reference image) were analyzed by the ZEBRA software using optical interferometric methods.

3 Results

All of the experimental runs showed a comparable evolution: (a) formation of an initial gentle depression without fracture formation in the central part of the bowl at the beginning of the water level decrease; (b) nucleation of small and isolated fractures after 5 min of the experiment run; (c) fracture coalescence to form a continuous ring fault with a geometry similar to the basement; (d) propagation of the fractures toward the external and/or internal part of the model. Differences in the model results depended mainly on the geometry of the diatreme.

Here, we discuss the results of two representative experiments. The first experiment discussed was of the type b in Fig. 3, parallel faults representing a diatreme with the shape of a dike. A central elliptical depression formed in the first 5 min, after 6 min of flow the parallel faults formed at the surface of the model. At minute 9 the parallel faults linked to form the ring fault with an elliptical shape. After 13 min the model ended its deformation; nevertheless some images were recorded after 30 min to corroborate that there was no further deformation. Figure 4 represents the evolution of the experiment in the digital elevation model, photographs with contour levels, and two topographic profiles obtained from the DEM's.

Subtle changes in the subsidence rate and geometry can be analyzed in the topographic profiles that were constructed using 25 images processed against a reference image of the

initial experimental conditions. Figure 5 shows differences in the evolution of subsidence in the profile parallel to the basement faults (RED) and the profile orthogonal to the parallel faults (BLUE). Localization of deformation occurs in the BLUE profile at an earlier time (500 s) and differences in compaction are attributable to the presence of the rigid basement and thinner sedimentary deposits. The inset of Fig. 5 (RED) shows the subsidence model obtained for the center of the model that follows a nearly lineal behavior.

The second experiment was designed to explore experimentally the influence of an asymmetric basement and the influence of tilted surfaces that could act as detachment planes during subsidence. The setup considered the structural differences observed in the crater (Aranda-Gómez and Carrasco-Nuñez, 2014; Cerca et al., 2014), mainly the formation of rollover folds along listric fault surfaces and the formation of shortening domes. This deformational features suggest an important lateral spreading of the lacustrine sequences during subsidence. Another experimental difference was that the second sedimentary layer was thicker than in other model runs. The overall evolution of the model was similar to the first experiment, but in this case the lateral sliding of the material formed extensional fractures in a zone of the ring fault and small folds toward the depocenter, implying that lateral displacements are important to accommodate deformation. Figure 6 shows a photographic sequence of the second experimental setup, the ellipse indicates the zone of folding and the nucleation of the fractures; as well as the initial depression formed.

Obtaining an accurate digital elevation model of the experiments allowed to perform further analysis of the data.

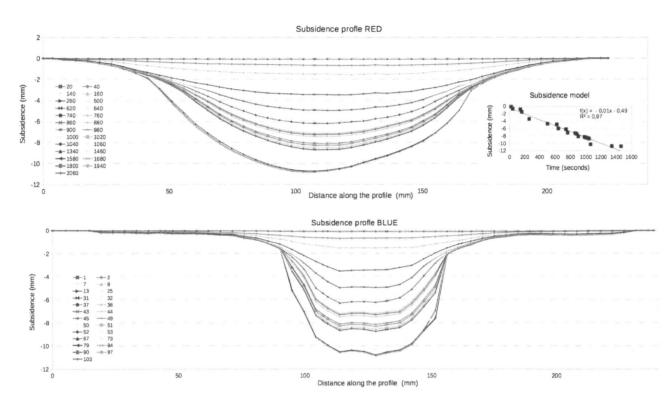

Figure 5. Topographic profiles obtained in the model in the direction: **(a)** parallel to the basement faults (RED) and, **(b)** the profile orthogonal to the parallel faults (BLUE).

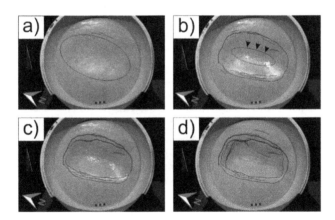

Figure 6. Photographic sequence of the second experiment showing the stages of deformation observed during desiccation of the model crater: **(a)** formation of a depression, **(b)** formation of fractures at the border of the depression and the displacement of material toward the depocenter, **(c)** growing and coalescence of faults, **(d)** desiccation of the model and formation of shortening structures in the depocenter.

For instance, the development of slopes after desiccation of the crater is important for the formation of extensional structures. In Fig. 7 we show the analysis of the subsidence and slope development of the images c and d of Fig. 6. The opening of the fractures and the formation of small folding along the axis of the elliptical depression suggest lateral spreading of the sediments in the late stages of deformation.

The comparison of model results with the deformation observed in the Rincón de Parangueo maar suggest that faults formed above the geometric borders of the rigid country rock. The ring fault pattern of faults in the mar suggest that the diatreme has a non uniform circular geometry. The presence of at least two major fault scarps in nature may indicate that the diatreme has two fault steps in the country rock.

4   Conclusion

Subsidence in the experiments formed patterns directly related to the geometry of the rigid country rock modeled below the sedimentary cover. Thus varying systematically the geometry of the diatreme and analyzing the results give important insights on the evolution of deformational features associated with land subsidence. Fractures and faults formed nearly above the border of the rigid basement, its position depending on the inclination of the border. The subsidence that can be generated in a crater is directly proportional to the thickness of the sedimentary lacustrine sequence. Groundwater extraction in the regional aquifer may trigger the deformation and needs to be regulated for a sustainable management of the crater lake. Finally, scaled physical experiments provided an important tool for understanding of subsidence mechanisms, for testing hypothesis of the causes of defor-

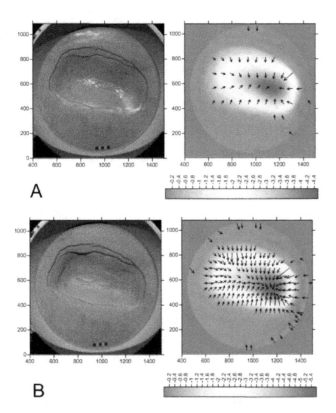

Figure 7. Interferometry analysis of selected photographs from Fig. 6 (c – growing of faults and d – dessication). (**a**) DEM of the image at the moment when the water level reached the bottom of the crater, color scale represent the subsidence in centimeters and the arrows indicate the magnitud and direction of the slope, the inplane scale is in pixels. (**b**) DEM of the image at a later desiccation time and deformation (color scale and arrows as above), note that the development of faulting has an effect in slope.

mation of the terrestrial surface, and for the development of monitoring techniques.

Acknowledgements. Ricardo Carrizosa of the Laboratorio de Mecánica de Geosistemas (LAMG) is thanked for his help with the experimental setup.

References

Aranda-Gómez, J. J. and Carrasco-Nuñez, G.: The Valle de Santiago maars, México: the record of magma-water fluctuations during the formation of a basaltic maar (La Alberca) and active post-desiccation subsidence at the bottom of a maar lake (Rincón de Parangueo): Intra-conference field trip guidebook, 5th International Maar Conference, 17–22 November, Centro de Geociencias, Universidad Nacional Autónoma de México, Querétaro, México, 30 pp., 2014.

Aranda-Gómez, J. J., Levresse, G., Pacheco-Martínez, J., Ramos-Leal, J. A., Carrasco-Núñez, G., Chacón-Baca, E., González-Naranjo, G., Chávez-Cabello, G., Vega-González, M., Origel-Gutiérrez, G., and Noyola-Medrano, C.: Active subsidence at

the bottom of a recently desiccated crater-lake and its environmental impact: Rincón de Parangueo, Guanajuato, México: Field trip guidebook, Eighth International Symposium on Land Subsidence, Universidad Nacional Autónoma de México, Centro de Geociencias, Querétaro, Qro., México, 1–48, 2010.

Aranda-Gómez, J. J., Levresse, G., Pacheco-Martínez, J., Ramos-Leal, J. A., Carrasco-Nuñez, G., Chacón-Baca, E., González-Naranjo, G. A., Chávez-Cabello, G., Vega-González, M., Origel, G., and Noyola-Medrano, C.: Active sinking at the bottom of the Rincón de Parangueo Maar (Guanajuato, México) and its probable relation with subsidence faults at Salamanca and Celaya, Boletín de la Sociedad Geológica Mexicana, 65, 169–188, 2013.

Barrientos, B., Cerca, M., García-Márquez, J., and Hernández-Bernal, C.: Three-dimensional displacement fields measured in a deforming granular-media surface by conbined frige projection and speckle photography, J. Opt. A-Pure Appl. Op., 10, 10 pp., 2008.

Cerca, M., Carreón-Freyre, D., Aranda-Gómez, J. J., and Rocha-Treviño, L.: GPR profiles for characterizing subsidence deformation in lake sediments within a mar crater, 15th International Conference on Ground Penetrating Radar GPR2014, Brussels, Belgium, 30 June–4 July, 4 pp., 2014.

Escolero-Fuentes, O. A. and Alcocer-Durand, J.: Desecación de los lagos cráter del Valle de Santiago, Guanajuato, in: El agua en México vista desde la Academia, edited by: Jiménez, B., Marín, L., Morán, D., Escolero, O., and Alcocer, J., Academia Mexicana de Ciencias, México, D.F., 99–116, 2004.

Kienel, U., Wulf Bowen, S., Byrne, R., Park, J., Böhnel, H., Dulski, P., Luhr, J. F., Siebert, L., Haug, G. H., and Negendank, J. F. W.: First lacustrine varve chronologies from Mexico: impact of droughts, ENSO and human activity since AD 1840 as recorded in maar sediments from Valle de Santiago, J. Paleolimnol., 42, 587–609, doi:10.1007/s10933-009-9307-x, 2009.

Lorenz, V.: On the growth of maars and diatremes and its relevance to the formation of tuff rings, B. Volcanol., 48, 265–274, 1986.

Pirrung, M., Büchel, G., Lorenz, V., and Treutler, H.-C.: Post-eruptive development of teh Ukinrek East Maar since its eruption in 1977 A.D. in the periglacial area of south-west Alaska, Sedimentology, 55, 305–334, 2008.

Suhr, P., Goth, K., Lorenz, V., and Suhr, S.: Long lasting subsidence and deformation in and above maar-diatreme volcanoes – a never ending story, Z. Dtsch. Ges. Geowiss., 157, 491–511, 2006.

29

Regional and local land subsidence at the Venice coastland by TerraSAR-X PSI

L. Tosi[1], T. Strozzi[2], C. Da Lio[1], and P. Teatini[1,3]

[1]Institute of Marine Sciences, National Research Council, Venice, Italy
[2]GAMMA Remote Sensing, Gümligen, Switzerland
[3]Dept. of Civil, Architectural and Environmental Engineering, University of Padova, Italy

Correspondence to: L. Tosi (luigi.tosi@ismar.cnr.it)

Abstract. Land subsidence occurred at the Venice coastland over the 2008–2011 period has been investigated by Persistent Scatterer Interferometry (PSI) using a stack of 90 TerraSAR-X stripmap images with a 3 m resolution and a 11-day revisiting time. The regular X-band SAR acquisitions over more than three years coupled with the very-high image resolution has significantly improved the monitoring of ground displacements at regional and local scales, e.g., the entire lagoon, especially the historical palaces, the MoSE large structures under construction at the lagoon inlets to disconnect the lagoon from the Adriatic Sea during high tides, and single small structures scattered within the lagoon environments. Our results show that subsidence is characterized by a certain variability at the regional scale with superimposed important local displacements. The movements range from a gentle uplift to subsidence rates of up to $35 \, \text{mm} \, \text{yr}^{-1}$. For instance, settlements of 30–$35 \, \text{mm} \, \text{yr}^{-1}$ have been detected at the three lagoon inlets in correspondence of the MoSE works, and local sinking bowls up to $10 \, \text{mm} \, \text{yr}^{-1}$ connected with the construction of new large buildings or restoration works have been measured in the Venice and Chioggia historical centers. Focusing on the city of Venice, the mean subsidence of $1.1 \pm 1.0 \, \text{mm} \, \text{yr}^{-1}$ confirms the general stability of the historical center.

1 Introduction

Cities in lowlying coastlands and deltaic regions are the sites most susceptible to land subsidence worldwide. The influence of land subsidence on these communities and environments can be very significant, increasing vulnerability to saltwater intrusion and flooding, threatening agriculture and ecological systems. Presently, land subsidence caused mainly by extraction of groundwater is threatening Jakarta (Ng et al., 2012), Ho Chi Minh City (Erban et al., 2014), Bangkok (Phien-wej et al., 2006), Shanghai (Dong et al., 2014), i.e. rapidly expanding urban areas which require huge amounts of water for domestic and industrial water supply.

Similarly to that occurring in these urban centres in developing countries, Venice experienced anthropogenic land subsidence due to aquifer overexploitation in the past (Carbognin et al., 2004). Although the effect of groundwater pumping ended a few decades ago, a reliable and detailed knowledge of land subsidence affecting the historical center

is even more important today as vulnerability has continuously increased over the years. The city of Venice and its surrounding lagoon is presently one of the sites most sensitive to land subsidence worldwide. Even a few $\text{mm} \, \text{yr}^{-1}$ loss of elevation with respect to the mean sea level can threaten the city's survival and irrevocably change the natural environment of the lagoon.

This is the reason why monitoring land subsidence in the Venice area at a very high accuracy has always been important. Recent studies have highlighted a significant heterogeneity of the subsidence at the local scale, both connected to the subsoil characteristics and human interventions (Strozzi et al., 2009; Tosi et al., 2012, 2013). Thus, in addition to the millimetric accuracy, a metric spatial resolution is also required for improving the understanding of the natural and human-induced components.

In this study, we use a long-term stack of TerraSAR-X scenes processed by PSI (Persistent Scatterer Interferome-

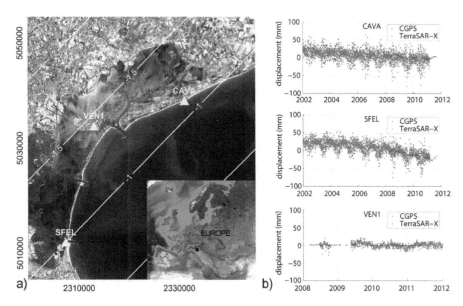

Figure 1. (a) Data calibration: plane model based on the Venice (VEN1), Chioggia (SFEL), and Cavallino (CAVA) CGPS displacements, used to correct the PSI outcome. **(b)** Comparison between the CGPS and the calibrated PS time series.

try) to monitor land subsidence at the Venice coastland from 2008 to 2011 and highlight the local movements of historical palaces and monuments in the Venice center, single small structures scattered within the natural lagoon environments, and the MoSE large structures under construction at the lagoon inlets.

2 Methods

The Interferometric Point Target Analysis (IPTA; Werner et al., 2003; Wegmüller et al., 2004; Teatini et al., 2005), a PSI implementation chain, is applied on a stack of 90 TerraSAR-X stripmap images for the period 2008 to 2011 with a regular 11-day revisiting time. In order to correct the so-called flattening problem, i.e. the slight phase tilt resulting by the inaccuracy in estimation of the orbital baseline due to the imperfect knowledge of the satellite positions (Teatini et al., 2012a, b), IPTA results have been calibrated using three permanent GPS stations (CGPS) well distributed in the monitored area, i.e. VEN1 at Venice, CAVA on the northern Cavallino littoral strip, and SFEL in the nearby of Chioggia at the southern tip of the lagoon (Fig. 1a). Figure 1a shows the tilting plane used to calibrate the PSI solution on the CGPS records. Even if the slope is relatively small, the corrections between 1 to 2 mm yr^{-1} are significant for the Venice coastal area. Figure 1b points out the satisfactory match between the CGPS time series and the average time series of the persistent scatterers (PS) in the surrounding of the GPS stations once calibrated the PSI solution.

IPTA provides land displacements along the line-of-sight (LOS) between the satellite and the targets. Due to the small incidence angle of the radar signal, IPTA measurements are

less sensitive to the horizontal components of the displacement than to the vertical one. Nevertheless, the occurrence of differential horizontal displacements, for example because of the tectonics, would affect the SAR interferometry measurement. The network of the CGPS stations shows however that the Venice coastland is uniformly moving north-eastward (Teatini et al., 2005; Tosi et al., 2010). Therefore, the differential component of the horizontal displacement rate perpendicular to the TerraSAR-X inclination orbit is smaller than 1 mm yr^{-1}, and negligible with respect to the vertical displacement rate. In our results the difference between the LOS and the vertical components of the movement (about 15 % for TerraSAR-X, i.e. less than 1 mm yr^{-1} for the large majority of the PS) is neglected.

3 Results

3.1 Land subsidence at the regional scale

Figure 2a shows the PSI solution in terms of average displacement rates on the whole coastal area of Venice. An impressive number of 1 933 690 PS, i.e. 2275 km^{-2}, has been detected. The movement velocity ranges from a gentle uplift (~2 mm yr^{-1}) to large subsidence (up to 30 mm yr^{-1}).

Similarly to previous results obtained with ENVISAT images (e.g., Teatini et al., 2012a, b), the monitoring outcome reveals that this portion of the Northern Adriatic coastland is characterized by a certain variability in terms of land subsidence at the large (or regional) scale. A number of zones with different features in the displacement rates is distinguished:

– the north-western zone (A circle in Fig. 2a): this area is characterized by uniform stability. The average sub-

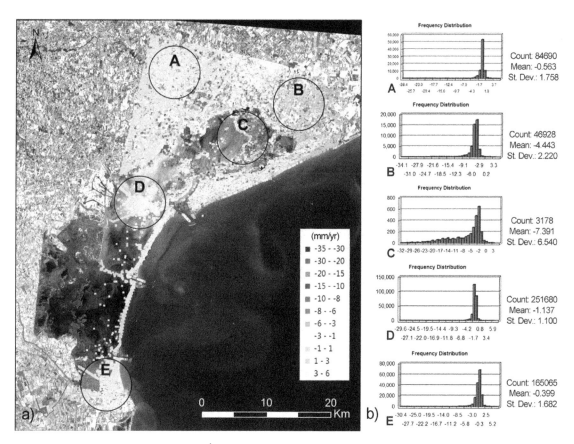

Figure 2. (a) Average land displacements (mm yr^{-1}) for the Venice coastland obtained by PSI on TerraSAR-X images acquired between March 2008 and August 2011. Positive values mean uplift, negative values land subsidence. A Landsat image obtained from the US Geological Survey – Earth Resources Observation and Science (EROS) Center is used as base map. (b) Frequency distribution of the displacement rates for the PS located in the A-E circular areas ($r = 3$ km) highlighted on the map.

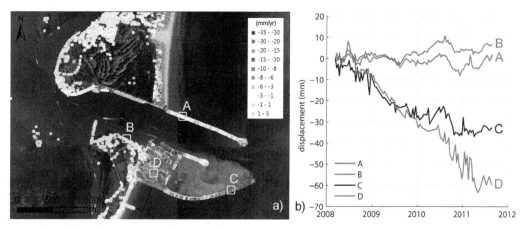

Figure 3. (a) Mean displacement rates between March 2008 and August 2011 from TerraSAR-X interferometry at Malamocco inlet. The image background is an aerophotograph acquired in 2007. Negative values indicate settlement, positive mean uplift. (b) Displacement history in the satellite line-of-sight direction for four scatterers (A, B, C, D). See location 1a in Fig. 1.

sidence η_r is less than 1 mm yr^{-1} and the standard deviation σ_η is about 1–2 mm yr^{-1} (Fig. 2b). Only local bowls of relatively large displacements caused by hu-

man activities (groundwater pumping, development of new urban and industrial zones) are detected;

– the north-eastern zone (B circle in Fig. 2a): a relatively uniform ($\sigma_\eta \approx 2$ mm yr^{-1}) land subsidence on the order

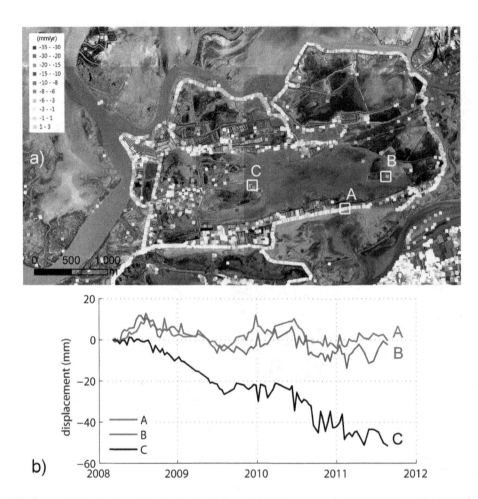

Figure 4. (a) Mean displacement rates between March 2008 and August 2011 from TerraSAR-X interferometry at Valle Paleazza, Northern Venice lagoon. The image background is an aerophotograph acquired in 2007. Negative values indicate settlement, positive mean uplift. **(b)** Displacement history in the satellite line-of-sight direction for three scatterers (A, B, C). See location 2a in Fig. 1.

of 4–6 mm yr^{-1} is threatening this part of the coastal area. Land subsidence due to aquifer exploitation occurs mainly along coastal sector; also processes related to the recent land reclamation (the zone was reclaimed between 1900 and 1930) likely contribute to the total movements;

– the northern lagoon (C circle in Fig. 2a): it is the zone characterized by the highest variability ($\sigma_\eta \approx 6$ mm yr^{-1}) with land displacements ranging between −1 and −30 mm yr^{-1}. The largely heterogeneous nature of the shallow Holocene deposits strongly impacts the movements;

– Venice and the central lagoon (D circle in Fig. 2a): the historical centre and the surrounding lagoon is relatively stable. On the average $\eta_r = 1$ mm yr^{-1}, with a very narrow v ariability of $\sigma_\eta \approx 1$ mm yr^{-1}. The Holocene deposits are relatively thin (about 5 m) and groundwater withdrawals have been precluded since the early 1970s;

– Chioggia and the southern coastland (E circle in Fig. 2a): moving to the south, land subsidence remains generally small. The city of Chioggia is stable ($\sigma_\eta < 1$ mm yr^{-1}), and a certain subsidence occurs mainly along the coastline in the soundings of the mouths of the main rivers (the Brenta and Adige rivers).

3.2 Ground displacements at the local scale

A significant local scale variability of the land displacements is superposed on the regional trend of land subsidence. In the following, a few cases showing the typical local movements observed along the Venice coast are presented and discussed.

3.2.1 The MoSE constructions at the lagoon inlets

Starting from the middle 1990s, the jetties at the three inlets were reinforced, and since 2005, in the framework of the MoSE works (i.e. the project of mobile barriers for the temporarily closure of the lagoon to the sea), they have been strongly reshaped and supplemented by offshore breakwa-

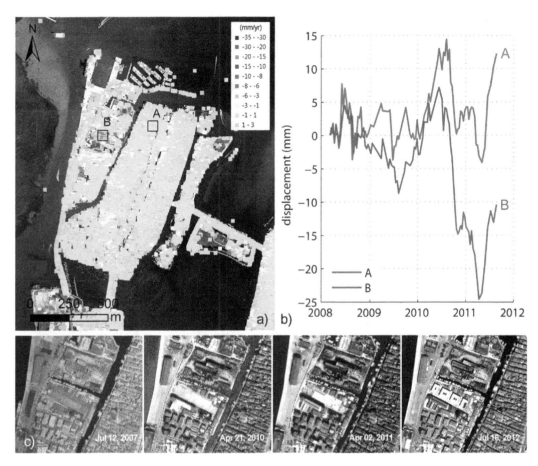

Figure 5. (a) Mean displacement rates between March 2008 and August 2011 from TerraSAR-X interferometry at Chioggia. The image background is an aerophotograph acquired in 2007. Negative values indicate settlement, positive mean uplift. (b) Displacement history in the satellite line-of-sight direction for two scatterers (A, B). (c) Satellite Google Earth images showing the building construction carried out at location B from 2010, cause of the large consolidation highlighted by PSI measurements in the zone from mid-2010 (Google Earth, data source: European Space Imaging, Image NASA; Digital Globe, Image). See location 3a in Fig. 1.

ters. Presently the works at the inlets are still ongoing and monitoring land displacements in these sectors, has a double aim, namely (i) the evaluation of possible effects of the works on the littoral environment, and (ii) the quantification of the consolidation of the new coastal structures. The former issue can support the public authorities in monitoring the environmental impacts of these giant works, the latter can help the engineering companies building the MoSE to check the absolute and differential displacements that can threaten the integrity and efficiency of the structures.

Figure 3a shows a large variability of the displacement velocities characterizing the different degree of the consolidation of the structures at the Malamocco inlet. Sinking rates less than 3 mm yr^{-1} characterize the parts of the jetties where older restoration works were carried out (A, B), while sinking rates up to 30 mm yr^{-1} are measured in the newer structures mainly built in the southern side (C, D). Figure 3b shows the displacement history of four scatterers selected in correspondence of old restoration works and new structures. The scatterers A and B highlight the ground stability after

the old restoration in these parts of the inlets. The scatterer C, located in the new structure of the outer jetty, sinks about 20 mm yr^{-1} in the whole monitored period, the lower values detected after 2010 show that the primary consolidation has likely been completed. Instead scatterer D, in the newest portion of the lock, is still experiencing primary consolidation due to the new load, with velocity rates of up to 30 mm yr^{-1}. Notice that the works at the inlet generally do not affect the land subsidence in the littoral surroundings.

3.2.2 The natural marshlands and the fish farms

Monitoring of land subsidence in natural marshlands and fish farms is a major challenge. The use of traditional methodologies (levelling, GPS) are impractical in these environment. SAR- based methodologies represent a unique approach in this context. Specifically, the X-band acquisitions, which are characterized by a higher resolution with respect to the C-band scenes, have demonstrated a high effectiveness. Figure 4 shows the impressive capability of IPTA to detect

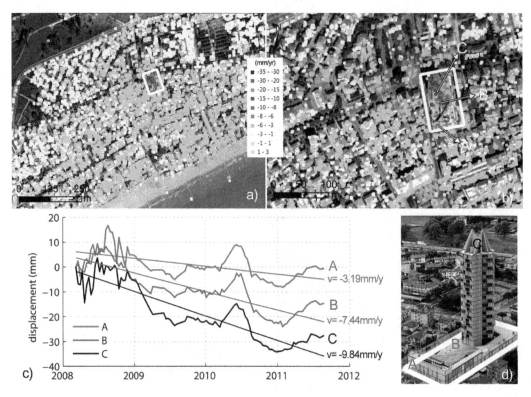

Figure 6. (a, b) Mean displacement rates between March 2008 and August 2011 from TerraSAR-X interferometry at Jesolo. The image background is an aerophotograph acquired in 2007. Negative values indicate settlement, positive mean uplift. **(c)** Displacement history in the satellite line-of-sight direction for three scatterers (A, B, C). **(d)** Photograph of Aquileia Tower, Jesolo, which is the cause of the largest movements. See location 4a in Fig. 1.

the displacements of thousands of radar targets in the Valle Paleazza, a fish farm in the northern Venice lagoon. The majority of the PS are located on artificial embankments bounding these lagoon zones whose water level is artificially managed, but several radar reflectors are also located within the fish farm and also on the natural marshland located to the west. The displacement rates are extremely variable, with the largest subsidence usually detected in the inner zones while the fish farm embankments, built up in the previous centuries, are more stable.

3.2.3   The lagoon and coastal urban areas

Apart from Venice, which has been the focus of several studies on land subsidence most recently (Tosi et al., 2013), other important cities and tourist villages are located in the area. Two main tourist centres are Chioggia and Jesolo at the southern and northern tip of the Venice lagoon, respectively.

The subsidence maps detected by PSI for these two cities are provided in Figs. 5 and 6, respectively. Chioggia is almost stable and Jesolo subsides at an average rate equal to $3\,\text{mm}\,\text{yr}^{-1}$. However, local bowls of large subsidence, up to $10\,\text{mm}\,\text{yr}^{-1}$, have been detected in both urban areas. Detailed investigations point out that these occurrences are usually connected with the construction of new large buildings that

exert new heavy loads on the land surface, leading to consolidation of the shallow subsoil below the foundations.

4   Conclusions

PSI on a large stack of stripmap TerraSAR-X images from 2008 to 2011 has proven particularly effective for monitoring land subsidence in the Venice coastland. The interferometric processing of this dataset allows for the characterization of the land movements at both the regional and local scales. The sufficiently large area covered by the images allows the use of a few CGPS stations to calibrate the radar solution. The high resolution and acquisition frequency guarantees the possibility of accurately detecting local-scale short-time anomalies in the subsidence pattern. Our results clearly show the large variability characterizing the consolidation of the structures at the inlets; hence, the effect of the primary consolidation is clearly visible close to the construction of new buildings due to the ongoing urbanization of the coastline. Venice is still experiencing natural land subsidence with rates in the range $1.1 \pm 1.0\,\text{mm}\,\text{yr}^{-1}$, mainly because of the heterogeneous nature and age of the lagoon subsoil.

TerraSAR-X acquisitions over the Venice coastland continued at a regular interval of 11 days after August 2011. We plan to update the PSI analysis using 49 new images acquired through November 2013. In the newest PSI analysis particular attention will be given to short observation periods, the occurrence of appearing and disappearing scatterers, and to the thermal dilation component of displacement (Monserrat et al., 2011) for a better characterization of the behaviour of large structures.

Acknowledgements. This work has been developed in the framework of the Action 2 (SP3-WP1) funded by the Flagship Project RITMARE – The Italian Research for the Sea – coordinated by the Italian National Research Council and funded by the Italian Ministry of Education, University and Research within the National Research Program 2011–2013. Data courtesy: (1) TerraSAR-X, Project COA0612© DLR "Assessing vertical movements of natural tidal landforms and anthropogenic structures at the Venice Lagoon inlets" and (2) CAVA and SFEL CGPS time series, Magistrato alle Acque (Venice Water Authority) through its concessionary Consorzio Venezia Nuova, and (3) VEN1 time series, Nevada Geodetic Laboratory (NGL), obtained from http://geodesy.unr.edu/.

References

Carbognin, L., Teatini, P., and Tosi, L.: Relative land subsidence in the lagoon of Venice, Italy, at the beginning of the new millennium, J. Mar. Sys., 51, 345–353, 2004.

Dong, S., Samsonov, S., Yin, H., and Ye, S.: Time-series analysis of subsidence associated with rapid urbanization in Shanghai, China measured with SBAS InSAR method, Environ. Earth Sci., 72, 677–691, 2014.

Erban, L. E., Gorelick, S. M., and Zebker, H. A.: Groundwater extraction, land subsidence, and sea-level rise in the Mekong Delta, Vietnam, Environ. Res. Lett., 9, 084010, doi:10.1088/1748-9326/9/8/084010, 2014.

Monserrat, O., Crosetto, M., Cuevas, M., and Crippa, B.: The Thermal Expansion Component of Persistent Scatterer Interferometry Observations, Geosci. Remote Sens. Lett., 8, 864–868, doi:10.1109/LGRS.2011.2119463, 2011.

Ng, A. H.-M., Ge, L., Li, X., Abidin, H. Z., Andreas, H., and K. Zhang: Mapping land subsidence in Jakarta, Indonesia using persistent scatterer interferometry (PSI) technique with ALOS PALSAR, Int. J. Appl. Earth Obs., 18, 232–242. 2012.

Phien-wej, N., Giao, P. H., and Nutalaya, P.: Land subsidence in Bangkok, Thailand, Eng. Geol., 82, 187–201, 2006.

Strozzi, T., Teatini, P., and Tosi, L.: TerraSAR-X reveals the impact of the mobile barrier works on Venice coastland stability, Remote Sens. Environ., 113, 2682–2688, 2009.

Teatini, P., Tosi, L., Strozzi, T., Carbognin, L., Wegmüller, U., and Rizzetto, F.: Mapping regional land displacements in the Venice coastland by an integrated monitoring system, Remote Sens. Environ., 98, 403–413, 2005.

Teatini, P., Tosi, L., Strozzi, T., Carbognin, L., Cecconi, G., Rosselli, R., and Libardo, S.: Resolving land subsidence within the Venice Lagoon by persistent scatterer SAR interferometry, Phys. Chem. Earth. Pt. A/B/C, 40, 72–79, 2012a.

Teatini, P., Tosi, L., and Strozzi, T.: Comment on "Recent subsidence of the Venice Lagoon from continuous GPS and interferometric synthetic aperture radar" by Y. Bock, S. Wdowinski, A. Ferretti, F. Novali, and A. Fumagalli., Geochem. Geophys. Geosys. 13, Q07008, doi:10.1029/2012GC004191, 2012b.

Tosi, L., Teatini, P., Strozzi, T., Carbognin, L., Brancolini, G., and Rizzetto, F.: Ground surface dynamics in the northern Adriatic coastland over the last two decades, Rend. Fis. Acc. Lincei, 21, S115–S129, 2010.

Tosi, L., Teatini, P., Bincoletto, L., Simonini, P., and Strozzi, T.: Integrating geotechnical and interferometric SAR measurements for secondary compressibility characterization of coastal soils, Surv. Geophys., 33, 907–926, 2012.

Tosi, L., Teatini, P., and Strozzi, T.: Natural versus anthropogenic subsidence of Venice, Nat. Sci. Reports, 3, 2710, doi:10.1038/srep02710, 2013.

Wegmüller, U., Werner, C., Strozzi, T., and Wiesmann, A.: Multitemporal interferometric point target analysis, Ser. Remote Sens., 3, 136–144, 2004.

Werner, C., Wegmüller, U., Strozzi, T., and Wiesmann, A.: Interferometric Point Target Analysis for deformation mapping. Proc. of the Geoscience and Remote Sensing Symposium (IGARSS 2003), IEEE International, 7, 4362–4364, 2003.

30

An analysis on the relationship between land subsidence and floods at the Kujukuri Plain in Chiba Prefecture, Japan

Y. Ito[1], H. Chen[2], M. Sawamukai[3], T. Su[1], and T. Tokunaga[1]

[1]Graduate School of Frontier Sciences, The University of Tokyo, 5-1-5 Kashiwanoha, Kashiwa City, Chiba, 277-8563, Japan
[2]School of Environmental Science and Engineering, Zhejiang Gongshang University, Hangzhou, 310018, China
[3]VisionTech Inc., Tsukuba City, Ibaraki, 305-0045, Japan

Correspondence to: Y. Ito (yu_ito@geoenv.k.u-tokyo.ac.jp)

Abstract. Surface environments at the Kujukuri Plain in Chiba Prefecture, Japan, in 1970, 2004, and 2013, were analyzed and compared to discuss the possible impact of land subsidence on the occurrence of floods. The study area has been suffered from land subsidence due to ground deformation from paleo-earthquakes, tectonic activities, and human-induced subsidence by groundwater exploitation. Meteorological data, geomorphological data including DEM obtained from the airborne laser scanning (1-m spatial resolution), leveling data, and the result of our assessment map (Chen et al., 2015) were used in this study. Clear relationship between floods and land subsidence was not recognized, while geomorphological setting, urbanization, and change of precipitation pattern were found to contribute to the floods. The flood prone-area is distributed on the characteristic geomorphological setting such as floodplain and back swamp. It was revealed that the urban area has been expanded on these geomorphological setting in recent years. The frequency of hourly precipitation was also shown to be increased in the past ca. 40 years, and this could induce rapid freshet and overflow of small- and medium-sized rivers and sewerage lines. The distribution of depression areas was increased from 2004 to 2013. This change could be associated with the ground deformation after the Tohoku earthquake ($M_w = 9.0$) in 2011.

1 Introduction

Land subsidence causes several problems which include changes in elevation, groundwater salinization in coastal area, structural damage, and increase in the potential for flooding (Galloway et al., 1999; Ng et al., 2015). Land subsidence has occurred at many areas in Japan. The Kujukuri Plain, Chiba Prefecture, Japan (Fig. 1), is one of the areas experiencing land subsidence which is caused by ground deformation from paleo-earthquakes, tectonic activities, and human-induced subsidence by groundwater abstraction. The maximum value of accumulated subsidence is ca. 1 m at the Mobara City during the period from 1969 to 2014 (Chiba Prefecture, 2014). This coastal area is relatively flat with its elevation lower than 10 m. The lowland consists of three main landforms (Fig. 1) (Moriwaki, 1979): (1) beach ridge;

(2) sand dune; and (3) back swamp and flood plain. Therefore, the impact of land subsidence on the surface environment has been concerned in this area; one of its effects may include the increase of the frequency of flooding. In the south Kujukuri plain area, floods have repeatedly occurred in the past. There were at least 16 large flood events in this area during the period between 1970 and 2013. The frequency of large flood events had relatively low before 1970. Figure 2 shows the flood inundation areas of the 1 July 1970, 8–11 October 2004, 16 October 2013.

In this study, we analyzed and compared the factors of the floods in 1970, 2004, and 2013 by using GIS to assess the possible impact of land subsidence on the occurrence of floods. The three periods were selected because of the following reasons; 1970 is after huge land modification, 2004 is

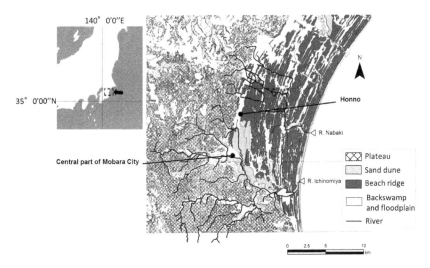

Figure 1. Geomorphological map of the Kujukuri Plain. Study area is Nabaki and Ichinomiya Rivers watershed.

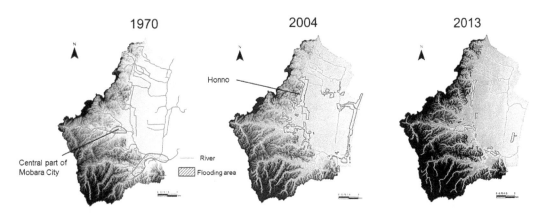

Figure 2. Flood inundation area in 1970, 2004, and 2013. Flood area was digitalized from the flood report of Chiba prefecture (1971, 2005, and 2015).

prior to the 2011 off the Pacific coast of Tohoku Earthquake ($M_{\mathrm{w}} = 9.0$), and 2013 is after the earthquake.

2 Analyses of the factors related to floods

Three major factors which were related to floods were analyzed – climatic, geomorphologic, and human factors.

2.1 Climatic factor

The intensity and frequency of precipitation may affect the frequency and magnitude of floods (Fitzpatrick et al., 1999). Spatial distribution of the Radar/raingauge-Analyzed precipitation data obtained from Japan Meteorological Agency in 2004 and 2013, indicate that the spatial rainfall pattern could affect the flooding area. The flooding area in 2004 was mainly recognized in coastal area and the central part of the Mobara city (Fig. 2) due to high intensity rainfalls in the lower reach of the Nabaki River and the higher tidal level. In 2013, the upper reaches of the Ichinomiya River and the central part of the Mobara city were flooded (Fig. 2) because the rainfall was concentrated in the upper reaches of the Ichinomiya River. The lower reaches were not flooded because the tidal level was low and the rainfall was not strong. The results of our assessment maps in 2004 and 2013 (Chen et al., 2015) also suggested that the flood areas were affected by the spatial rainfall pattern.

Frequency of hourly precipitation of more than $50\,\mathrm{mm\,h^{-1}}$ is shown to be increased during the past ca. 40 years within the watersheds, and this could induce rapid freshet and overflow of small- and medium-sied rivers and sewerage lines.

2.2 Geomorphological factor

Alluvial and coastal lowlands have a potential floods risk (Umitsu, 2012). Flood inundation area is mainly recognized on floodplains, back swamp, Holocene terrace, and reclaimed land in three periods (Fig. 3 and Table 1). Coastal plain and beach ridges are also flooded.

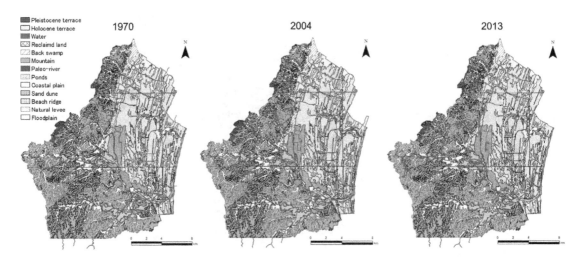

Figure 3. Geomorphological classification map in 1970, 2004, and 2013 based on land condition map (1 / 25 000) of Geospatial Information Authority of Japan. Red line indicates flood inundation area.

Table 1. Ratio of flood inundation area by geomorphological condition.

Geomorphological classification	1970	2004	2013
Holocene terrace	29.01 %	5.74 %	13.74 %
Flood plain	23.06 %	16.42 %	31.74 %
Natural levee	17.34 %	1.88 %	4.99 %
Beach ridge	6.92 %	25.94 %	5.07 %
Reclaimed land	7.57 %	9.19 %	10.22 %
Water	5.60 %	2.88 %	5.11 %
Sand dune	4.15 %	11.66 %	5.00 %
Ponds	2.40 %	0.82 %	0.52 %
Coastal plain	2.29 %	11.91 %	3.95 %
Back swamp	–	12.69 %	14.82 %
Mountain	0.94 %	0.41 %	3.27 %
Paleo-river	0.62 %	0.36 %	1.39 %
Pleistocene terrace	0.09 %	–	–
Others	–	0.11 %	0.17 %

Table 2. Impermeable area ratios in each land cover.

Land cover classification	1970	2004	2013
Urban	0.20	0.30	0.30
Building	0.14	0.25	0.21
Farmland	0.12	0.18	0.11
Paddy	0.09	0.09	0.10
Sands	0.12	0.09	0.07
Grassland	0.07	0.09	0.10
Forest	0.07	0.07	0.08
Water	0.08	0	0.03
Bareland	0.15	–	–

change might be the influence of the ground deformation after the $M_\mathrm{w} = 9.0$ earthquake in 2011.

2.3 Human factor

Urban development of the Mobara city and its surroundings was quite rapid during 1970–2013 (Fig. 5). The ratio of urban and building estate areas were 15.4 % in 2004 and 17.4 % in 2013, respectively, while 9.5 % in 1970. Farm lands and paddy areas were decreased during the period from 1970 to 2013. Impermeable area ratios in each land cover also change with land cover changing (Table 2). The area ratio of urban and building have increased from 0.20 and 0.14 in 1970 to 0.30 and 0.21–0.25 in 2004–2013, respectively.

The Urban areas have spread into lowland area such as flood plains and back swamps since 1970. This is clearly reflected in the relationship between the urbanization area and the landforms. The results of the analysis on the relationship between built-up area and the geomorphological changes (Fig. 6) indicate that the built-up area have expanded on vulnerable landforms with respect to floods, i.e., flood

In particular, in the central part of the Mobara City which is flood-prone area (Fig. 2), the floods occur on the flood plains surrounded by slightly highland (relative elevation: +1–2 m) such as natural levee and Holocene terraces. Honno, are the flood-prone area, is also situated on back swamp and is often flooded even with a light rainfall due to the area is surrounded by sandbars or beach ridges. Therefore, the flood-prone area may depend heavily on the characteristics of the geomorphological setting.

Local topographic depressions were extracted from the 1-m mesh resolution DEM. The results showed that the distribution of depressions was mostly unchanged from 1970 to 2004, however, was changed in 2013, particularly in the Nabaki River mouth and its surroundings (Fig. 4). This

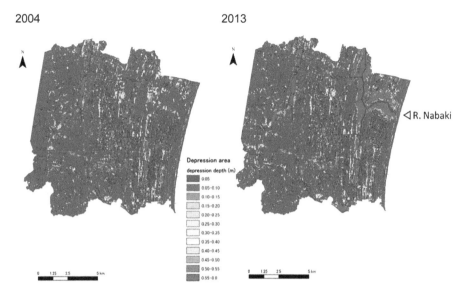

Figure 4. Distribution of depressions in 2004 and 2013. 1-m mesh digital elevation model obtained from the airborne laser scanning survey was used for the analysis.

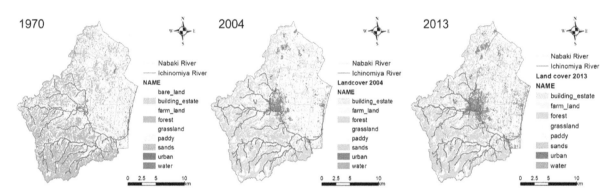

Figure 5. Land cover changes in 1970, 2004, and 2013. Land cover derived from Landsat-MSS and TM was classified by methods of maximum likelihood.

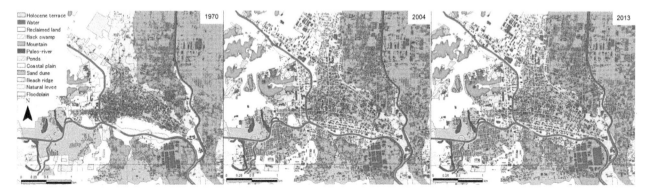

Figure 6. Expansion of the built-up area and geomorphological classification map in the central Mobara City. Red features indicate the building.

plains and paleo-channels. Majority of the flood plains in the central Mobara City was also changed to reclaimed lands since 1970 by land development. The buildings were located on the relatively less vulnerable land and ground conditions in 1970, as a result of urban development and the expansion to the flood plains and paleo-channels since 1970, the one might be changed to be prone to floods.

3 Summary

The visible change of the spatial distribution of hazard potential areas with land subsidence were not clearly recognized whereas the influence of artificial land modification and the natural geomorphological condition are considered to be the factors that explain the spatial distribution and the temporal changes of the vulnerable areas to floods. Back swamp and flood plains are surrounded by beach ridges in the center of the Mobara City and its surrounding area; its geomorphological condition suggests that water tends to accumulate in the area. Honno area could be flood even with a light rainfall because of the back swamp area surrounded by the sand bars. Urbanizations have expanded the urban area into lowlands such as flood plains since 1970, and the function to control the floods might be reduced in the area because flood plains were changed to the reclaimed land.

It is important to consider natural factors such as changes of the rainfall patterns and geomorphological factor, artificial factors such as urbanization, as well as the influence of the land subsidence all together to discuss the spatial distribution of the flood-prone area in its temporal changes.

Acknowledgements. We would like to express our sincere gratitude to the Keiyo Natural Gas Association for supporting this study. Residential maps provided by ZENRIN CO., LTD are used as the CSIS Joint Research (No. 524) using spatial data provided by Spatial Information Science, The University of Tokyo.

References

Chiba prefecture: Report of Chiba prefecture flooding, 1971 (in Japanese).

Chiba prefecture: Report of Chiba prefecture flooding, 2005 (in Japanese).

Chiba Prefecture: Report of current state of land subsidence in Chiba prefecture, 2014 (in Japanese).

Chiba Prefecture: Documents of Ichinomiya River watershed meeting, 2015 (in Japanese).

Chen, H. L., Ito, Y., Sawamukai, M., and Tokunaga, T.: Flood hazard assessment in the Kujukuri Plain of Chiba Prefecture, Japan, based on GIS and multicriteria decision analysis, Nat. Hazards, 78, 105–120, doi:10.1007/s11069-015-1699-5, 2015.

Fitzpatrick, F. A., Knox, J. C., and Whitman, H. E.: Effects of historical land-cover changes on flooding and sedimentation, North Fish Creek, Wisconsin, USGS Water-resources investigations report, 99–4083, 1–12, 1999.

Galloway, D., Jones, D. R., and Ingebritsen, S. E.: Land subsidence in the United States, U.S. Geological survey circular 1182, USA, 1999.

Moriwaki, H.: The landform evolution of the Kujukuri Coastal Plain, Central Japan, Quaternary Res., 18, 1–16, 1979 (in Japanese).

Ng, A. h.-M., Ge, L., and Li, X.: Assessments of land subsidence in the Gippsland Basin of Australia using ALOS PALSAR data, Remote Sens. Environ., 159, 86–101, 2015.

Umitsu, M.: Alluvial lowlands and floods, Geo-environment of alluvial and coastal lowlands, Kokon-Shoin Publishers, Tokyo, 47–48, 2012 (in Japanese).

31

Characterization of earth fissures in South Jiangsu, China

S. Ye[1], Y. Wang[1], J. Wu[1], P. Teatini[2], J. Yu[3], X. Gong[3], and G. Wang[3]

[1]School of Earth Sciences and Engineering, Nanjing University, Nanjing, China
[2]Department of Civil, Environmental and Architectural Engineering, University of Padova, Padova, Italy
[3]Key Laboratory of Earth Fissures Geological Disaster, Ministry of Land and Resources, (Geological Survey of Jiangsu Province), Nanjing, China

Correspondence to: S. Ye (sjye@nju.edu.cn)

Abstract. The Suzhou-Wuxi-Changzhou (known as "Su-Xi-Chang") area, located in the southern part of Jiangsu Province, China, experienced serious land subsidence caused by overly exploitation of groundwater. The largest cumulative land subsidence has reached 3 m. With the rapid progress of land subsidence since the late 1980s, more than 20 earth fissures developed in Su-Xi-Chang area, although no pre-existing faults have been detected in the surroundings. The mechanisms of earth fissure generation associated with excessive groundwater pumping are: (i) differential land subsidence, (ii) differences in the thickness of the aquifer system, and (iii) bedrock ridges and cliffs at relatively shallow depths. In this study, the Guangming Village Earth Fissures in Wuxi area are selected as a case study to discuss in details the mechanisms of fissure generation. Aquifer exploitation resulted in a drop of groundwater head at a rate of 5–$6\,\mathrm{m\,yr}^{-1}$ in the 1990s, with a cumulative drawdown of 40 m. The first earth fissure at Guangming Village was observed in 1998. The earth fissures, which developed in a zone characterized by a cumulative land subsidence of approximately 800 mm, are located at the flank of a main subsidence bowl with differential subsidence ranging from 0 to 1600 mm in 2001. The maximum differential subsidence rate amounts to $5\,\mathrm{mm\,yr}^{-1}$ between the two sides of the fissures. The fissure openings range from 30 to 80 mm, with a cumulative length of 1000 m. Depth of bed rock changes from 60 to 140 m across the earth fissure. The causes of earth fissure generation at Guangming Village includes a decrease in groundwater levels, differences in the thickness of aquifer system, shallow depths of bedrock ridges and cliffs, and subsequent differential land subsidence.

1 Introduction

Excessive exploitation of groundwater in Suzhou-Wuxi-Changzhou (known as "Su-Xi-Chang") area, located in the southern part of Jiangsu Province, China, started in the 1970s. It has resulted in serious land subsidence and earth fissure hazard that caused significant damage and heavy economic losses. The first earth fissure case in the Su-Xi-Chang area was reported in 1989 at Henglin town of Changzhou City. Twenty earth fissures occurred in the Su-Xi-Chang area between 1990 and 2000. The rapid development of earth fissures from 1990 to 2000 destroyed many buildings and the local governments decided to stop groundwater pumping in 2000 to alleviate land subsidence and earth fissures.

Currently, a total of 25 earth fissures have been found in Su-Xi-Chang area, mainly located in the Wujin County in Changzhou and Xishan County in Wuxi (Fig. 1) (Liu et al., 2004). According to geological surveys, the orientation of the earth fissure in this area are mainly NE (NNE), EW and SN direction (Fig. 1), showing no obvious unified direction in the region. All the fissures are in a dense banded distribution and their orientation are in perfect accordance with the underlying bedrock. The width of the earth fissure zone generally ranges from 30 to 100 m, and the length varies between 200 and 600 m, and several fissures can reach lengths of 1000 m (Yu et al., 2004).

A significant amount of research has been done to study the occurrence of earth fissures in this area. Liu et al. (2004)

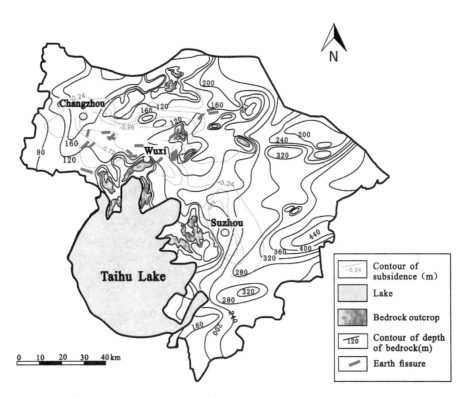

Figure 1. The distribution of earth fissures in Su-Xi-Chang area. The bedrock outcrops and depth (m from the mean sea level) and the cumulative land displacement from 1980 to 2000 are mapped (after Liu et al., 2004).

Figure 2. The ground rupture (a) and damaged house (b) at Guangming Village (red line in (a) shows the direction of the earth fissure).

identified the factors related to the development of earth fissure including bedrock hills, buried terraces, Karst collapse, structural differences of the aquifer system, and over-exploitation of groundwater. Wang et al. (2009, 2010) reported earth fissures triggered by excessive groundwater withdrawal and the variability of bedrock depth in Hetang town, Jiangyin City. Yu et al. (2006) analyzed the mechanisms leading to occurrence of some typical earth fissures using three-dimensional seismic exploration techniques. Based on the inversion of the 3-D seismic data, the sedimentary structure and morphology of the bedrock surface were obtained. Yu et al. (2004) reported that earth fissures in the Su-Xi-Chang area develop under the combined effects of the

visco-elasto-plastic deformation of some sedimentary layers, buried bedrock surface morphology, thickness differences of some highly compressible sedimentary layers, and excessive exploitation of groundwater.

In Su-Xi-Chang area, the second confined aquifer was the main over-exploited unit. Groundwater pumping has formed a large, wide ranging cone of land subsidence because of long term over-exploitation of groundwater. The groundwater levels in the drawdown cone declined to −80 m in 1990s, and the largest cumulative subsidence was close to 2000 mm (Zhang et al., 2008). The groundwater levels recovered gradually from the previous lowest value to about −75 m in 1994 and −58 m in 2003, since the implementation of severe restrictions of groundwater exploitation by the Jiangsu government in 1995 (Zhang et al., 2008). The Jiangsu government finally decided to prohibit all groundwater exploitation in the Su-Xi-Chang area from 2000. As a result, land subsidence became controlled in some areas, and the rate of land subsidence decreased from 110 mm yr^{-1} in 1994 to 10–0 mm yr^{-1} in 2006 (Wang et al., 2009). The movement rates for these fissures slowed gradually, and the number of earth fissure are no longer increasing except for a few active earth fissures. The most active fissures in Su-Xi-Chang area are located in the Guangming Village, Xishan District of Wuxi City, with the maximum subsidence rate of about 5 mm yr^{-1}.

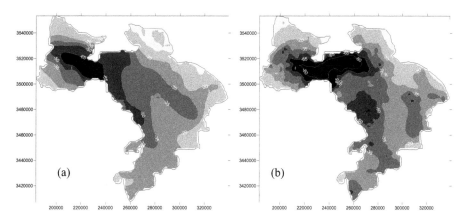

Figure 3. Groundwater level (m above mean sea level) in the second confined aquifer in 1995 (**a**) and 2001 (**b**). The red line represents the earth fissure at Guangming Village.

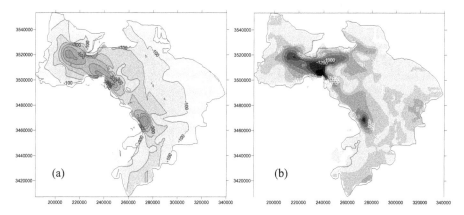

Figure 4. Cumulative land subsidence (mm) from 1980 to 1995 (**a**) and 2001 (**b**). The red line represents the earth fissure at Guangming Village.

2 Earth fissures at Guangming Village

The first earth fissure at Guangming Village was discovered in 1998. Earth fissuring advanced significantly since 2000, with tilting farmlands and ground fractures at the early stages of development. Fissuring increased drastically since 2007 with more than 400 newly damaged houses (Fig. 2). The orientation of the earth fissures at Guangming Village are mostly in NE direction which is along the ridge tip.

Regional and local groundwater levels

The second confined aquifer is the main withdrawn aquifer in the Su-Xi-Chang area. Groundwater was overly exploited in the 1990s, and the total pumped groundwater reached 380 million m^3 in 1995. The groundwater levels in the second confined aquifer dropped from -35 to -75 m, with values between -60 and -55 m at Guangming Village (Fig. 3a). The total pumping rate decreased from 1996 to 2000 to 250 million m^3 in 2000. The groundwater exploitation was prohibited after 2001 in the Su-Xi-Chang area. However, the groundwater levels in some areas still continued to drop. For example, the groundwater levels in the second confined aquifer dropped to -95 m in the center of the drawdown cone and the groundwater level at Guangming Village ranged from -75 to -70 m (Fig. 3b).

Regional and local cumulative land subsidence

The lowering of the groundwater levels resulted in a severe land subsidence. The cumulative land subsidence in 1995 reached to 700 mm at the center of the subsidence bowl surrounding Guangming Village, and between 400–500 mm at Guangming Village (red line in Fig. 4a). The cumulative land subsidence in 2001 peaked to 1600 mm at the center of subsidence bowl located about 20 km to the east of the Guangming Village, and to 700–800 mm at Guangming Village (Fig. 4b). From the regional cumulative land subsidence maps, one can see that earth fissures in Guangming Village are in the middle of the subsiding area.

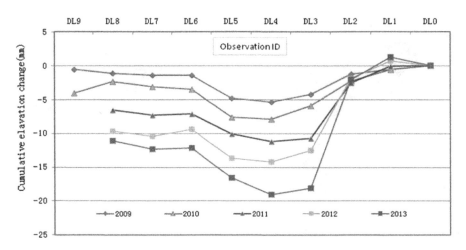

Figure 5. Cumulative elevation changes from 2009 to 2013 at the ten leveling benchmarks established across the fissure at Guangming Village. DL0 is used as reference point. The distance between DL0 and DL9 is approximately 80 m.

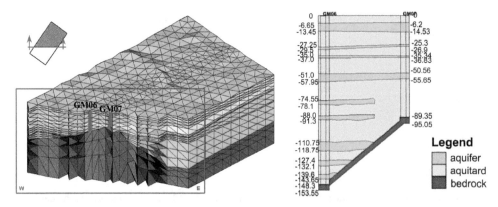

Figure 6. Perspective view of the Finite-Element model developed by Franceschini et al. (2015) to simulate the earth fissure generation at Guangming Village (**a**). The geological setting is highlighted in various colours: rock in grey, aquifers in yellow, and aquitards in blue). Hydrogeological profile between boreholes GM06 and GM07 (**b**). The numbers provide the buried depths of the top of each layer.

Local differential land subsidence

The farmland is obviously tilting at Guangming Village, which signifies the occurrence of differential land subsidence. Ten leveling points were established in 2007 along a trail that crosses one of the earth fissures. Because no deep stable benchmark was set at Guangming Village, the leveling benchmark DL0 is used as a reference point. The other 9 levelling points show a significant different elevation compared to DL0 (Fig. 5) over the monitoring interval from 2009 to 2013. Benchmark DL4 always shows the highest land subsidence values, and the average differential land subsidence is 5 mm yr^{-1}.

Geological setting with buried rock ridge

The thickness of the aquifer system in the Su-Xi-Chang area is not uniform because of buried rock ridges whose depth varies significantly in some local areas. Due to this peculiar geological setting, the thickness of the aquifer system clearly

differs across the rock ridges, with thicker sediments further away from ridge (Fig. 6a). For example, the thickness of the aquifer system at borehole GM07 is 95 m and that at GM06, only 95 m apart, is 153 m (Fig. 6b). The hydrostratigraphic section in Fig. 6b also shows that there are more aquifers at GM06 than at GM07.

3 Conclusions

Earth fissures are serious problems in Su-Xi-Chang area. They are distributed along the buried rock ridge primarily in the main NE direction. Excessive groundwater exploitation triggers earth fissuring. Earth fissuring at Guangming Village typically demonstrates the causes and mechanism of earth fissure generation and development, which includes groundwater level decrease, buried rock ridges with differing depths, the heterogeneous aquifer system, and subsequent differential land subsidence. A numerical model to simulate the generation and development of earth fissures at Guangming Village is being developed (Franceschini et al., 2015).

Acknowledgements. Funding supported by the Special Project for the Public Scientific Research of the Ministry of Land and Resources of China No. 201411096, KLEFGD No. 201403, NSFC No. 41272259 and NSF of Jiangsu Province No. BK2012730 is appreciated. Pietro Teatini was partially supported by the University of Padova, Italy, within the 2014 International Cooperation Programme. The international collaboration is supported by UNESCO IGCP Project No. 641 (M3EF3 – Deformation and fissuring caused by exploitation of subsurface fluids).

References

Franceschini, A., Teatini, P., Janna, C., Ferronato, M., Gambolati, G., Ye, S., and Carreon-Freyre, D., Modelling ground rupture due groundwater withdrawal: applications to test cases in China and Mexico, Proceeding of the Ninth International Symposium on Land Subsidence, Nagoya, November 2015.

Liu, C., Yuan, X. J., and Zhu, J. Q.: Earth fissures in Su-Xi-Chang area, China University of Geosciences Press, 2004 (in Chinese).

Wang, G. Y., You, G., Shi, B., Yu, J., Li, H. Y., and Zong, K. H. : Earth fissures triggered by groundwater withdrawal and coupled by geological structures in Jiangsu Province, China, Environ. Geol., 57, 1047–1054, 2009.

Wang, G. Y., You, G., Shi, B., Qiu, Z. L., Li, H. Y., and Tuck, M.: Earth fissures in Jiangsu Province, China and geological investigation of Hetang earth fissure, Environ. Earth Sci., 60, 35–43, 2010.

Yu, J., Wang, X., Su, X., and Yu, Q.: The mechanism analysis on ground failure disaster formation in Suzhou- Wuxi – Changzhou area, Journal of Jilin University (Earth Science Edition), 34, 236–241, 2004 (in Chinese).

Yu, J., Chen, G, Wang, X. M., He, H. S., Xu, X. L., and Wu, J. H.: Formation mechanisms analysis on ground failure based on 3D seismic exploration data in Yinguo'an area, Hydrogeology & Engineering Geology, 34, 117–123, 2006 (in Chinese).

Zhang, Y., Xue, Y. Q, Wu, J. C., Yu, J., Wei, Z. X., and Li, Q. F.: Land subsidence and earth fissures due to groundwater withdrawal in the Southern Yangtse Delta, China, Environ. Geol., 55, 751–762, 2008.

32

Potential of Holocene deltaic sequences for subsidence due to peat compaction

32

Potential of Holocene deltaic sequences for subsidence due to peat compaction

32

Potential of Holocene deltaic sequences for subsidence due to peat compaction

E. Stouthamer and S. van Asselen

Dept. of Physical Geography, Faculty of Geosciences, Utrecht University, Utrecht, the Netherlands

Correspondence to: E. Stouthamer (e.stouthamer@uu.nl)

Abstract. Land subsidence is a major threat for the livability of deltas worldwide. Mitigation of the negative impacts of subsidence, like increasing flooding risk, requires an assessment of the potential of the deltas' subsurfaces for subsidence. This enables the prediction of current and future subsidence and optimization of sustainable management strategies. In this paper we present a method to determine the amount of compaction within different Holocene deltaic peat sequences based on a case study from the Rhine-Meuse delta, the Netherlands, showing the potential of these sequences for subsidence due to peat compaction.

1 Introduction

Land subsidence is a major threat to hundreds of millions of people living in deltas worldwide. It leads to increasing flooding risk, damage to buildings and infrastructure, intrusion of salt water, and hence, has a negative impact on the livability in deltas. Natural subsidence rates, caused by tectonics, isostasy, sediment compaction and oxidation of organic material are generally in the order of $mm\,yr^{-1}$ (van Asselen et al., 2011). Human-induced subsidence rates, due to extraction of groundwater, oil and gas, drainage for land reclamation, and loading by for example roads and buildings are commonly an order of magnitude higher ($cm\,yr^{-1}$; e.g. van Asselen, 2011; Erban et al., 2014). Consequently, in many populated deltas the contribution of subsidence to relative sea-level rise is much larger than eustatic (absolute) global sea-level rise, which is in the order of $mm\,yr^{-1}$ (Church et al., 2013).

The potential of delta sequences for both natural and human-induced subsidence due to compaction and oxidation of peat, and hence subsidence magnitude and rate, is highly variable in time and space, mainly depending on the composition of the subsurface. Subsurface composition and characteristics are determined by the palaeogeographical development (i.e. the spatial and temporal distribution of depositional environments and hence deposits), post-depositional processes of erosion (i.e. preservation) and loading, and physical, geochemical and biological processes, including

compaction and soil formation. Based on the 3-D distribution of specific sequences within a delta and their physical properties, the potential for subsidence due to peat compaction of these sequences at the spatial scale of an entire delta can be determined.

To mitigate the negative impacts of human-induced subsidence it is essential to know the potential of the subsurface for compaction and oxidation. This enables the prediction of current and future subsidence under different delta management strategies and to develop and optimize sustainable management strategies. In this paper we present a method to determine the amount of peat compaction for typical Holocene distal deltaic organo-clastic sequences based on a case study from the Rhine-Meuse delta, the Netherlands, showing the potential of these sequences for subsidence due to peat compaction.

2 Subsidence potential due to peat compaction

Holocene delta sequences are usually very heterogeneous, and composed of alternating layers consisting of mixtures of sand, silt, clay and peat. The main factors influencing the potential for subsidence due to peat compaction are (1) the thickness of organic, mainly peat and clay, layers in the subsurface, (2) the organic matter content of these layers, and (3) the effective stress, which is a function of the weight of the overburden and pore water pressures (van Asselen et al.,

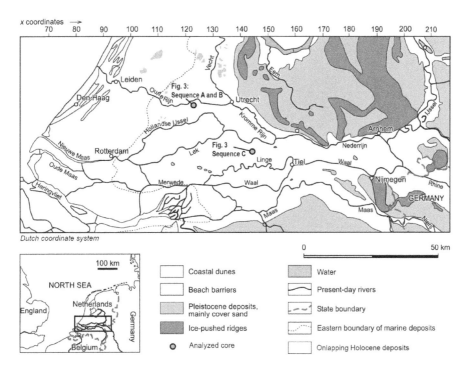

Figure 1. The Holocene Rhine-Meuse delta, the Netherlands; rivers run E–W. *x* coordinates according to the Dutch map coordinate system (in kilometers, after Stouthamer and Berendsen, 2001 and Stouthamer et al., 2011).

2010). In addition, the Holocene sedimentation and groundwater table history is extremely important, because this determines if organic layers have experienced past compaction and/or oxidation, and hence their present geotechnical properties. Lowering of the phreatic groundwater level, for e.g. agricultural purposes, leads to a drop in hydrostatic pressure, causing diminishing pore water pressures and consequently an increase in effective stress (= total weight – pore water pressure; Terzaghi, 1943) and compaction of the peat sequence. At the depth interval between the former, higher, and new, lower, groundwater level, oxidation of organic material occurs. This leads to a decrease in volume and collapse of the original soil structure that existed before groundwater level was lowered. Both compaction and oxidation contribute to total subsidence. Sequences that are composed of thick high-organic peat layers, especially in the upper meters of the sequence have the highest potential for compaction and oxidation and hence subsidence. Compaction rates will be much lower if a peat layer has already experienced compaction by loading. Rates of compaction due to loading are highest within decades to a few centuries after loading (van Asselen et al., 2011).

3   Study area: Rhine-Meuse delta, the Netherlands

The Rhine-Meuse delta is located in the Netherlands, in the tectonically subsiding Southern North Sea Basin. The estimated Holocene-averaged basin subsidence rate is 0.1–

0.3 mm yr^{-1} (Cohen, 2005). The Holocene delta was jointly formed by the rivers Rhine and Meuse. The delta apex is located near the Dutch-German border (East). The Rhine and Meuse debouch into the North Sea (West) (Fig. 1). The distance between the modern delta apex and the Rhine-Meuse river mouth is ca. 150 km. The Holocene sequence is approximately 2 m thick near the delta apex and just over 20 m at the present river mouth and coastal barrier. The upstream part of the delta is confined by higher topography of sandy Pleistocene deposits (ice-pushed pre-Saalian glacial sandy Rhine and Meuse deposits) and relatively narrow; 15–25 km wide. Downstream the delta widens to 60 km and grades laterally into tidal-estuary and lagoonal coastal plain environments. The southern and central-northern part of the delta are bordered by higher topography formed by Late Glacial cover sand deposits. Between 1100 and 1300 AD all Rhine and Meuse distributaries in the delta were embanked, stopping flood basins from receiving sediments (van de Ven, 2003). Flooding and channel-migration activity has since remained restricted to the embanked flood plains.

The Holocene subsurface of the delta, fed with sediment by the mixed-load rivers Rhine and Meuse, is notoriously heterogeneous. Channel-belt sand bodies dissect areas of overbank deposits (dominantly clayey, but holding sand and silt too). Multiple generations of channel belts avulsed adding further complexity (Stouthamer and Berendsen, 2000). Downstream in the delta, the overbank clastics alternate with peat layers (Fig. 2). The distal Holocene coastal,

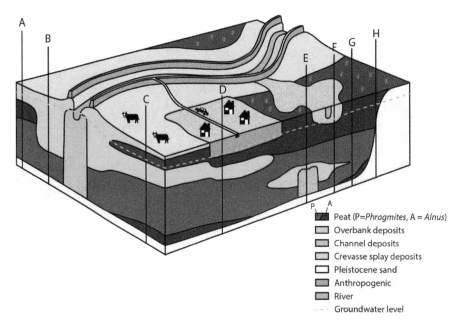

Figure 2. Schematic representation of the lithological subsurface composition of the distal (western) part of the Holocene Rhine-Meuse delta. The vertical lines (A–H) indicate different sequences with a varying potential for subsidence due to compaction and oxidation depending on the subsurface composition and characteristics.

tidal, estuarine and fluvial deposits are onlapping the Pleistocene cover sand deposits in the North and the South and braided river deposits in the central part of the delta (Fig. 1).

4 Methods: determining the potential for compaction

The Holocene sequence composition of the Rhine-Meuse delta is well-known from previous studies (e.g. Berendsen and Stouthamer, 2000; Stouthamer, 2001; Cohen, 2003; Gouw and Erkens, 2007; Gouw, 2008; Bos, 2010; Stouthamer et al., 2011). These studies allow to distinguish key clastic-organic sequences (Fig. 2), which have different potentials for subsidence due to peat compaction.

The degree of peat compaction has been determined at different locations in the organo-clastic central part of the Rhine-Meuse delta (van Asselen, 2011; Van Asselen, unpublished work; here we present three sequences A–C in Figs. 1 and 3), by comparing the dry bulk density of compacted and uncompacted peat of similar organic-matter content (Bird et al., 2004; van Asselen et al., 2009; van Asselen, 2011). The dry bulk density and organic-matter content of compacted peat in a delta sequence are measured from sediment cores at 5-cm intervals using a 1 cm × 1 cm × 5 cm sampler. The cores are extracted with a 100 cm × 6 cm wide gouge auger, and subsequently cut in half lengthways using a thin stretched wire, to sample of the inner least disturbed part. The dry bulk density of each 5 cm³ peat sample has been determined by drying it at 105 °C and weighing it on an electronic scale with an accuracy of 0.001 g

(dry bulk density = weight / 5 cm³). The organic-matter content has been determined by loss-on-ignition (LOI = ((dried weight − ashed weight) / dried weight) × 100 %; cf. Heiri et al., 2001). The dried peat samples were heated at 550 °C for 4 h and subsequently weighted to determine the ashed weight.

Uncompacted peat samples were obtained from the Biebrza National Park (Poland) and the Rhine-Meuse delta (van Asselen, 2011). The Biebrza National Park is a wetland that has experienced minimal human disturbance, where similar peat types occur as in the Holocene Rhine-Meuse delta. The fresh, uncompacted peat has been sampled using a special device that was designed for this purpose (van Asselen and Roosendaal, 2009). The dry bulk density and LOI of uncompacted peat samples were determined following the same procedure as for compacted 5 cm³ samples. Based on these data an equation for calculating the dry bulk density of uncompacted peat of any LOI has been constructed:

$$\rho_{\text{dry, uncompacted}} = a - ce^{-(b/\text{LOI})}, \tag{1}$$

in which a, b and c are fitted parameters (van Asselen, 2011). The calculated dry bulk density was subsequently used to calculate the decompacted thickness of each 5 cm³ compacted peat sample:

$$h_{\text{decomp}} = \frac{\rho_{\text{dry, comp}}}{\rho_{\text{dry, uncomp}}} \times 5. \tag{2}$$

The percentage of compaction is expressed as the ratio between the volume reduction

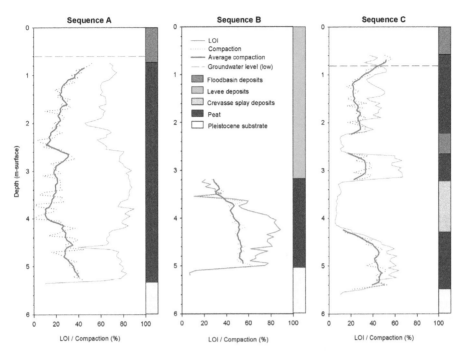

Figure 3. Lithological composition, LOI and amount of compaction of cores OR-I (sequence B), OR-II (sequence A), and CB-I (sequence C) (after van Asselen, 2011). These sequences are representative for the sequences A, B and C in Fig. 2.

$(v_{\mathrm{red}} = (1 \times 1 \times h_{\mathrm{decomp}}) - (1 \times 1 \times 5))$ and the calculated decompacted volume of a $5\,\mathrm{cm}^3$ peat sample $(v_{\mathrm{decomp}} = 1 \times 1 \times h_{\mathrm{decomp}})$:

$$\mathrm{compaction} = \frac{v_{\mathrm{red}}}{v_{\mathrm{decomp}}} \times 100\,\%. \qquad (3)$$

5 Results: compaction of sequences A–C

Sequence B is located close to the river (Fig. 2) and is characterized by approximately 3 m of natural levee deposits on top of a 2-m-thick peat layer between 3–5 m below the surface that is situated on the incompressible Late-Pleistocene sand substrate (Fig. 3). Sequence A and C are located at further distance from the river (Fig. 2). Sequence A is characterized by a 60 cm thick layer of flood basin deposits on a 4.5 m thick peat layer on the Late-Pleistocene sand substrate (Fig. 3). Sequence C comprises alternating layers of peat and fluvial deposits; 50 cm of flood basin deposits at the top, 30 cm flood basin deposits at 2.20–2.40 m depth and 1 m of crevasse-splay deposits at 3.20–4.20 m depth.

Fluctuating LOI-values within the sequences can be explained by the varying clastic content; deposits with a high clastic content show low LOI-values (e.g. flood basin deposits, clayey peat intervals), deposits with a low clastic content show high LOI-values (e.g. peat). For sequences B and C, in which peat layers are considerably compacted, the amount of peat compaction is positively related to LOI (van Asselen, 2011). For example, this is observed in the com-

pacted peat layer in sequence B, where lower compaction values are observed in the low-organic top of the peat layer.

Relatively high peat compaction values occur at the top of the peat layer in sequence A at 120–80 cm depth and at the top of the upper peat layer in sequence C at 120–70 cm depth (Fig. 3). This can be explained by artificial lowering of the phreatic groundwater level in this agricultural area, which causes compaction and oxidation that both lead to an increase in the dry bulk density in the zone above and just below groundwater level (see Sect. 2). In sequence B the peat layer is situated too deep for groundwater level lowering to have any effect. Furthermore, high values occur beneath the natural levee and crevasse-splay deposits in sequence B and C respectively, due to the overburden weight. Also, at the base of sequence A the amount of compaction increases, which may be attributed to palaeo-groundwater level fluctuations, (auto)compaction and/or fluctuations in the hydrostatic pressure (Fig. 3).

6 Discussion: assessing the potential of Holocene deltaic sequences for subsidence due to compaction

The calculated compaction values for the sequences A–C indicate the potential for future subsidence due to compaction for comparable delta settings. For example, strongly compacted peat beneath thick natural levee deposits (Fig. 2, sequence B) is less vulnerable for future subsidence due to compaction than an uncompacted thick peat sequence with-

188

Hydrology: The Scientific Study of Water

Table 1. Relative potential of key sequences for peat compaction, based on lithological composition. For the lithological built-up of the sequences see Fig. 2.

Sequence	Compaction potential
G	High
A	
F	
E	
C	
D	
H	
B	Low

out clastic overburden (Fig. 2, sequence G). Based on the lithological sequence composition and its properties, the relative potential for future subsidence due to peat compaction can be estimated (Table 1). In addition, the depth of the groundwater level can affect peat compaction: a drop in the pore water pressure following groundwater level lowering may induce additional compaction, and oxidation, of the peat sequence (van Asselen et al. 2009, 2010). This should be taken into account in assessing the compaction potential at locations where the groundwater table has been significantly lowered.

In summary, assessing the relative potentials for peat compaction of Holocene deltaic sequences requires determining the (1) lithological composition of the sequence, (2) the geotechnical properties of the different lithologies, which is predominantly a function of organic-matter and clastic content, and post-depositional processes of loading, and physical, geochemical and biological processes, determining the present degree of compaction and geotechnical properties, and (3) the effective stress, which is a function of the weight of the overburden and pore water pressure that is influenced by respectively the built-up of the sequence and its properties and fluctuations in groundwater level or the absence thereof. Based on this information, the future potential for additional subsidence due to peat compaction can be estimated.

7 Conclusion

The potential for subsidence due to peat compaction of Holocene deltas is mainly determined by the 3-D distribution of different lithologies, and associated geotechnical properties, in the subsurface. In this study the amount of compaction within 3 sequences that are representative for distal deltaic settings is determined to assess the potential of comparable sequences in other deltas. In all sequences A–C the peat layers are compacted. Relatively high compaction values occur in peat layers within approximately 120 cm below surface, which are subjected to compaction and oxidation following groundwater level lowering. Furthermore, high amounts of compaction occur in peat layers beneath nat-

ural levee and crevasse-splay deposits, and at the base of the Holocene sequences, due to the weight of the overburden. In compacted peat layers, high-organic peat generally compacts more than low-organic peat.

References

Berendsen, H. J. A. and Stouthamer, E.: Late Weichselian and Holocene palaeogeography of the Rhine–Meuse delta, the Netherlands, Palaeogeogr. Palaeocl., 161, 311–335, 2000.
Bos, I. J.: Architecture and facies distribution of organic-clastic lake fills in the fluviodeltaic rhine-meuse system, the Netherlands, J. Sediment. Res., 80, 339–356, 2010.
Bird, M. I., Fifield, L. K., Chua, S., and Goh, B.: Calculating sediment compaction for radiocarbon dating of intertidal sediments, Radiocarbon 46, 421–435, 2004.
Church, J. A., Clark, P. U., Cazenave, A., Gregory, J. M., Jevrejeva, S., Levermann, A., Merrifield, M. A., Milne, G. A., Nerem, R. S., Nunn, P. D., Payne, A. J., Pfeffer, W. T., Stammer, D., and Unnikrishnan, A. S.: Sea Level Change, in: Climate Change 2013: The Physical Science Basis. Contribution of Working Group I to the Fifth Assessment Report of the Intergovernmental Panel on Climate Change, edited by: Stocker, T. F., Qin, D., Plattner, G.-K., Tignor, M., Allen, S. K., Boschung, J., Nauels, A., Xia, Y., Bex, V., and Midgley, P. M., Cambridge University Press, Cambridge, UK, New York, NY, USA, 2013.
Cohen, K. M.: Differential subsidence within a coastal prism: late-Glacial – Holocene tectonicsin The Rhine Meuse delta, the Netherlands, PhD thesis Utrecht University, Utrecht, the Netherlands, Geographical Studies, 316, 2003.
Cohen, K. M.: 3D Geostatistical interpolation and geological interpretation of paleo-groundwater rise in the Holocene coastal prism in the Netherlands, in: River Deltas – Concepts, models, and examples, edited by: Giosan, L. and Battacharya, J. P., SEPM Special Publication, 83, 341–364, 2005.
Erban, L. E., Gorelick, S. M., and Zebker, H.: Groundwater extraction, land subsidence, and sea-level rise in the Mekong Delta, Vietnam, Environ. Res. Lett., 9, 1–6, 2014.
Gouw, M. J. P.: Alluvial architecture of the Holocene Rhine-Meuse delta, the Netherlands, Sedimentology, 55, 1487–1516, 2008.
Gouw, M. J. P. and Erkens, G.: Architecture of the Holocene Rhine-Meuse delta (the Netherlands) – A result of changing external controls, Neth. J. Geosci., 86, 23–54, 2007.
Heiri, O., Lotter, A. F., and Lemcke, G.: Loss on ignition as a method for estimating organic and carbonate content in sediments: Reproducibility and comparability of results, J. Paleolimnol., 25, 101–110, 2001.
Stouthamer, E.: Sedimentary products of avulsions in the Holocene Rhine-Meuse Delta, the Netherlands, Sediment. Geol., 145, 73–92, 2001.
Stouthamer, E. and Berendsen, H. J. A.: Factors controlling the Holocene avulsion history of the Holocene Rhine-Meuse delta (the Netherlands), J. Sediment. Res. A, 70, 1051–1064, 2000.
Stouthamer, E., Cohen, K. M., and Gouw, M. J. P.: Avulsion and its implications for fluvial-deltaic architecture: insights from the Holocene Rhine-Meuse delta. in: From River to Rock Record: The preservation of fluvial sediments and their subsequent interpretation, edited by: Davidson, S. K., Leleu, S., and North, C. P., Society for Sedimentary Geology, 11, 215–232, 2011.

Terzaghi, K.: Theoretical Soil Mechanics, John Wiley and Sons, New York, USA, 510 pp., 1943.

van Asselen, S.: The contribution of peat compaction to total basin subsidence: Implications for the provision of accommodation space in organic-rich deltas, Basin Res., 23, 239–255, 2011.

van Asselen, S. and Roosendaal, C.: A new method for determining the bulk density of uncompacted peat from field settings, J. Sediment. Res., 79, 918–922, 2009.

van Asselen, S., Stouthamer, E., and Van Asch, Th. W. J.: Effects of peat compaction on delta evolution: A review on processes, responses, measuring and modelling, Earth-Sci. Rev., 92, 35–51, 2009.

van Asselen, S., Stouthamer, E., and Smith, N. D.: Factors controlling peat compaction in alluvial floodplains – A case study in the cold-temperate Cumberland Marshes, Canada, J. Sediment. Res., 80, 155–166, 2010.

van Asselen, S., Karssenberg, D., and Stouthamer, E.: Holocene peat compaction in fluvial lowlands: Implications to subsidence within deltas, Geophys. Res. Lett., 38, L24401, doi:10.1029/2011GL049835, 2011.

van de Ven, G. P.: Leefbaar laagland. Geschiedenis van de waterbeheersing en landaanwinning in Nederland, Utrecht: Uitgeverij Matrijs, 456 pp., 2003.

33

Continuous monitoring of an earth fissure in Chino, California, USA – a management tool

M. C. Carpenter

formerly at: U.S. Geological Survey, Tucson, Arizona, USA

Correspondence to: M. C. Carpenter (mccarp@dakotacom.net)

Abstract. Continuous measurements of deformation have been made in Chino, California across an earth fissure and nearby unfissured soil since 2011 in two buried, horizontal, 150 mm pipes, 51 m long, which are connected by sealed boxes enclosing vertical posts at mostly 6 m intervals. Horizontal displacements and normal strain are measured in one line using nine end-to-end quartz tubes that are attached to posts and span fissured or unfissured soil. The free ends of the tubes are supported by slings and move relative to the attachment post of the next quartz tube. Linear variable differential transformer (LVDT) sensors measure the relative movements. Five biaxial tilt sensors were also attached to selected posts in that line. Relative vertical movement was measured at nine locations along the line in the second pipe using low-level differential pressure sensors. The second pipe is half full of water giving a free water surface along its length. Data are recorded on a Campbell CR10 using multiplexers.

The quartz-tube horizontal extensometers have exhibited more than 3 mm of predominantly elastic opening and closing in response to about 32 m of seasonal drawdown and recovery, respectively, in an observation well 0.8 km to the south. The nearest production well is 1.6 km to the west. The horizontal strain was 5.9×10^{-5} or 30 % of the lowest estimate of strain-at-failure for alluvium. Maximum relative vertical movement was 4.8 mm. Maximum tilt in the fissure zone was 0.09 arcdeg while tilt at a separate sensor 100 m to the east was 0.86 arcdeg, indicating a wider zone of deformation than is spanned by the instrumentation. High correlation of horizontal displacements during drawdown, and especially recovery, with change in effective stress supports differential compaction as the mechanism for earth-fissure movement.

The continuous measurements of horizontal strain coupled with water-level fluctuations and vertical borehole extensometry can provide a real-time adaptive management tool for restricting pumping if strain approaches the lower limit of strain-at-failure or a stress-strain curve deviates from the previous mostly elastic regimen.

1 Introduction

A zone of earth fissures opened as early as 1974 on the California Institute for Men prison in Chino, California. Renewed fissuring occurred in 1991 on prison grounds and northward for 1.6 km in a zone as much as 150 m wide (Stewart et al., 1998). In the 1990s, the fissure split a house and drained a liquid manure pond at a dairy, possibly directly contaminating the shallow aquifer. The house was condemned and destroyed. As a result of management of groundwater pumping by the Chino Basin Watermaster, fissuring has been minimal since the 1990s and limited to asphalt parking lots and vacant properties. The structural setting of the Chino earth fis-

sure is a north-south, strike-slip, fault-zone, groundwater barrier that bounds a subsidence bowl to the west and concentrates differential subsidence over a limited horizontal distance (Wildermuth Environmental, 2007).

2 Instrumentation

In October 2011, instruments were installed in a west to east line at 5500 Daniels St., Chino, California, in a vacant lot, which had been the location of the dairy and destroyed house (Fig. 1). The instruments provide a continuous record of horizontal displacements, differential vertical movement, and tilt across the earth fissure and nearby unfissured soil. Two buried, horizontal, 150 mm pipes, 51 m long, are connected by sealed boxes enclosing vertical, 3-m, grounding rod posts driven into the ground at mostly 6 m intervals. Data are recorded on a Campbell CR10 datalogger using multiplexers.

Horizontal displacements are measured in one line using nine end-to-end quartz tubes that are attached to posts and span fissured or unfissured soil. Tubes are 15 mm OD, 13 mm ID. Trans-Tek ± 6 mm, 0242-0000 linear variable differential transformer (LVDT) sensors measure the relative displacements. The free ends of the tubes are supported by beaded-chain slings and move relative to the attachment post of the next quartz tube. A tube is attached to a post by means of an aluminum strip that extends behind the end of its attachment post, and the LVDT is attached to a nonmagnetic stainless steel strip that extends beyond the end of its post. The length of aluminum times its coefficient of expansion $(2.4 \times 10^{-5}\,^\circ\mathrm{C}^{-1})$ plus the length of stainless steel times its coefficient of expansion $(1.2 \times 10^{-5}\,^\circ\mathrm{C}^{-1})$ equals the length of quartz times its coefficient of expansion $(5.7 \times 10^{-7}\,^\circ\mathrm{C}^{-1})$. This compensating design reduces temperature effects by as much as an order of magnitude.

Relative vertical movement was measured at nine locations along the line in the second pipe using Honeywell Micro Switch 24PCEFA6D ± 34 hPa $(\pm 0.5$ psi) low-level differential pressure sensors. Differential silicon-strain-gage pressure sensors exhibit a voltage shift in response to a pressure change common to both ports. For low-level sensors used for high precision, a secondary common-mode calibration must be done over the range of barometric-pressure change. An adjustment can be applied or sensors can be selected with identical shifts. The second pipe is half full of water giving a free water surface along its length.

Five Applied Geomechanics (now Jewell Instruments) 902-TH biaxial tilt sensors were attached to selected posts in the quartz-tube line. Temperature sensors incorporated in the tilt sensors provided the temperature measurements for the instrument enclosures. Biaxial tilt sensors provide the easiest information for the extent to which deformation is restricted to a vertical plane orthogonal to the fissure.

3 Results

From west to east the quartz-tube spans are designated Q1–Q3, Q11, and Q4–Q8 (Fig. 2). Q11 spans the historical fissure zone. Curves represent cumulative displacement along the line. For example, Q8 is the sum of all displacements Q1–Q8. The quartz-tube horizontal extensometers have exhibited more than 3 mm of predominantly elastic opening and closing in response to about 32 m of seasonal drawdown and recovery, respectively, in an observation well 0.8 km to the south. The nearest production well is 1.6 km to the west, screened in a deep confined aquifer at depths of 91–

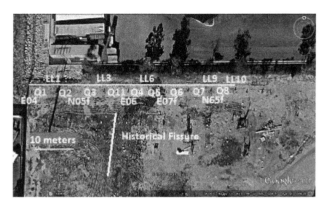

Figure 1. Instrument layout at 5500 Daniels horizontal extensometer site, Chino, California. Q1–Q8 are quart-tube horizontal extensometers. LL1-LL10 are liquid-level sensors. E04-N65f are biaxial tilt sensors.

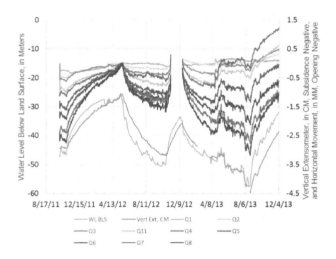

Figure 2. Quartz-tube horizontal movement compared with water-level fluctuation in an observation well and deep-extensometer compaction 0.8 km to the south.

140 m and 152–298 m. The horizontal strain was 5.9×10^{-5} or 30 % of the lowest estimate of strain-at-failure for alluvium (Jachens and Holzer, 1982).

Coefficients of determination for the pairs water-level decline and deep extensometer, water-level decline and horizontal displacements (Q8), and deep extensometer and horizontal displacements (Q8) range from 0.89 to 0.99 (Table 1). The lower correlation between horizontal displacements and both water-level decline and the deep extensometer during drawdown 2013 reflects closing of and major deviation of span Q4 during that period.

The deviation of span Q4 began during drawdown 2012 when that span closed while spans to the east of Q5 opened uniformly. That deviation indicates concentration of fissure opening across span Q5. A much larger closing across Q4 occurred during drawdown beginning in April 2013. That closing, little movement across spans Q1 and Q2, and fairly

Table 1. Coefficients of determination for periods of recovery and drawdown for water-level fluctuations, deep extensometer compaction, and horizontal movements.

	Water-level decline and deep extensometer	Water-level decline and horizontal movements	Deep extensometer and horizontal movements
Recovery winter 2011–2012	0.93	0.95	0.99
Drawdown 2012	0.95	0.97	0.93
Drawdown 2013	0.95	0.89	0.89
Recovery winter 2013	0.98	0.97	0.98

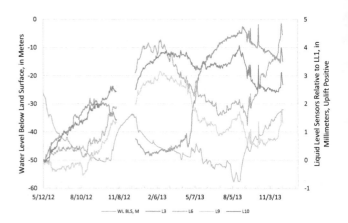

Figure 3. Vertical movement of liquid-level sensors with respect to LL1 compared with water-level fluctuation in observation well 0.8 km to the south. Sensor locations are in Fig. 1.

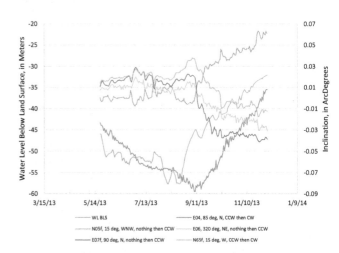

Figure 4. Biaxial tilt compared with water-level fluctuation in observation well 0.8 km to the south. Sensor locations are in Fig. 1. The legend gives the azimuth of the vertical plane of tilt, viewing direction, and sense of rotation for two segments of each graph. CW is clockwise; CCW is counterclockwise. Response to water-level fluctuation is delayed about one month and sensors on opposite sides of the fissure zone respond with the opposite sense. Only E04 and E07f exhibited tilt in the vertical plane orthogonal to the fissure.

uniform opening and closing across spans Q3, Q11, and Q5–Q8 indicates that most fissure movement was concentrated across span Q5, at least 6 m east of the historic fissure zone, Q11.

The liquid-level, vertical-movement instrumentation is a work in progress. Sensors were built using the wire bundle of the cable as the vent tube. This technique had worked exceptionally well in the past, but excessive noise existed in the liquid-level data, presumably because of tight bends in the cable restricting pressure equalization in the reference side of the sensors. Venting the pipe to the atmosphere, piercing the cable, and inserting a syringe needle with a compliant diaphragm (sandwich bag) improved five of the sensors, LL1, LL3, LL6, LL9, and LL10. Movement of the other sensors is with respect to LL1 (Fig. 3).

The relation between liquid-level sensors and water-level fluctuation is not clearly defined. Maximum relative vertical movement was 4.7 mm at LL3. During drawdown 2012 all sensors exhibited uplift with respect to LL1, and LL3 reversed during late drawdown. During drawdown 2013, LL6, LL9, and LL10 exhibited similar reversing movement during early drawdown followed by decline during late drawdown, and LL3 experienced uplift delayed by more than 4 months. In contrast with horizontal strain, vertical movement in the liquid level sensors is delayed with respect to water-level fluctuations. Sensors on opposite sides of the fissure generally exhibit opposite movement.

Tilt sensors exhibited response to water-level fluctuation that was delayed by about 1 month (Fig. 4). E04 and N05f on the west side tracked water level fluctuation, while E06, E07f, and N65f exhibited opposite movement. Only tilts at E04 and E07f are in the vertical plane orthogonal to the fissure. Tilts at N05f, E06 and N65f were 75, 40, and 75 arcdeg off axis, respectively.

Maximum tilt in the fissure zone was 0.09 arcdeg while tilt at a separate sensor 100 m to the east was 0.86 arcdeg, indicating a wider zone of deformation than is spanned by the instrumentation (Fig. 5). Tilt at Station 23 tracks water-level fluctuation with a 1-month delay similar to the sensors near the fissure zone.

4 Fissure mechanisms

Among several mechanisms proposed for earth fissuring, differential subsidence caused by differential aquifer compaction at depth remains the most substantiated (Fig. 6). Lateral change in thickness of the compressible aquifer-aquitard system, lateral change in compressibility within the system, or abrupt lateral change in water-level decline can cause fissuring. The conventional model treats the sediments overlying the zone of differential compaction as a bending plate. In that model, maximum horizontal strain is at the surface (Jachens and Holzer, 1982). Field evidence at many fissures indicates that fissures can have greater horizontal strain at

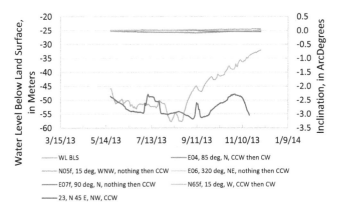

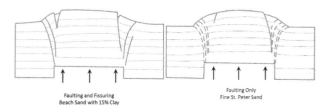

Figure 7. Results of sandbox experiments demonstrating fissuring and faulting consistent with shear and tensile failure of an elastic body exhibiting differential offset at depth (modified from Sanford, 1959 by Carpenter, 1993).

Figure 5. Biaxial tilt compared with water-level fluctuation in nearby observation well. Sensor locations are in Fig. 1. Station 23 is 100 m east of the fissure. The legend gives the azimuth of the vertical plane of tilt, viewing direction, and sense of rotation for two segments of each graph. CW is clockwise; CCW is counterclockwise. Deformation extends at least 100 m east of the fissure zone, and tilt at Station 23 is approximately ten times greater than tilt in the fissure zone.

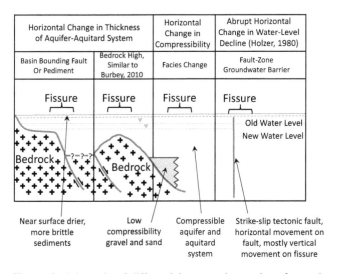

Figure 6. Schematic of differential compaction settings for earth fissures.

depth than at the surface (Carpenter, 1999). An earlier model of upward propagation of stress and strain in an elastic body (Sanford, 1959) solves this dilemma. In addition to the elastic model, Sanford performed physical experiments that demonstrate the patterns of deformation that can develop at the land surface in response to differential offset at depth (Fig. 7). High correlation of horizontal strain with change in effective stress supports differential compaction as the mechanism for earth-fissure movement.

Horizontal seepage stress is another mechanism proposed for earth fissuring (Helm, 1994). In Carpenter (1993), water-level fluctuations in pairs of piezometers on opposite sides of a fissure compared with horizontal strain across the fissure

enabled a test of seepage stresses as a fissure mechanism. The coefficients of determination between water-level fluctuation in the piezometers (change in effective stress) and horizontal strain ranged from 0.83 to 0.85, while the coefficients of determination for horizontal seepage stresses and horizontal strain ranged from 0.03 to 0.17. In this study, coefficients of determination between water-level fluctuations and horizontal strain range from 0.89 to 0.97 (Table 1). During water-level recovery, fissures close (Fig. 2). Unless proponents of seepage stresses have a mechanism of water flowing back uphill during recovery dragging aquifer matrix, the mechanism of seepage stresses should be dismissed.

5 Discussion

The precise measurements of this study are incorporated in a larger monitoring network of semi-annual levelling and EDM with monument spacing of more than 400 m. With error bars of ±3 mm horizontal, the seasonal elastic deformation of about 30 % of strain-at-failure would be invisible in that network. This difficulty illustrates the need for closer monument spacing and greater precision in zones of known or suspected fissuring. Monument spacing of 12 m would enable delimitation of a zone to be instrumented, and infilling of monuments with 2 m spacing would enable spacing of quartz-tube monuments.

At present, the horizontal quartz tubes are the best established technique for continuous monitoring (Davis et al., 1969; Wolff, 1970). Displacement resolution can be finer than 0.1 μm. Biaxial tilt sensors (Haneberg and Friesen, 1995) are the easiest to install but, as point measurements, may be the least definitive technique. The liquid-level technique adapted after Riley (1970) remains unverified by precise levelling but has great promise.

Continuous measurements of horizontal displacements using quartz tubes presents a unique method for real-time adaptive management to prevent fissure opening or reopening. Municipal groundwater pumping can be restricted if horizontal strain approaches the lower limit of strain-at-failure or, coupled with water-level fluctuations and vertical borehole extensometry, stress-strain curves deviate from the previous mostly elastic regimen. Fissures have caused catastrophic damage to buildings and infrastructure. A municipal water agency that actively produces groundwater from fine-grained aquifer units runs the risk of causing such damage. This monitoring tool can be used to keep their operations safely below renewed fissuring.

The clarity of response of the horizontal-displacement data demonstrates the superiority of resolution of quartz tubes over EDM surveys and the superiority of continuous measurements over periodic surveys. The seasonal elastic deformation, which constitutes a significant percentage of strain at failure would be lost in the error bars of repeated surveys.

6   Recommendations

Wherever both horizontal and vertical deformation have been made at earth fissures, they have been of comparable magnitude (Carpenter, 1993, 1999). Fissures termed surface faults by Holzer (1980) that have significant vertical offset can pose particular hazards. Fissure opening of as much as 25 mm has been documented in a few hours during major frontal storms (Carpenter, 1999). A sudden vertical offset of 50–75 mm in a concrete highway during a storm with limited visibility could cause a multi-vehicle accident resulting in several deaths. Repeated precise levelling and EDM surveys where known or suspected fissures cross highways would indicate locations where GPS receivers could be placed along the highway on opposite sides of the fissure. Pairs of receivers could be monitored automatically for a threshold change that would trigger a warning sign or highway closing.

Fissures have been known to close over many years. That is, monuments on opposite sides have converged in repeated surveys in addition to infilling of the void space by collapse and deposition. In many cases the closure has been accompanied by water-level recovery (Carpenter, 1999). This presents the intriguing possibility of fissure mitigation by "wait and see" or determination of a rate of water-level decline that could be sustained without causing renewed fissuring. T. Holzer's old USGS project and the old USGS Subsidence Research Project EDM and levelling measurements at numerous fissures in the Western United States could be remeasured using original instruments and state of the art instruments, located by precise GPS, and compared with long-term water-level fluctuations. Application of a Bingham-substance model with appropriate plastic, elastic, and viscous material properties could lead to a greater understanding of long-term fissure behaviour (LSCE et al., 2014).

Fissure systems are varied and complex. Two sets of published records of continuous measurements of fissure movement have been made to date (Carpenter, 1993, and this study). As droughts lengthen and basins become more depleted, fissuring will increase. Several more sets of continuous measurements would add immensely to the understanding of fissures and increase the possibility of finding mitigating measures. In particular, comparison of fissures associated with fault-zone groundwater barriers, fissures associated with known subsurface structure, and fissures without known subsurface structure would be desirable.

A study would begin with repeated EDM and levelling over monuments with 40 m spacing on a line 500 m long (LSCE et al., 2014). Deformation within that line would determine the locations of more closely spaced surveying monuments that would be used for selection of quartz-tube posts and liquid-level stations with 6-m spacing. Diameter of the quartz tubes should be increased to 18 mm OD, 15 mm ID for improved rigidity. Permanent instrument and target monuments of 100 mm pipe crash posts for precise EDM would be installed beside the quartz-tube posts. Semiannual repeated precise EDM (Leica TC2003 or TDM5000) and levelling (Leica DNA03) would be compared with continuous quartz-tube horizontal displacements, liquid-level vertical movements, and tilt. Precise surveying would extend over greater lengths than the instruments providing continuous record. Surveying and continuous measurements would provide cross checks at the limit of resolution of the surveying instruments.

Acknowledgements. Funding for this work is provided by Wildermuth Environmental, Incorporated under the auspices of the Chino Basin Watermaster. Richard Wilson and Eric Lindberg provided field assistance and reviews of this paper. George Pardo also provided field assistance.

References

Burbey, T. J.: Mechanisms for earth fissure formation in heavily pumped basins, Land Subsidence, Associated Hazards and the Role of Natural Resources Development, Proceedings of EISOLS 2010, 17–22 October 2010, Queretaro, Mexico, IAHS Publ. 339, 3–8, 2010.

Carpenter, M. C.: Earth-fissure movements associated with fluctuations in ground-water levels near the Picacho Mountains, south-central Arizona, 1980-84, U.S. Geological Survey Professional Paper 497-H, 49 pp., available at: http://pubs.usgs.gov/pp/0497h/report.pdf, last access: 15 October 2015, 1993.

Carpenter, M. C., South-central Arizona, in: Land Subsidence in the United States, edited by: Galloway, D., Jones, D., and Ingebritsen, S., U.S. Geological Survey Circular 1182, 65–78, available at: http://pubs.usgs.gov/circ/circ1182/pdf/09Arizona.pdf, last access: 15 October 2015, 1999.

Davis, S. N., Peterson, F. L., and Halderman, A. D.: Measurement of small surface displacements induced by fluid flow, Water Resour. Res., 5, 129–138, 1969.

Haneberg, W. C. and Friesen, R. L.: Tilts, strains, and ground-water levels near an earth fissure in the Mimbres Basin, New Mexico, Geol. Soc. Am. Bull., 107, 316–326, 1995.

Helm, D. C.: Hydraulic forces that play a role in generating fissures at depth, Bulletin of the Association of Engineering Geologists, 31, 293–304, 1994.

Holzer, T. H.: Faulting caused by groundwater level declines, San Joaquin Valley, California, Water Resour. Res., 16, 1065–1070, 1980.

Jachens, R. C. and Holzer, T. L.: Differential compaction mechanism for earth fissures near Casa Grande, Arizona, Geol. Soc. Am. Bull., 93, 998–1012, 1982.

LSCE (Luhdorff & Scalmanini Consulting Engineers), Borchers, J. W., and Carpenter, M. C.: Land subsidence from groundwater use in California, California Water Foundation, 151 pp., available at: http://californiawaterfoundation.org/uploads/1397858208-SUBSIDENCEFULLREPORT_FINAL.pdf, last access: 15 October 2015, 2014.

Riley, F. S.: Land-surface tilting near Wheeler Ridge, Southern San Joaquin Valley, California, U.S. Geological Survey Professional Paper 497-G, 29 pp., 1970.

Sanford, A. R.: Analytical and experimental study of simple geologic structures: Geol. Soc. Am. Bull., 70, 19–52, 1959.

Stewart, C. A., Colby, N. D., Kent, R. T., Eagan, J. A., and Hall, N. T.: Earth fissuring, ground-water flow, and water quality in the Chino basin, California, in: Dr Joseph F. Poland symposium, Land subsidence case studies and current research: proceedings of the Dr Joseph F Poland symposium on land subsidence; 1995, Sacramento, CA, edited by: Borchers, J. W., Association of Engineering Geologists Special Publication 8, 195–205, 1998.

Wildermuth Environmental: MZ-1 Subsidence Management Plan, variously paginated, Chino Basin Watermaster, Rancho Cucamonga, California, USA, available at: http://ayala.wildermuthenvironmental.com:8888/AyalaPark/documents/20071017_MZ1_Plan.pdf, last access: 15 October 2015, 2007.

Wolff, R. G.: Relationship between horizontal strain near a well and reverse water level fluctuation, Water Resour. Res., 6, 1721–1728, 1970.

34

Improving predictions of the effects of extreme events, land use, and climate change on the hydrology of watersheds in the Philippines

Rubianca Benavidez[1]**, Bethanna Jackson**[1]**, Deborah Maxwell**[1]**, and Enrico Paringit**[2]

[1]School of Geography, Environment and Earth Sciences, Victoria University of Wellington, Wellington, 6012, New Zealand
[2]Disaster Risk and Exposure Assessment for Mitigation Program, Quezon City, 1101, Philippines

Correspondence to: Rubianca Benavidez (rubianca.benavidez@vuw.ac.nz)

Abstract. Due to its location within the typhoon belt, the Philippines is vulnerable to tropical cyclones that can cause destructive floods. Climate change is likely to exacerbate these risks through increases in tropical cyclone frequency and intensity. To protect populations and infrastructure, disaster risk management in the Philippines focuses on real-time flood forecasting and structural measures such as dikes and retaining walls. Real-time flood forecasting in the Philippines mostly utilises two models from the Hydrologic Engineering Center (HEC): the Hydrologic Modeling System (HMS) for watershed modelling, and the River Analysis System (RAS) for inundation modelling. This research focuses on using non-structural measures for flood mitigation, such as changing land use management or watershed rehabilitation. This is being done by parameterising and applying the Land Utilisation and Capability Indicator (LUCI) model to the Cagayan de Oro watershed (1400 km^2) in southern Philippines. The LUCI model is capable of identifying areas providing ecosystem services such as flood mitigation and agricultural productivity, and analysing trade-offs between services. It can also assess whether management interventions could enhance or degrade ecosystem services at fine spatial scales. The LUCI model was used to identify areas within the watershed that are providing flood mitigating services and areas that would benefit from management interventions. For the preliminary comparison, LUCI and HEC-HMS were run under the same scenario: baseline land use and the extreme rainfall event of Typhoon Bopha. The hydrographs from both models were then input to HEC-RAS to produce inundation maps. The novelty of this research is two-fold: (1) this type of ecosystem service modelling has not been carried out in the Cagayan de Oro watershed; and (2) this is the first application of the LUCI model in the Philippines. Since this research is still ongoing, the results presented in this paper are preliminary. As the land use and soil parameterisation for this watershed are refined and more scenarios are run through the model, more robust comparisons can be made between the hydrographs produced by LUCI and HEC-HMS and how those differences affect the inundation map produced by HEC-RAS.

1 Introduction

The Philippines is regularly struck by tropical cyclones. An annual average of twenty enter the region, of which about nine make landfall and often cause destructive floods (Lasco et al., 2009). In 2011, Typhoon Washi caused heavy flooding and destruction in the Cagayan de Oro region, leaving 1268 casualties and PHP 2 billion in damages ($\sim$ USD 46 million in 2011) (National Disaster Risk Reduction Management Council, 2012a). Since the Philippines is located within the typhoon belt, the country is vulnerable to the effects of climate change due to possible increases in typhoon frequency and intensity (Intergovernmental Panel on Climate Change, 2012). The vulnerability of the country to tropical cyclones and climate change underscores the need for more proactive disaster risk management, which to date is mainly carried out through flood forecasting and structural flood protection projects.

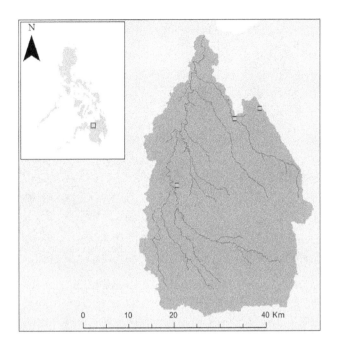

Figure 1. The Cagayan de Oro watershed and rain gauges relative to the Philippines.

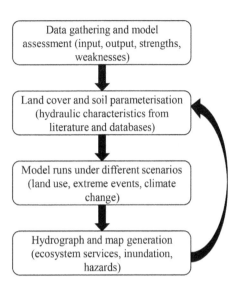

Figure 2. Methodology of the study.

The official programme of disaster mitigation in the Philippines is Project NOAH (Nationwide Operational Assessment of Hazards). This programme mainly uses two models from the United States Hydrologic Engineering Center (HEC): the Hydrologic Modeling System (HMS) for watershed modelling, and the River Analysis System (RAS) for inundation modelling (Brunner, 2010; Scharffenberg, 2013). Project NOAH uses these models for real-time flood forecasting and produces flood hazard maps available on a public website (Santillan et al., 2013). These models have also been used to analyse the watershed response to extreme events and changing land use, estimating peak flow and occurrence time, and modelling the floodplain inundation (Mabao and Cabahug, 2014; Santillan et al., 2011).

Although these models can assist in decision-making through running different land use scenarios, they do not identify areas to target for potential management interventions. Reforestation activities in the Cagayan de Oro (CDO) watershed (Fig. 1) tend to occur in easily-accessible areas, which may not be the areas that would optimally benefit from management interventions (Center for Environmental Studies and Management, 2014). This research applies a model developed to aid in scientifically-sound decision-making regarding land management: the Land Utilisation and Capability Indicator (LUCI). This is being used to identify areas providing existing ecosystem services, areas where management interventions can enhance or degrade ecosystem services, trade-offs between services, and impacts of changing land cover (Jackson et al., 2013). This type of ecosystem services modelling has not been carried out in the Philippines to

date, and the LUCI model can aid in more sustainable land management in the CDO watershed. This research is also the first application of LUCI in a tropical catchment, which will allow testing of how well the model is able to represent hydrological processes in such regions.

This research is still on-going, so the results presented in this paper are preliminary. As the soil and land use parameterisation is refined, the models will be run under more scenarios of changing land use, extreme events, and extreme events affected by climate change. The main aims of this research are: (1) to apply the LUCI model to the Cagayan de Oro catchment to help identify priority areas for land management; (2) assess how the LUCI model fits into the existing disaster risk management framework in the Philippines, and (3) assess how the hydrological response of the watershed changes under different scenarios of land use, extreme events, and climate change.

2 Methodology

An overview of the methodology is shown in Fig. 2. First, thematic datasets of land cover, soil, and topography were gathered from various sources and at different scales (local, national, global). Next, land cover and soil parameterisation for the LUCI model was carried out through correlating local soil names with the United States Department of Agriculture (USDA) soil taxonomy, and using the Soil Water Characteristics (SWC) model to estimate the soils' hydraulic characteristics (Saxton and Rawls, 2006). The watershed and floodplain model in HMS and RAS were made and parameterised by the Disaster Risk and Exposure Assessment for Mitigation Program (DREAM) (2015).

In LUCI, the land cover and soil dataset was used to produce ecosystem services maps in order to identify which ar-

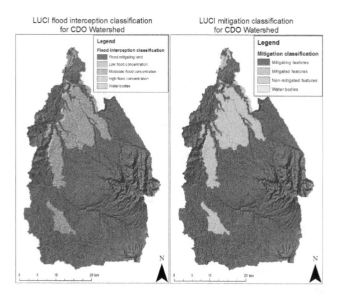

Figure 3. Flood interception and mitigation classification for the CDO watershed.

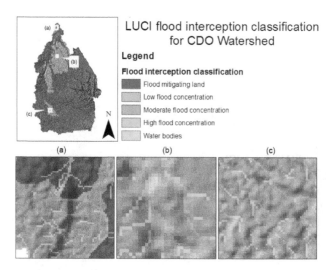

Figure 4. A zoom of some areas of the watershed to show flood interception at a smaller scale.

eas could be targeted for management interventions to improve flood mitigation.

Although this study aims to run the models under different scenarios, this paper presents one extreme event scenario: Typhoon Bopha (local name: Pablo). In December 2012, Typhoon Bopha hit the region and caused a total of 1067 casualties and estimated PHP 36 billion in damages ($\sim$ USD 870 million in 2012) (NDRRMC, 2012b). In this scenario, the land cover and soil dataset was the same dataset used to construct the current flood forecasting model for the CDO watershed. The rainfall data was taken from three different rain gauges: Bubunawan, Libona, and Talakag, and was provided by the DREAM Program (2015).Using the Typhoon Bopha rainfall data, rainfall-runoff modelling was performed in both LUCI and HEC-HMS to produce hydrographs for comparison. Both hydrographs were then used as input into HEC-RAS for inundation modelling, and the resulting inundation maps were prepared in ArcMap 10.3.1.

3 Results and discussion

This section presents the preliminary results of this research: the initial LUCI results for the CDO watershed, the hydrographs from LUCI and HEC-HMS, and the resulting inundation maps from HEC-RAS.

3.1 Initial LUCI results

The LUCI model is used to assess the existing and potential impacts of changes in land use management on the following ecosystem services: habitat connectivity, flood mitigation, erosion/sediment delivery, carbon sequestration, and agricultural productivity (Jackson et al., 2013). The flood

mitigation service is the main focus for this paper, but this LUCI application aims to model the rest of the ecosystem services and trade-offs as the land cover and soil parameterisation is further refined. The model makes use of a "traffic light" system for its output maps: red means to stop or that changes in land use may degrade the ecosystem service, yellow/orange means that careful management interventions can be carried out, and green areas are priority areas since management interventions have the potential to improve the service (Jackson et al., 2013).

Figure 3 shows the initial LUCI results for the study site for the ecosystem service of flood mitigation. For the flood interception map, red areas are those providing good flood mitigation and any land management changes are likely to adversely affect its capability for flood mitigation. Conversely, green areas are areas of high and moderate flood concentration and can be targeted for management interventions such as rehabilitation or reforestation. The yellow/orange areas of flood interception are areas of low flood concentration, but can be targeted for management interventions since the LUCI algorithm identifies them as non-mitigated features.

At the watershed scale, the green areas or areas that are possible targets for management interventions are not immediately obvious. Figure 4 shows that the model has identified areas for management interventions at the smaller scale, identifying areas where flow accumulates before reaches streams and rivers. Based on the initial LUCI results, small-scale and targeted management efforts have the potential to be more useful to improving flood mitigation services than choosing large portions of land for reforestation. This is important because limited funding for watershed rehabilitation was one of the problems identified in the development of the watershed management plan, so a smaller-scale but more targeted approach may be more beneficial (CESM, 2014). As

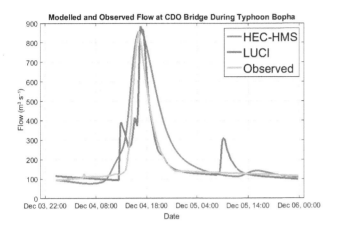

Figure 5. The modelled hydrographs from HEC-HMS and LUCI compared to the observed flow during Typhoon Bopha.

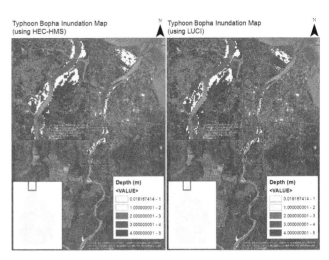

Figure 6. Inundation maps produced in HEC-RAS using the HEC-HMS hydrograph and the LUCI hydrograph, with the inset showing the flood water encroaching on the river banks and buildings.

the soil and land cover parameterisation in LUCI is refined and iterative model runs are done, this application can produce a much clearer picture of priority rehabilitation areas in the watershed and be used to assess the potential result of the local government's plans for rehabilitation or utilisation.

3.2 Typhoon Bopha results: LUCI, HEC-HMS, and HEC-RAS

Under the scenario of Typhoon Bopha, the hydrographs of HEC-HMS and LUCI compare reasonably well with the observed flow at the Cagayan de Oro Bridge (Fig. 5). The peak flows were: $864\,679\,\mathrm{m^3\,s^{-1}}$ (HEC-HMS), $883\,963\,\mathrm{m^3\,s^{-1}}$ (LUCI), and $858.3\,\mathrm{m^3\,s^{-1}}$ (observed). Both hydrographs from LUCI and HEC-HMS show a similar peak flow volume and occurrence time, as well as a smaller peak after the main flood event. The LUCI hydrograph shows a flashier watershed response and a later peak flow occurrence time compared to the HEC-HMS hydrograph. However, the LUCI hydrograph shows a more similar falling limb when compared to the observed flow. Another similarity between the LUCI hydrograph and the observed flow is the flood volume, as LUCI is able to capture the more abrupt rise and fall of the flood peak, while the HEC-HMS hydrograph shows a more gradual rise and fall, implying an over-prediction of the flood volume.

When both hydrographs are run through HEC-RAS, the produced inundation maps are relatively similar (Fig. 6). The maximum inundation depths were: 4.82 (HEC-HMS) and 4.63 m (LUCI). Although the inundation extent using the HEC-HMS hydrograph is larger compared to the inundation extent produced by the LUCI hydrograph, the inundation occurs in the same areas: at the mouth of the river and areas where the river meanders. Based on this initial comparison, LUCI is able to predict the watershed's response with similar performance to the HEC-HMS model, and the inundation

map based on the LUCI hydrograph is also similar to the inundation map based on the HEC-HMS hydrograph.

LUCI, HEC-HMS, and HEC-RAS can be used in combination to complement each other's strengths and weaknesses. For example, LUCI has the capability to go down to significantly finer spatial scales than HEC-HMS. Given its fine resolution, LUCI can run reasonably efficiently at the watershed scale, which has advantages for flood forecasting and uncertainty estimation. However, HEC-HMS takes less processing time to run over the entire watershed. The current configuration of HEC-HMS used by the disaster risk programme uses the curve number method to determine the sub-basin runoff. This method uses one composite number to represent the different soil and land use combinations present within a sub-basin (Scharffenberg, 2013). While this allows the programme to take less processing time, it cannot capture the explicit spatial configuration of elements within the watershed. This makes it less useful for assisting in decision-making regarding land use management. By contrast, the LUCI model uses the soil and land use data to account for the storage and permeability of the different elements within the watershed, which allows the model to perform watershed modelling and represent possible management impacts at fine spatial scales (Jackson et al., 2013). This method allows LUCI to take spatial configuration into account and identify areas that would either benefit from management interventions, or where changing land use would degrade an ecosystem service. This would make the LUCI model more useful to stakeholders and local government making management decisions, or choosing areas for rehabilitation or development.

The models address different areas of disaster risk management: the HEC models for short-term flood forecasting and LUCI for long-term land use planning. The expedience of the HEC models allows the DREAM Program to automate the process of watershed modelling and inundation modelling at 10-min intervals to produce maps that are accessible through a public website (Santillan et al., 2013). The HEC models have also been well-applied in the Philippines as this automated system is being used for flood forecasting in eighteen major river basins in the country (Lagmay, 2012). Through running scenarios of local government plans for rehabilitation and development, the LUCI model can complement the existing system by assessing the potential impacts of land use changes, and identifying priority areas for management interventions. By coupling LUCI with HEC-RAS, the impacts of management decisions can be elucidated through inundation modelling and comparing the resulting maps with maps from the baseline scenarios.

4 Conclusions and further research

Given the vulnerability of the Philippines to typhoons and their resulting floods, it is important to invest in proactive measures for disaster risk mitigation. Three of the ways this can be done is through robust hydrological modelling for flood forecasting, structural measures for flood protection, and through making environmentally-sound land management decisions within the watershed. The initial LUCI application in the Cagayan de Oro watershed seeks to address the component of land management decisions through modelling the ecosystem service of flood mitigation in the CDO watershed. The model was able to identify areas to target for management interventions to improve the ecosystem service of flood mitigation. The priority areas are areas before the flow reaches streams and rivers, suggesting that the small-scale targeted rehabilitation approach may be more useful to improving flood mitigation compared to reforesting large areas within the watershed. Under the Typhoon Bopha scenario, LUCI and HEC-HMS produced similar hydrographs in terms of peak flow and occurrence time, and similar inundation maps. The next steps for this research are to refine the soil and land cover parameterisation to produce LUCI maps of the different ecosystem services, identify trade-offs, and to run the model under different land cover scenarios. After that, the rainfall-runoff modelling capability of the LUCI model can be compared with the HEC models to assess how the LUCI model is able to accurately represent watershed hydrological response.

Data availability

This research used secondary data provided by the DREAM Program (https://dream.upd.edu.ph/).

Acknowledgements. R. Benavidez acknowledges the support and guidance of her supervisors: B. Jackson, D. Maxwell, and E. Paringit. Thanks to the fellow hydrology researchers at the Victoria University of Wellington and the University of the Philippines.

Thanks to the DREAM Program for providing thematic datasets, rainfall datasets, and the basin and floodplain model. Thanks to the Cagayan de Oro City Hall, the Department of Public Works and Highways, the Regional Development Council, the Department of Agriculture, and the Philippine Atmospheric, Geophysical and Astronomical Services Administration for providing datasets and watershed reports.

References

Brunner, G.: HEC-RAS User's Manual, 2010.

Center for Environmental Studies and Management: Formulation of an Integrated River Basin Management and Development Master Plan for Cagayan de Oro river Basin, San Juan City, Metro Manila, Philippines, 2014.

Disaster Risk and Exposure Assessment for Mitigation Program: HEC-RAS and HEC-HMS Model Parameterised for Cagayan de Oro, Philippines, University of the Philippines Diliman, Quezon, City, 2015.

Intergovernmental Panel on Climate Change: Managing the Risks of Extreme Events and Disasters to Advance Climate Change Adaptation, Cambridge, UK, and New York, USA, 2012.

Jackson, B., Pagella, T., Sinclair, F., Orellana, B., Henshaw, A., Reynolds, B., and Eycott, A.: Polyscape: A GIS mapping framework providing efficient and spatially explicit landscape-scale valuation of multiple ecosystem services, Landscape Urban Plan., 112, 74–88, doi:10.1016/j.landurbplan.2012.12.014, 2013.

Lagmay, A. M. F.: Disseminating near real-time hazards information and flood maps in the Philippines through Web-GIS, DOST-Project NOAH Open-File Reports, 1, 28–36, ISSN 2362-7409, 2012.

Lasco, R. D., Pulhin, F. B., Jaranilla-Sanchez, P. A., Delfino, R. J. P., Gerpacio, R., and Garcia, K.: Mainstreaming adaptation in developing countries: The case of the Philippines, Climate and Development, 1, 130, doi:10.3763/cdev.2009.0009, 2009.

Mabao, K. and Cabahug, R. G.: Assessment and Analysis of the Floodplain of Cagayan De Oro River Basin, Mindanao Journal of Science and Technology, 12, 147–170, 2014.

National Disaster Risk Reduction and Management Council: Final Report on the Effects and Emergency Management re Tropical Storm "SENDONG" (Washi), 2012a.

National Disaster Risk Reduction and Management Council: Update re Effects of Typhoon "PABLO" (Bopha), 2012b.

Santillan, J. R., Makinano, M., and Paringit, E.: Integrated Landsat Image Analysis and Hydrologic Modeling to Detect Impacts of 25-Year Land-Cover Change on Surface Runoff in a Philippine Watershed, Remote Sensing, 3, 1067–1087, doi:10.3390/rs3061067, 2011.

Santillan, J. R., Ramos, R. V., Recamadas, S., David, G., Paringit, E. C., Espanola, N. C., and Alconis, J.: Use of geospatial technologies and numerical modeling to monitor and forecast flooding along Marikina River, Philippines, Proceedings of the 12th SEASC: Geospatial Cooperation Towards a Sustainable Future, 2013.

Saxton, K. E. and Rawls, W. J.: Soil Water Characteristic Estimates by Texture and Organic Matter for Hydrologic Solutions, Soil Sci. Soc. Am. J., 70, 1569, doi:10.2136/sssaj2005.0117, 2006.

Scharffenberg, W. A.: HEC-HMS: User's Manual, Washington, D.C., 2013.

35

Inflow forecasting using Artificial Neural Networks for reservoir operation

Chuthamat Chiamsathit, Adebayo J. Adeloye, and Soundharajan Bankaru-Swamy

Institute for Infrastructure and Environment, Heriot-Watt University, Edinburgh, EH14 4AS, UK

Correspondence to: Adebayo J. Adeloye (a.j.adeloye@hw.ac.uk)

Abstract. In this study, multi-layer perceptron (MLP) artificial neural networks have been applied to forecast one-month-ahead inflow for the Ubonratana reservoir, Thailand. To assess how well the forecast inflows have performed in the operation of the reservoir, simulations were carried out guided by the systems rule curves. As basis of comparison, four inflow situations were considered: (1) inflow known and assumed to be the historic (Type A); (2) inflow known and assumed to be the forecast (Type F); (3) inflow known and assumed to be the historic mean for month (Type M); and (4) inflow is unknown with release decision only conditioned on the starting reservoir storage (Type N). Reservoir performance was summarised in terms of reliability, resilience, vulnerability and sustainability. It was found that Type F inflow situation produced the best performance while Type N was the worst performing. This clearly demonstrates the importance of good inflow information for effective reservoir operation.

1 Introduction

The planning of reservoirs for various purposes including flood and drought control relies on the historic inflow data at the reservoir site. Due to natural variability and other factors (e.g. climate and land-use changes), however, the inflow situation when the reservoir is being operated will be different. It is therefore important that reservoirs are properly operated so that they continue to perform satisfactorily during changing hydro-climatology.

Reservoir operation concerns taking decisions on water release from a reservoir based on the amount of water available vis-à-vis the demand placed on the system. The available water is the sum of starting period storage and the inflow expected during the period. Consequently, effective reservoir operation relies on reliable forecast of the inflow into the reservoir. Traditional forecasting methods using hydrologic, hydraulic and time-series models require specification of the functional relationship of the model which can be problematic (Zhang et al., 1998), which is why focus has recently shifted to the use of data-driven techniques that do not require knowledge of this functional relationship. In particular, artificial neural networks (ANN) have been widely used to forecast reservoir inflows (see e.g. Edossa and Ba-

bel, 2012; Mohammadi et al., 2005) due to their effectiveness and flexibility and have been proven to be superior to other approaches such as regression-based and time series models.

The aim of this study is to apply multi-layer perceptron (MLP)-ANN for the one-month-ahead inflow forecasting for the Ubonratana reservoir, Thailand. To investigate the effect of the forecasts on reservoir operation performance, four situations were considered for the one-month-ahead inflow: (1) inflow is known and assumed to be the historic (Type A); (2) inflow is known and assumed to be the ANN forecast (Type F); (3) inflow is known and assumed to be the historic average for the given month (Type M); and (4) inflow is not known and the release decision is conditioned only on the starting reservoir storage (Type N). Simulations of the Ubonratana reservoir were then carried out with these alternative inflow scenarios and the resulting reservoir performance was summarised in terms of reliability, resilience, vulnerability and sustainability.

In the next section, further details about the methodology will be given. This is then followed by the presentation of the case study. Next the results are presented and discussed and finally, the main conclusions are given.

2 Methodology

2.1 Artificial neural networks modelling

The theory and mathematical basis of ANN have been described excellently by Shamseldin (1997). Essentially, the structure of ANN comprises an input layer, an output layer and one or more hidden layers as illustrated in Fig. 1. The schematic in Fig. 1 has a single hidden layer which is generally sufficient to approximate any complex, non-linear function (Mulia et al., 2015). The layers contain nodes or neurons which are connected by weights. Determining optimal values for these weights and other parameters of the network is the purpose of the ANN training exercise.

For a given problem, the number of nodes in the output layer is fixed by the problem, e.g. in the current work, it is the 1-month ahead inflow forecast. The input nodes must be determined by the factors known to affect the output variable and this has been achieved through an examination of the cross-correlation matrix (see Adeloye and De Munari, 2006). The number of neurons in the hidden layer is much more difficult to arrive at and is normally determined as part of the training by trial and error as described by Adeloye and De Munari (2006).

Training is often improved through the use of early-stop-rule (ESR) that helps to avoid over-fitting. In ESR, the available data are divided into three parts: (i) a training set, used to determine the network weights and biases, (ii) a validation set, used to estimate the network performance and decide when the training should be stopped, and (iii) a test set, used to verify the effectiveness of the stopping criterion and to estimate the expected performance in the future.

The tested ANN architectures (in trying to arrive at the best value for the number of hidden neurons) were compared using the correlation coefficient (R) criterion, i.e.:

$$R = \frac{\sum y_{sim}y_{obs} - \frac{\sum y_{sim}\sum y_{obs}}{N}}{\sqrt{\left(\sum y_{sim} - \frac{(\sum y_{sim})^2}{N}\right)\left(\sum y_{obs}^2 - \frac{(\sum y_{obs})^2}{N}\right)}} \tag{1}$$

where y_{sim} and y_{obs} are respectively the simulated and observed values of the output variable and N is the number of exemplars used.

2.2 Reservoir performance simulation

Reservoir behaviour simulation employed the mass balance equation (McMahon and Adeloye, 2005):

$$S_{t+1} = S_t + Q_t - D_t' - E_t \tag{2}$$

subject to the operational policy for the reservoir, where S_t and S_{t+1} are respectively storage at the beginning and end of time t; Q_t is the inflow to the reservoir during t; E_t is the net evaporation (evaporation minus direct rainfall) in period

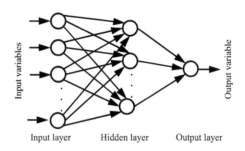

Figure 1. Schematic of artificial neural network.

t; D_t' is the total water release towards meeting the target demand of D_t during t.

As noted previously, the water available for allocation during t, WA_t, is:

$$WA_t = S_t + Q_t \tag{3}$$

and assumes that the inflow is known at the start of the month when making the release decision. In practice, however, this is not the case and assumptions about the size of the anticipated inflow must be made. If the actual inflow turns out to be exactly the same as the assumed inflow, then the end of period storage will be exactly as given by Eq. (2). If, however, there is a discrepancy, the actual end of period storage will be different from Eq. (2).

Let the actual end-of-period storage be $S_{end,t}$, the relationships between this and S_{t+1} for each of the assumed inflow knowledge assumptions become:

1. Type A: $WA_t = S_t + Q_t$ and $S_{end,t} = S_{t+1}$
2. Type F: $WA_t = S_t + Q_t'$ and $S_{end,t} = S_{t+1} + Q_t - Q_t'$
3. Type M: $WA_t = S_t + \overline{Q}_t$ and $S_{end,t} = S_{t+1} + Q_t - \overline{Q}_t$
4. Type N: $WA_t = S_t$ and $S_{end,t} = S_{t+1} + Q_t$

where Q_t is the observed (correct) inflow during time t, Q_t' is the corresponding forecast inflow, $\overline{Q}_t$ is the historic mean flow for the month of time t, and $S_{end,t}$ is the adjusted end-of-period storage.

With the available water determined, release then takes place guided by the rule curves as follows:

Case 1: For $WA_t \geq URC_m$ this is the excess operation case, i.e., $D_t' \geq D_t$

$$D_t' = S_t + Q_t - E_t - URC_m \tag{4}$$
$$Y_t = D_t' - D_t \tag{5}$$

Case 2: For $LRC_m < WA_t < URC_m$ this is the normal operation case, i.e., $D_t' \leq D_t$

$$Y_t = 0 \tag{6}$$
If $WA_t - D_t \geq LRC_m$, $D_t' = D_t$ \tag{7}
If $WA_t - D_t < LRC_m$, $D_t' = WA_t - LRC_m$ \tag{8}

Case 3: For $WA_t \leq LRC_m$ this is the deficit operation case, i.e., $D_t' = 0$ (No water released), where URC_m is the upper rule curve during month $m\,(=1, 2, 3, \ldots, 12)$ of the year; LRC_m is the lower rule curve during month m; Y_t is the excess water released during period t. In general, $t = 12(y-1) + m$ for years $y = 1, 2, 3, \ldots, n$, where n is the number of years in the data record.

Once the simulation is complete, performance indices are then evaluated as follows (McMahon and Adeloye, 2005):

i. Time-based Reliability (R_t): $R_t = N_s/N$, where N_s is the total number of intervals out of N that the demand was met.

ii. Volume-based Reliability (R_v): $R_v = \sum\limits_{t=1}^{N} D'_t / \sum\limits_{t=1}^{N} D_t$, $\forall D'_t \le D_t$.

iii. Resilience: $\phi = 1/\left(\frac{f_d}{f_s}\right) = \frac{f_s}{f_d}$; $0 < \phi \le 1$, where ϕ is resilience, f_s is number of continuous sequences of failure periods and f_d is the total duration of the failures, i.e. $f_d = N - N_s$.

iv. Vulnerability: $\eta = \dfrac{\sum_{k=1}^{f_s}\left(\frac{\max(sh_k)}{D_k}\right)}{f_s}$, where $\max(sh_k)$ is the maximum water shortage in failure sequence k and D_k corresponding demand.

v. Sustainability index (Sandoval-Solis et al., 2011): $\lambda = (R_t\phi(1-\eta))^{1/3}$, where there are multiple users or sectors, each of the above indices will be evaluated for each sector and these can later be combined to determine a weighted group (or global) index. This was done for the sustainability index λ using:

$$\lambda_G = \sum_{j=1}^{M} w_j \lambda_j \qquad (9)$$

where w_j is a weight, given by (Sandoval-Soils et al., 2011):

$$w_j = \frac{DS^j}{\sum\limits_{j=1}^{M} DS^j} \qquad (10)$$

and λ_G is the group sustainability; λ_j is the sustainability for users category j; w_j is the weighting for user j; M is the total number of users sectors and DS^j is the average annual water demand for users sector j.

3   Study area and data

The Ubonratana reservoir is the largest, single multi-purpose reservoir in the upper Chi River Basin in north-eastern Thailand. The dam provides water for consumptive uses (domestic, industrial, irrigation), Pong River in-stream flow augmentation as well as flood control (EGAT, 2002). However, the

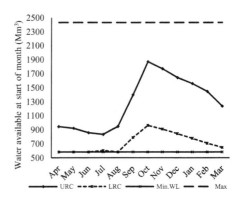

Figure 2. Rule curves for Ubonratana reservoir.

water deliveries first pass through turbines for power generation (installed capacity $= 25.2$ MW) before being allocated to the other uses. The release is prioritised in the order of public (i.e. domestic and industrial), instream flow augmentation and irrigation. The maximum storage capacity of the reservoir is 2431 Mm^3 at elevation of 182 m above mean sea level (m a.m.s.l). Minimum water level of the reservoir is 175 m a.m.s.l. or 581.67 Mm^3 which has been prescribed for the purpose of hydropower generation.

Data collected for the study included daily reservoir inflows, evaporation, area-height-storage relationship, weekly and monthly water requirements and operating rule curves for the reservoir. The observed monthly inflow from April 1970 to March 2012 and rainfall from April 1981 to March 2012 were provided by the Electricity Generating Authority of Thailand (EGAT) and the Royal Irrigation Department (RID). The analysis, however, used the overlapping period of April 1982 to March 2012 (i.e. 360 months) for which the rainfall and runoff data were complete. Data on historical water releases to the various sectors were also provided by the RID. The gross water requirements for the analysis period were 28 952 Mm^3, i.e. average monthly of: 0.98 Mm^3 for public (municipal and industrial) demands; 18.83 Mm^3 for downstream requirements; and 60.6 Mm^3 for irrigation. The original rule curves were also provided by the EGAT; the improved versions of these (see Fig. 2) developed by Chiamsathit et al. (2014) were used in the current study.

4   Results and discussion

4.1   ANN inflow forecasts

Based on extensive testing involving the examination of the auto-correlation function (acf – Fig. 3a), partial-autocorrelation function (pcf – Fig. 3b) and cross-correlation function (ccf – Fig. 3c), six input variables (i.e. current month historic mean inflow, lagged inflows $(t-1, t-2, t-3)$, and lagged rainfall $(t-1, t-2)$) were used for the ANN modelling. The acf (Fig. 3a) shows infinite attenuation with only the first three lags of inflow being significant. Additionally,

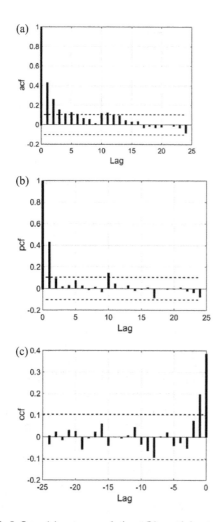

Figure 3. Inflow (**a**) auto-correlation, (**b**) partial autocorrelation functions, and (**c**) inflow-rainfall cross-correlation function for Ubonratana system.

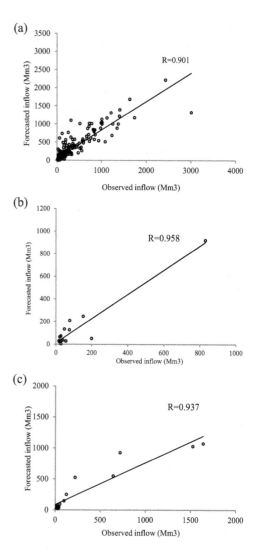

Figure 4. Comparing the 1-month ahead observed and forecast inflow during (**a**) training, (**b**) validation, and (**c**) testing.

the ccf in Fig. 3c indicates that the first two lags of the rainfall are significant. With these, the functional form of the forecast model becomes:

$$Q_t = f(Q_{t-1}, Q_{t-2}, Q_{t-3}, R_{t-1}, R_{t-2}, \overline{Q}_t) \qquad (11)$$

where Q_t is the one-month ahead inflow forecast; Q_{t-1}, Q_{t-2} and Q_{t-3} are lagged inflows of one-month, two-month and three-month, respectively; R_{t-1} and R_{t-2} are lagged rainfall of one-month and two-month, respectively; and $\overline{Q}_t$ is historic mean inflow for the current month.

The ESR was used for the ANN training and for this the 360 months of data were split into three $(90:5:5)$ for training, validation and testing, respectively. The number of hidden neurons was varied between 1 and 35 and based on the R criterion the best architecture had 33 neurons in the hidden layer. Indeed, the final model performed very well with the R exceeding 0.9 in each of the training, validation and testing. Figure 4a, b and c compare the predicted and observed inflow during training, validation and testing, respectively

and further confirm the good performance of the forecasting model. The time series of the forecast inflows (April 1982 to March 2012, i.e. 360 months) are also compared in Fig. 5 and this together with the estimated Nash–Sutcliffe efficiency (NSE) of 0.75 is further evidence of the efficacy of the forecasting model. Additionally, the fact that the NSE was higher than zero is an indication that the model has been a better predictor than the mean value of the observed time series.

4.2 Reservoir performance evaluation

The results of the performance evaluation are summarised in Table 1. For convenience, the operating policy with Type A, Type F, Type M and Type N are denoted by P-A, P-F, P-M and P-N, respectively.

As seen in Table 1, in terms of the total amount of water released, P-A, P-F and P-M were significantly better than P-N, which is not surprising given that P-N did not have any

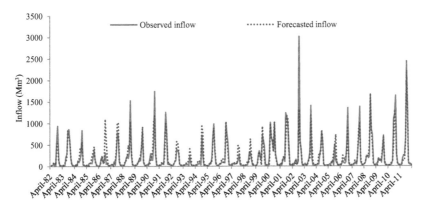

Figure 5. Time series of 1-month ahead of observed and forecast inflows for the complete data record.

Table 1. Summary of evaluated reservoir performance indices for Ubonratana reservoir.

Policy	Water user	Total water shortage (Mm³)	Excursions of $S_{end,t}$ below LRC	f_d	f_s	Reliability (%)		ϕ	η	λ_{user}	λ_G
						R_t	R_v				
P-A	Domestic	0.0	8.0	0	0	100.00	100.00	–	0.000	1.000	0.557
	Downstream	0.5		1	1	99.72	99.99	1	0.026	0.990	
	Irrigation	309.4		15	3	95.83	98.58	0.200	0.626	0.415	
P-F	Domestic	0.0	14.0	0	0	100.00	100.00	–	0.000	1.000	0.655
	Downstream	0.0		0	0	100.00	100.00	–	0.000	1.000	
	Irrigation	244.5		10	4	97.22	98.88	0.400	0.591	0.542	
P-M	Domestic	0.0	16.0	0	0	100.00	100.00	–	0.000	1.000	0.464
	Downstream	0.0		0	0	100.00	100.00	–	0.000	1.000	
	Irrigation	166.8		6	1	98.33	99.24	0.167	0.853	0.289	
P-N	Domestic	3.2	4.0	6	5	98.33	99.09	0.833	1.000	0.000	0.543
	Downstream	132.7		10	9	97.22	98.04	0.900	0.770	0.586	
	Irrigation	1062.6		28	15	92.22	95.13	0.536	0.684	0.539	

additional water from inflows. In terms of reliability (R_t and R_v), the P-F was marginally better than using P-A and significantly better than P-N; P-F was, however, inferior to P-M. A possible reason for this is that in some of the months, the historic monthly mean and forecast inflows were higher than the actual inflows, implying that more water will be released in those months with P-M and P-F than with the other two inflow situations. However, the net effect of such large releases (based on the upwardly-biased inflow forecasts) is the increased number of excursions of the end-of-period storage ($S_{end,t}$) into the region below the LRC as shown in Table 1 for both the P-F and P-M.

The other performance indices reported in Table 1 all reveal the superiority of P-F relative to the other inflow situations. For example, the group sustainability index for P-F was the highest of all four; indeed, the same better performance of P-F was recorded across all three (public, instream and irrigation) demand sectors supplied by the reservoir. As

expected, the conservative nature of P-N resulted in the least number of excursions below the LRC. This is likely to benefit the hydro-power generation potential of the reservoir albeit, as revealed by this study, at the expense of its performance in meeting the consumptive demands.

5 Conclusion

This study has developed MLP-ANN model to forecast one-month-ahead inflow for the Ubonratana reservoir in north-eastern Thailand. Extensive testing of the model showed that it was able to provide inflow forecasts with reasonable accuracy. The performance of the ANN forecasts was tested against those of three other inflow scenarios and the reservoir simulation results showed that the ANN forecasts produced superior reservoir performance. The worst performing inflow situation was when there was complete lack of knowledge about the inflow and release decision was based on the starting storage alone. All this represents an objective demonstration of good inflow forecast knowledge for effective reservoir operation.

Acknowledgements. This study formed part of the PhD research undertaken by the first author with PhD Scholarship provided by the Royal Thai Government. We also thank officials of EGAT for providing the data and other information used for the study.

References

Adeloye, A. J. and De Munari, A.: Artificial neural network based generalized storage-yield-reliability models using Levenberg-Marquardt algorithm, J. Hydrol., 362, 215–230, 2006.

Chiamsathit, C., Adeloye, A. J., and Soundharajan, B.: Assessing competing policies at Ubonratana reservoir, Thailand, Proceedings ICE (Water Management), 167(WM10), 551–560, 2014.

Edossa, D. C. and Babel, M. S.: Forecasting Hydrological Droughts Using Artificial Neural Network Modeling Technique, South Africa: University of Pretoria, Proceedings of 16th SANCIAHS National Hydrology Symposium, 1–3 October 2012, Pretoria, 2012.

EGAT: Improved Rule Curve, Procedure of the Ubonratana reservoir operation: Electricity Generating Authority of Thailand (EGAT) in the Ubonratana dam, EGAT, Khon Kaen, Thailand, 2002.

McMahon, T. A. and Adeloye, A. J.: Water resources yield, Water Resources Publications, LLC, Colorado, USA, 2005.

Mohammadi, K., Eslami, H. R., and Dardashti, S. D.: Comparison of Regression, ARIMA and ANN Models for Reservoir Inflow Forecasting using Snowmelt Equivalent (a Case study of Karaj), J. Agric. Sci. Technol., 7, 17–30, 2005.

Mulia, E. I., Asano, T., and Tkalich, P.: Retrieval of missing values in water temperature series using a data-driven model, Earth Science Informatics, 8, 787–798, 2015.

Sandoval-Soils, S., Mckinney, D. C., and Loucks, D. P.: Sustainability index for water resources planning and management, Water Resources Planning and Management, ASCE, 137, 381–389, 2011.

Shamseldin, A. Y.: Application of Neural Network Technique to Rainfall-Runoff Modelling, Hydrol. J., 199, 272–294, 1997.

Zhang, G., Patuwo, B. E., and Hu, Y. M.: Forecasting with artificial neural networks: The state of the art, Int. J. Forecasting, 14, 35–62, 1998.

Permissions

List of Contributors

T. L. Yang, X. X. Yan, H. M. Wang, X. L. Huang and G. H. Zhan
Shanghai Institute of Geological Survey, Shanghai, 200072, China
Key Laboratory of Land Subsidence Monitoring and Prevention,Ministry of Land and Resources, Shanghai, 200072, China
Shanghai Engineering Research Center of Land Subsidence, Shanghai, 200072, China

A. A. Malinowska and R. Hejmanowski
AGH University of Science and Technology, Cracow, Poland

G. Auvinet, E. Méndez-Sánchez and M. Juárez-Camarena
Instituto de Ingeniería, UNAM, Mexico

C. H. Lu and C. F. Ni
Graduate Institute of Applied Geology, National Central University, Zhongli District, Taoyuan City, Taiwan

C. P. Chang
Center for Space and Remote Sensing Research, National Central University, Zhongli District, Taoyuan City, Taiwan
Institute of Geophysics, National Central University, Zhongli District, Taoyuan City, Taiwan

J. Y. Yen
Department of Natural Resources and Environmental Studies, National Dong Hwa University, Shoufeng, Hualien, Taiwan

W. C. Hung
Green Environmental Engineering Consultant Co. Ltd., Hsinchu, Taiwan

M. Hernandez-Marin, J. Pacheco-Martinez, J. A. Ortiz-Lozano and G. Araiza-Garaygordobil
Departamento de Geotecnia e Hidráulica, Universidad Autónoma de Aguascalientes, Aguascalientes, Mexico

A. Ramirez-Cortes
Doctorado en Ciencias de los Ámbitos Antrópicos, Universidad Autónoma de Aguascalientes, Aguascalientes, México

H. Guo, G. Cheng and Z. Zhang
China Institute of Geo-Environment Monitoring, Beijing, 100081, China

L. Wang
Beijing Geo-Environmental Monitoring Station, Beijing, 100195, China

H. Z. Abidin, H. Andreas and I. Gumilar
Geodesy Research Group, Faculty of Earth Science and Technology, Institute of Technology Bandung, Bandung 40132, Indonesia

J. J. Brinkman
Deltares, Delft, Netherlands

G. Erkens and E. H. Sutanudjaja
Deltares Research Institute, Utrecht, the Netherlands
Utrecht University, Utrecht, the Netherlands

J. Pacheco-Martínez and J. A. Ortiz-Lozano
Departamento de Construcción y Estructuras, Universidad Autónoma de Aguascalientes, Aguascalientes, Mexico

S. Wdowinski, T. Oliver-Cabrera, D. Solano-Rojas and E. Havazli
Department of Marine Geosciences, University of Miami, Miami, Florida, USA

E. Cabral-Cano
Unidad de Geomagnetismo y Exploración, Universidad Nacional Autónoma de Mexico, D.F., Mexico, USA

M. Hernández-Marín
Departamento de Geotécnia e Hidráulica, Universidad Autónoma de Aguascalientes, Aguascalientes, Mexico

K. Furuno
IUGS-GEM Japan branch, 6-41-8, Kotehashidai, Hanamigawa-ku, Chiba 262-0005, Japan

A. Kagawa and O. Kazaoka
Research Institute of Environmental Geology, Chiba, Japan

T. Kusuda
Chiba Environmental Foundation, Chiba, Japan

H. Nirei
The Geo-pollution Control Agency, Chiba, Japan

K. Lei, G. Guo, Y. Yang, Y. Luo and R. Wang
Beijing Institute of Hydrogeology and Engineering Geology, Beijing 100195, China

B. Chen
College of Resources Environment and Tourism, Capital Normal University, Beijing 100048, China

M. Guo
Beijing Bureau of Geology and Mineral Exploration and Development, Beijing 100195, China

J. A. de Waal, A. G. Muntendam-Bos and J. P. A. Roest
State Supervision of Mines, The Hague, the Netherlands

A. Franceschini, P. Teatini, C. Janna, M. Ferronato and G. Gambolati
Department ICEA, University of Padova, Padova, Italy

S. Ye
School of Earth Sciences and Engineering, Nanjing University, Nanjing, China

D. Carreón-Freyre
Laboratorio de Mecanica de Geosistemas, Mexican National University, Queretaro, Mexico

M. Sneed and J. T. Brandt
US Geological Survey, 6000 J Street, Placer Hall, Sacramento, CA 95819, USA

A. Kouda
Teikoku International Corporation, 2-8 Hashimoto-cho, Gifu, Japan

K. Nagata
Geospatial Information Authority of Japan, 2-5-1 Sannomaru, Naka-ku, Nagoya, Japan

T. Sato
Dept. of Civil Engineering, Gifu University, 1-1 Yanagido, Gifu, Japan

I. Martinez-Noguez and R. Hinkelmann
Water Resources Management and Modeling of Hydrosystems, TU Berlin, Berlin, Germany

A. Zhao and A. Tang
School of Civil Engineering, Harbin Institute of Technology, Harbin, China
School of Civil Engineering, Heilongjiang University, Harbin, China

M. González-Hernández, R. Gutierrez-Calderon and W. Flores-Garcia
Centro de Evaluación de Riesgos Geológicos CERG, Iztapalapa, Mexico City, Mexico

M. Cerca and D. Carreón-Freyre
Centro de Geociencias de la UNAM, Juriquilla, Querétaro, Mexico

P. A. Fokker, J. Gunnink, O. Leeuwenburgh and E. F. van der Veer
TNO, Utrecht, the Netherlands

G. de Lange
Deltares, Utrecht, the Netherlands

K. Yasuhara
Institute for Global Change Adaptation Science, Ibaraki University, 2-1-1 Bunkyo, Mito, Ibaraki, 310-8512, Japan

M. Kazama
Graduate School of Engineering, Tohoku University, 2-1-1 Katahira Aoba-ku Sendai-shi, Miyagi, 980-8577, Japan

E. Glowacka, O. Sarychikhina, F. A. Nava, F. Farfán and M. A. García Arthur
Centro de Investigacion Cientifica y Educacion Superior de Ensenada, Ensenada, Mexico

V. H. Márquez Ramírez
UNAM Campus, Centro de Geociencias, Juriquilla, Querétaro, Mexico

B. Robles
Instituto Mexicano de Tecnología de Agua, Jiutepec, Morelos, Mexico

H. L. Chen
School of Environmental Science and Engineering, Zhejiang Gongshang University, Hangzhou, 310018, China

Y. Ito, T. Su and T. Tokunaga
Department of Environment Systems, School of Frontier Sciences, The University of Tokyo, Tokyo, 277-8563, Japan

M. Sawamukai
VisionTech Inc., Tsukuba-city, Ibaraki, 305-0045, Japan

E. Luna-Sánchez
Posgrado en Ciencias de la Tierra, Universidad Nacional Autónoma de México (UNAM), Mexico City, Mexico

F. A. Centeno-Salas
Posgrado en Ciencias de la Tierra, Universidad Nacional Autónoma de México (UNAM), Mexico City, Mexico
Centro de Evaluación de Riesgo Geológico (CERG), Delegación Iztapalapa del Distrito Federal, Mexico

D. Carreón-Freyre
Laboratorio de Mecánica Geosistemas (LAMG), Centro de Geociencias, UNAM, Querétaro, Mexico

W. A. Flores-García and R. I. Gutiérrez-Calderón
Centro de Evaluación de Riesgo Geológico (CERG), Delegación Iztapalapa del Distrito Federal, Mexico

O. Kazaoka, M. Morisaki, A. Kagawa, T. Yoshida, M. Kimura, Y. Sakai, T. Ogura, T. Kusudaa and K. Furunoa
Research Institute of Environmental Geology, Chiba (RIEGC), Japan

S. Kameyama
Environmental Protection division of Chiba Prefectural Government, Japan

K. Shigeno and Y. Suzuki
Meiji Consultante Co., Ltd, Japan

D. Carreón-Freyre, S. Solís-Valdéz, M. Vega-González and M. Cerca
Centro de Geociencias de la UNAM Juriquilla. Queretaro, Mexico

M. González-Hernández and R. Gutiérrez-Calderón
Centro de Evaluación de Riesgos Geológicos CERG, Iztapalapa, Mexico City, Mexico

D. Martinez-Alfaro and F. Centeno-Salas
Posgrado en Ciencias de la Tierra, UNAM Juriquilla, Queretaro, Mexico

B. Millán-Malo
Centro de Física Aplicada y Tecnología Avanzada, UNAM Juriquilla, Queretaro, Mexico

T. Bucx, G. de Lange and J. Lambert
Deltares Research Institute, Utrecht, the Netherlands

G. Erkens
Deltares Research Institute, Utrecht, the Netherlands
Utrecht University, Utrecht, the Netherlands

R. Dam
WaterLand Experts, Amsterdam, the Netherlands

T. Toulkeridis, D. Simón Baile, F. Rodríguez and R. Salazar Martínez
Universidad de las Fuerzas Armadas ESPE, Sangolquí, Ecuador

N. Arias Jiménez
Empresa Metropolitana de Alcantarrillado y Agua Potable de Quito, Quito, Ecuador

D. Carreon Freyre
Universidad Nacional Autónoma de México, Queretaro, Mexico

M. Cerca, D. Carreón-Freyre and J. Aranda
Centro de Geociencias, Universidad Nacional Autónoma de México, Campus Juriquilla, Querétaro, Qro., 76230, México

L. Rocha
Posgrado en Ciencias de la Tierra, Centro de Geociencias, Universidad Nacional Autónoma de México, Campus Juriquilla, Querétaro, Qro., 76230, México

L. Tosi and C. Da Lio
Institute of Marine Sciences, National Research Council, Venice, Italy

T. Strozzi
GAMMA Remote Sensing, Gümligen, Switzerland

P. Teatini
Institute of Marine Sciences, National Research Council, Venice, Italy
Dept. of Civil, Architectural and Environmental Engineering, University of Padova, Italy

Y. Ito, T. Su and T. Tokunaga
Graduate School of Frontier Sciences, The University of Tokyo, 5-1-5 Kashiwanoha, Kashiwa City, Chiba, 277-8563, Japan

H. Chen
School of Environmental Science and Engineering, Zhejiang Gongshang University, Hangzhou, 310018, China

M. Sawamukai
VisionTech Inc., Tsukuba City, Ibaraki, 305-0045, Japan

S. Ye, Y. Wang and J. Wu
School of Earth Sciences and Engineering, Nanjing University, Nanjing, China

P. Teatini
Department of Civil, Environmental and Architectural Engineering, University of Padova, Padova, Italy

J. Yu, X. Gong and G. Wang
Key Laboratory of Earth Fissures Geological Disaster, Ministry of Land and Resources, (Geological Survey of Jiangsu Province), Nanjing, China

E. Stouthamer and S. van Asselen
Dept. of Physical Geography, Faculty of Geosciences, Utrecht University, Utrecht, the Netherlands

M. C. Carpenter
U.S. Geological Survey, Tucson, Arizona, USA

Rubianca Benavidez, Bethanna Jackson and Deborah Maxwell
School of Geography, Environment and Earth Sciences, Victoria University of Wellington, Wellington, 6012, New Zealand

Enrico Paringit
Disaster Risk and Exposure Assessment for Mitigation Program, Quezon City, 1101, Philippines

Chuthamat Chiamsathit, Adebayo J. Adeloye, and Soundharajan Bankaru-Swamy
Institute for Infrastructure and Environment, Heriot-Watt University, Edinburgh, EH14 4AS, UK

Index

Printed in the USA
CPSIA information can be obtained
at www.ICGtesting.com
JSHW052022301024
72690JS00004B/134